Lecture Notes of the Institute for Computer Sciences, Social and Telecommunications E

Konstantina S. Nikita
James C. Lin
Dimitrios I. Fotiadis
Maria-Teresa Arredondo Waldmeyer (Eds.)

Wireless Mobile Communication and Healthcare

Second International ICST Conference
MobiHealth 2011
Kos Island, Greece, October 5-7, 2011
Revised Selected Papers

 Springer

Volume Editors

Konstantina S. Nikita
National Technical University of Athens
15773 Athens, Greece
E-mail: knikita@ece.ntua.gr

James C. Lin
University of Illinois at Chicago
Chicago, IL 60607, USA
E-mail: lin@ece.uic.edu

Dimitrios I. Fotiadis
University of Ioannina
45110 Ioannina, Greece
E-mail: fotiadis@cs.uoi.gr

Maria-Teresa Arredondo Waldmeyer
Polytechnic University of Madrid
28040 Madrid, Spain
E-mail: mta@lst.tfo.upm.es

ISSN 1867-8211   e-ISSN 1867-822X
ISBN 978-3-642-29733-5   e-ISBN 978-3-642-29734-2
DOI 10.1007/978-3-642-29734-2
Springer Heidelberg Dordrecht London New York

Library of Congress Control Number: 2012937079

CR Subject Classification (1998): J.3-4, K.4, C.2, H.1.2, H.2.7-8, D.2

Typesetting: Camera-ready by author, data conversion by Scientific Publishing Services, Chennai, India

Printed on acid-free paper

Springer is part of Springer Science+Business Media (www.springer.com)

Preface

2nd International ICST Conference on Wireless Mobile Communication and Healthcare – MobiHealth 2011

The Second International ICST Conference on Wireless Mobile Communication and Healthcare—MobiHealth 2011—took place on the island of Kos, Greece, during October 5–7, 2011. MobiHealth 2011 was held in parallel with the 10th International Workshop on Biomedical Engineering providing the opportunity to their more than 150 scientific participants to have many fruitful discussions and exchanges that contributed to the success of the events. Kos, the birthplace of the "father" of medicine, Hippocrates, was a delightful venue for the conference.

The MobiHealth International conference series started in Cyprus last year. This year the number of submissions doubled. More than 80 high-quality papers were received. Each paper was carefully evaluated by at least two independent experts. The final program featured 60 papers presented in ten sessions and two workshops with topics covering: intrabody communications, chronic disease monitoring and management, ambient assistive technologies, implantable and wearable sensors, emergency and disaster applications. Invited and contributed papers showed in a unique manner the rapidly changing face and context of healthcare delivery services facilitated by the advances in wireless communications, mobile computing and sensing technologies. Participants from 27 countries worldwide made the conference truly international in scope.

Apart from invited and contributed presentations and workshop papers, during MobiHealth 2011, participants also had the privilege to attend timely keynote lectures by four leading experts, which motivated vigorous discussions.

In his keynote lecture, Yadin David outlined recent progress in telehealth and telemedicine systems emphasizing how on-demand communication of clinical information can be achieved. He reviewed the forces and barriers to mainstream deployment of telehealth systems, and concluded that a good understanding of the forces of quality, cost, home care support and remote management is required by developers, integrators, users and payers.

Nikos Bourbakis addressed the issue of security and protection of private patient data in continuous monitoring applications in terms of safe exchange of health information, as well as secure authentication/authorization access of these valuable data. In his keynote lecture, he presented a mobile-health monitoring system/prototype (Prognosis) and a secure access protection mechanism for information exchange based on strong compression-encryption-hiding mechanisms for information protection and biometrics for information access.

In his keynote lecture, Yang Hao focused on body-centric wireless communications with the use of wearable and implantable wireless sensors. He provided a comprehensive review of recent developments in monitoring behavior related to human physiological response and presented background information on the use of wireless technology and sensors to develop wireless physiological measurement systems. Emphasis was given to recent progress in non-invasive detection of vital signs for chronic disease management.

Sergio Guillen, in his keynote presentation, outlined the concept of "Ambient Assisted Living (AAL)" as a strategy based on the use of information and communications technologies to develop and to provide applications and services that enable older people to live longer independently, while reducing the dependency time. He discussed the interconnection between AAL and "ambient intelligence," and emphasized that although AAL is still more a vision than a reality, in its completion, each user could set the AAL solutions according to individual needs, tastes and economic ability, as they would do with the furnishings and equipment in their homes.

Current and emerging developments in wireless communications integrated with advances in pervasive and wearable technologies are having a radical impact on healthcare delivery systems. The contributions presented in Mobihealth 2011 represented some of these recent developments and illustrated the multi-disciplinary nature of this important and emerging concept. A number of very timely mobile communication systems that can be used for patient monitoring and healthcare delivery, as well as several information and communication technology platforms for chronic disease management and support of the ageing population were presented. The field of wireless medical devices was also explored. Novel implantable and wearable sensory and monitoring devices were proposed and the performance of protocols that are widely used for biomedical telemetry was investigated. Mobile and wireless technologies for healthcare delivery and emergency as well as ambient assistive technologies for pervasive healthcare services were also investigated.

Furthermore, sessions in Mobihealth 2011 explored, examined and debated the ways in which mobile technology developments are expected to transform healthcare delivery, research, business and policy for the twenty-first century. Several open issues and technical challenges were identified as key factors for invigorating healthcare delivery and assisting the shift toward preventive, personalized and people-centered care. Innovative solutions for remote management of diseases, treatment and rehabilitation, outside hospitals and care centers would be based on closed-loop approaches and would integrate components into wearable, portable or implantable devices coupled with appropriate platforms and services. Emphasis would be placed on less obtrusive, self-calibrating and energy-efficient pervasive devices with multi-sensing, advanced on-board processing, communication and actuation capabilities. Context-aware, multi-parametric monitoring of health parameters, lifestyle, activity, ambient environment parameters becomes of utmost importance. Analysis, interpretation and use of the acquired multi-parametric data, in conjunction with established or newly created medical knowledge, are expected to revolutionize medical decision making and action.

Mobihealth 2011 would not have been possible without the dedicated work of many people. Many thanks go to our special session organizers and to the members of the Technical Program Committee, who did an excellent job to uphold quality and promote academic excellence in the review process.

We would also like to extend our sincere thanks to our Local Arrangements Chair, Yiannis Gkialas, and all the members of the Organizing Committee, Asimenia Kiourti, Evi Tripoliti and Maria Christopoulou, who contributed to this venture with great energy and enthusiasm. Our work was made as easy as it could be through the exceptional professionalism of our Conference Coordinator, Justina Senkus.

We are grateful to the Institute for Computer Sciences, Social Informatics and Telecommunications Engineering, and the European Alliance for Innovation for sponsoring this event. Furthermore, generous support for the conference was provided by the Institute of Communication and Computer Systems, the National Technical University of Athens, the IEEE Greece Section and the IEEE EMBS Greece Chapter. We gratefully acknowledge the technical co-sponsorship and endorsement provided by the IEEE EMBS and the IFMBE. Finally, we would like to thank the Hellenic Ministry of Culture, the Prefecture of South Aegean and the Local Authorities of Kos for their sponsorships.

The papers included in these proceedings are the end result of a tremendous amount of creative work and a highly selective review process. We hope that they will serve as a valuable source of information on the state of the art in mobile health.

Konstantina S. Nikita
James C. Lin
Dimitrios I. Fotiadis
Maria-Teresa Arredondo

Organization

Steering Committee

Steering Committee Chair

Imrich Chlamtac   CREATE-NET Research Consortium

Steering Committee Members

Dimitrios Koutsouris	National Technical University of Athens, Greece
James C. Lin	University of Illinois at Chicago, USA
Janet Lin	University of Illinois at Chicago, USA
Arye Nehorai	Washington University, St. Louis, USA
Konstantina S. Nikita	National Technical University of Athens, Greece
George Papadopoulos	University of Cyprus, Cyprus

Organizing Committee

General Chairs

James C. Lin	University of Illinois at Chicago, USA
Konstantina S. Nikita	National Technical University of Athens, Greece

Technical Program Chairs

Dimitrios I. Fotiadis	University of Ioannina, Greece
Maria Teresa Arredondo	Polytecnic University of Madrid, Spain

Local Chair

Ioannis Gkialas   University of the Aegean, Greece

PhD Forum Chairs

Ilias Maglogiannis	University of Central Greece, Greece
Andriana Prentza	University of Piraeus, Greece

Publications Chair

Maria Christopoulou   National Technical University of Athens, Greece

Publicity Chair

Thanos Kakarountas   University of Central Greece, Greece

Travel Grant Co-chairs

Guang-Zhong Yang   Imperial College London, UK
Surapa Thiemjarus   Thammasat University, Thailand

Tutorial Chair

Thanos Kakarountas   University of Central Greece, Greece

Web Chair

Asimina Kiourti   National Technical University of Athens,
                           Greece

Workshops Chair

Paolo Nepa   University of Pisa, Italy

Table of Contents

Session We.3: Advances in Wireless Implantable Devices (Special Session)

Session We.4: Mobile Devices for Patient Monitoring

Session Th.1: Healthcare Telemetry and Telemedicine

Session Th.2: ICT Platforms and Technologies for the Daily Management of Chronic Diseases and the Support of the Ageing Population (Special Session)

Session Th.3: Patient Monitoring and Management

Session Th.4: Mobile and Wireless Technologies for Healthcare Delivery and Emergency

Session Fr.1: Measurement and Monitoring Technologies

Session Fr.2: Ambient Assistive Technologies for Pervasive Healthcare Services (Special Session)

Workshop We.1: Electromagnetic Issues in Advanced Mobile Healthcare Applications

Workshop Fr.1: Mobile Systems and Technologies for Patient Safety, Guidance and Empowerment

A Distributed-Parameter Approach
to Model Galvanic and Capacitive Coupling
for Intra-body Communications

M. Amparo Callejón[1], Javier Reina-Tosina[3,2],
Laura M. Roa[1,2], and David Naranjo[2,1]

[1] Biomedical Engineering Group, University of Seville, Seville, Spain
[2] CIBER de Bioingeniería, Biomateriales y Nanomedicina (CIBER-BBN), Spain
[3] Dept. of Signal Theory and Communications, University of Seville, Seville, Spain
{mcallejon,jreina,lroa,dnaranjo}@us.es

Abstract. In this paper, we propose a simple, but accurate propagation
model through the skin based on a *RGC* distributed-parameter circuit
that leads to the obtaining of simple and general attenuation expressions
for both galvanic and capacitive coupling methods that could assist in
the design of Intra-body Communications (IBC) systems. The objective
of this model is to study the influence of the skin impedance in the
propagation characteristics of a particular signal. In order to depict that
skin impedance, the model is based on the major electro-physiological
properties of the skin, which also allows a personalized model. Simulation
results have been successfully compared with several published results,
thus showing the tuning capability of the model to different experimental
conditions.

Keywords: Attenuation, capacitive coupling, distributed parameter circuit, electrophysiological properties, galvanic coupling, intra-body
communication, skin admittance.

1 Introduction

Intrabody Communication (IBC) is a technique that uses the human body as a
transmission medium for electrical signals to connect wireless body sensors [1].
Therefore, it is a human-centric connectivity technology that exhibits interesting advantages over conventional RF standards such as Bluetooth and Zigbee.
Specifically, low frequency bands without large antennas can be used and there
is no need to transmit high-power signals, thus considerably reducing energy
consumption and the interference with external devices [2]. These features make
IBC a promising alternative for the communication among sensors in biomedical monitoring systems, where minimally-invasive, small size and power-saving
devices are required, which provides the basis of pervasive computing technologies with personal health systems [3]. These advantages have led many research
groups to improve their IBC prototypes [4], [5]. However, no criteria have been

K.S. Nikita et al. (Eds.): MobiHealth 2011, LNICST 83, pp. 1–8, 2012.
© Institute for Computer Sciences, Social Informatics and Telecommunications Engineering 2012

proposed yet that allow achieving an optimal electronic design for an IBC system (in terms of power consumption, data rate, carrier frequency, modulation scheme, etc.) probably due to the lack of knowledge about the IBC propagation mechanisms. For this reason, a model of the human body and in particular the skin (where the signal is mainly confined) as a transmission medium is needed. In the literature there are several modeling approaches: FDTD electromagnetic simulations [6], liquid and solid phantoms [7] and propagation theoretical models that use simple geometries such as cylinders and planes to model the human body [8], as well as circuital models that emulate it as a set of impedances. In [9], a four-port lumped circuit model of the IBC galvanic coupling up to tens of MHz is introduced. As frequencies increase, the previous model becomes more imprecise. In this sense, a distributed RC circuit model of the IBC capacitive coupling was proposed in [10] based on three T-shaped cylinders that simulate the trunk and the arms. Nevertheless, the human body impedance values were discrete and neither physiological characteristics of skin nor the frequency dependence of their dielectric properties were addressed. Only in [11] several electrophysiological considerations were taken into account, but the attenuation study was limited up to 1 MHz for the galvanic coupling technique. The authors of this work have already proposed in [12], [13] a general transmission model through the skin from the cascade, along a longitudinal axis, of basic electrical cells that emulate the skin transcutaneous admittance, forming a RGC distributed-parameter circuital structure, whose elements depend on the frequency up to 1 GHz. In this paper, we propose an improved version of this RGC distributed parameter model, which is able to account for the major electrophysiological properties of the skin, with the added value that it can easily reproduce both galvanic and capacitive coupling, by considering as many longitudinal axes as needed to reproduce the experimental conditions for both coupling techniques as well as the coupling capacitance with the air, which were considered negligible in the previous model. This paves the way for the comparison between both techniques, which has not been yet theoretically undertaken in the literature. Notice that IBC mechanism is based on the fact that the signal is confined to the body surface, with a negligible electromagnetic energy component radiated into the air. For this reason, the frequency of our study only ranges from the hundreds of kHz to 1 GHz, which is the frequency band in which most IBC applications have been developed. Nevertheless, we agree that the most suitable frequency band for IBC performance is below 100 MHz, because the human body does not act as an antenna and communication is limited to the human body surface, without radiation into the air [2]. The ultimate objective of the proposed model is to obtain some insight about the communications performance, through simple expressions of signal attenuation that could assist in the identification of the main IBC design parameters. The simulations have been successfully compared with several published results, thus showing the validity and tuning capability of the model to different experimental conditions.

2   Proposed Distributed RGC Model

The methodology used in this work consisted in obtaining a transmission model through the skin by means of the cascade of basic RGC blocks, along a longitudinal axis, which forms a distributed parameter circuital structure. The objective of this model is to study the influence of the skin cross-sectional admittance Y_{skin} and the skin longitudinal impedance Z_{skin} in the propagation characteristics of a particular signal. Specifically, Y_{skin} is a GC shunt circuit where the conductance G represents the conductive pathways of the skin, which are mainly the sweat glands and the ionic channels that cross the cell membrane, and the conductance C represents the keratinized cells of the stratum corneum (SC) and the lipid bilayer, which are respectively more or less negligible depending on the frequency range [14]. In addition, the impedance Z_{skin} is reduced to a resistance R parameter that emulates the signal propagation between the basic GC cells, that are repeated along the propagation axis, thereby obtaining a distributed parameter circuit model. These parameters were easily obtained from the equations $G = A\sigma/d$, $R = 1/G$ and $C = \varepsilon_r \varepsilon_0 G/\sigma$ where d and A are the skin depth and the cross-sectional area of the GC cell, respectively. The permittivity ε and the conductivity σ dielectric properties of the skin where taken from [15]. It can be noticed that this circuit configuration can correspond to the equivalent electrical circuit model of a lossy transmission line without the inductive element L, i.e $L = 0$. In this way, we can obtain the propagation constant γ and subsequently the attenuation constant α of the skin. Once we have modeled the skin as a transmission medium, we only have to introduce the electrodes configuration of both galvanic and capacitive coupling, as well as the coupling capacitance with the external ground in the latter case. Both circuital model schemes are shown in Fig 1. Z_e represents the electrode impedance, whose values were taken from [16] for two kinds of materials (AgCl and copper). R_L is the load impedance, which was set to $50\,\Omega$. In Figs. 1c-d, the C_a coupling capacitance with the air has been added in order to model the IBC capacitive coupling. Its value strongly depends on the environment, and it further increases with the presence of interfering devices. We have chosen values about tens of pF, as reported in [9]. C_{ad} is the distributed coupling capacitance that appears between each point of the skin and the nearby space towards the external ground. In order to estimate it, a 1m-distance between the person and the floor was considered.

The propagation constant γ of both models can be found by means of

$$\gamma = \sqrt{Z_{skin}Y'_{skin}} \tag{1}$$

where $Z_{skin} = R$ and Y'_{skin} depends on the coupling type.

-   For galvanic coupling,

$$Y'_{skin} = Y_{skin} = 2(G + j\omega C) \tag{2}$$

where (2) is multiplied by a constant factor equal to 2, due to the differential configuration of the galvanic coupling. Therefore, a virtual ground line

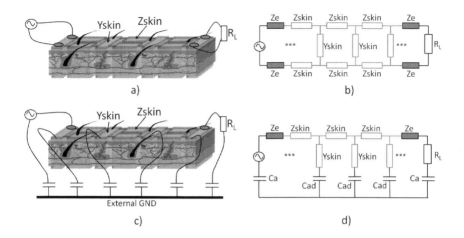

Fig. 1. a) Galvanic coupling IBC method. b) Galvanic coupling IBC equivalent circuital model. c) Capacitive coupling IBC method. d) Capacitive coupling IBC equivalent circuital model.

appears in the middle of the distributed circuit, thus dividing C into $2C$-value capacitors and G into $2G$-value conductances.

– For capacitive coupling, Y'_{skin} is the serie circuit of Y_{skin} and C_{ad},

$$Y'_{skin} = \frac{1}{\frac{1}{G+j\omega C} + \frac{1}{j\omega C_{ad}}} \tag{3}$$

Thus, this model allows to obtain the propagation constant γ related to both electrophysiological properties of the skin, which is modeled by means of R, G and C parameters; as well as the coupling type, through the two different expressions of Y'_{skin}. Once we have the propagation constant γ we can calculate its real part in order to find the α attenuation constant of the skin and predict the behavior differences between both capacitive and galvanic coupling approaches. Then, the electrodes effect have been introduced in order to obtain the total pathloss of the IBC system. Thus, the introduction of the electrodes might cause an impedance mismatch, represented by the reflection coefficient Γ_l,

$$L(\mathrm{dB}) = 20 \log_{10} \frac{1 + \Gamma_l e^{-2\gamma l}}{(1 + \Gamma_l) e^{-\gamma l}} \tag{4}$$

where l is the length between the electrodes.

3   Results

The attenuation constant of the skin was found by means of the real part of (1) for both coupling methods using RGC model parameters values detailed in Section 2. The result is shown in Fig. 2. It can be seen that there is a significant difference between both couplings techniques mainly at low frequencies. This behaviour could be explained by the fact that at low frequencies in the galvanic coupling approach the signal penetrates transversely across the skin into the muscle, therefore muscle conductivity is higher at such frequencies [17]. On the other hand, the capacitive coupling is based on the near-field coupling mechanism that ensures the signal to be confined in the body surface, mainly through the skin, which acts as a signal guide that couples the signal electrostatically without penetrating across it, and which does not cause losses as high as those of galvanic coupling. As a proof of the validity of the model, the IBC pathloss in (4) has been simulated and the theoretical results have been successfully compared with experimental measurements reported by other authors in the literature. We remark that the same parameters used for the simulation results presented in Fig. 1 have been applied, together with the inclusion of the corresponding electrode models. For the sake of representation, in Fig. 3 we have chosen an author that works with galvanic coupling with three different commercial electrodes [7] and two authors that work with capacitive coupling but with different material electrodes [10],[18] in Fig. 4, in order to illustrate the adaptability of the model to different frequency ranges, electrode types and coupling methods. In all these cases it is evidenced a good agreement between the attenuation predicted by the model and the measurement data, notwithstanding that there is a large variability with the values reported in the literature, as a consequence of the different test set-ups and measurement conditions under which the authors carried out their experiments. In addition, skin admittance varies considerably between different people and environmental conditions. Changes in hydration mechanisms due to sweat gland activity and temperature can be manifested in large variations of skin admittance [14]. Thus, we show the tuning capability of our model, which besides retains an electrophysiological significance.

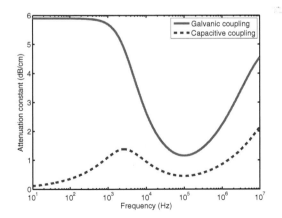

Fig. 2. Attenuation constant (dB/cm) for both galvanic and capacitive coupling between 10 Hz and 10 MHz

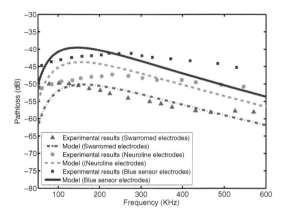

Fig. 3. Pathloss results for galvanic coupling and three different types of commercial electrodes. Comparison between simulated results from model and experimental results in [7].

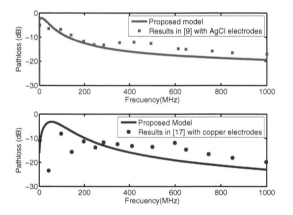

Fig. 4. Pathloss results for capacitive coupling and two different electrodes material: copper and Agcl. Comparison between simulated results from model and experimental results in [10] and [18].

4   Summary and Conclusion

We have proposed a simple but accurate model based on an RGC distributed parameter circuit that combines its simplicity with the flexibility to approximately match diverse experimental results, regarding both IBC galvanic and capacitive coupling methods. Notice that model validation is not aimed at an accurate prediction of the characteristics of attenuation, but to approach the trends and behavior observed in practice, because of the large variability of the measurement conditions under which IBC results have been carried out. In conclusion, a helpful tool to guide IBC designs has been obtained. In addition, we accounted for the main electrical properties of the skin. Moreover, RGC parameters can be derived from bioimpedance measurements, thus obtaining a personalized model. Our next work will be focused on the practical implementation of both IBC coupling techniques in order to obtain different experimental measurements with which the model data could be fitted.

Acknowledgments. The authors are grateful to M.A Estudillo and G. Barbarov for their helpful comments. This work was supported in part by the Consejería de Economía, Innovación y Ciencia, Government of Andalucía, under Grant P08-TIC-04069 and in part by the Fondo de Investigaciones Sanitarias, Instituto de Salud Carlos III, under Grant PI082023.

References

1. Zimmerman, T.G.: Personal area networks. Ph.D. thesis, Massachusetts Inst. Technol., Cambridge (1995)
2. Baldus, H., Corroy, S., Fazzi, A., Klabunde, K., Schenk, T.: Human-centric connectivity enabled by body-coupled communications. IEEE Commun. Mag. 47, 172–178 (2009)
3. Maglaveras, N., Bonato, P., Tamura, T.: Guest Editorial Special Section on Personal Health Systems. IEEE Trans. Inf. Technol. Biomed. 14, 360–363 (2010)
4. Cho, N., Yan, L., Bae, J., Yoo, H.-J.: A 60 kb/s-10 Mb/s Adaptive Frequency Hopping Transceiver for Interference-Resilient Body Channel Communication. IEEE J. Solid-State Circuits 44, 708–717 (2009)
5. Sasaki, A.-I., Shinagawa, M., Ochiai, K.: Principles and Demonstration of Intrabody Communication With a Sensitive Electrooptic Sensor. IEEE Trans. Instrum. Meas. 58, 457–466 (2009)
6. Xu, R., Hongjie Zhu, H., Yuan, J.: Electric-Field Intrabody Communication Channel Modeling With Finite-Element Method. IEEE Trans. Biomed. Eng. 58, 705–712 (2011)
7. Wegmueller, M.S., Oberle, M., Felber, N., Kuster, N., Fichtner, W.: Signal Transmission by Galvanic Coupling Through the Human Body. IEEE Trans. Instrum. Meas. 59, 963–969 (2010)
8. Pun, S.H., Gao, Y.M., Mou, P.A., Mak, P.U., Vai, M.I., Du, M.: Multilayer limb quasi-static electromagnetic modeling with experiments for Galvanic coupling type IBC. In: 2010 Annual International Conference of the IEEE Engineering in Medicine and Biology Society, pp. 378–381 (2010)
9. Song, Y., Qun Hao, Q., Zhang, K., Wang, M., Chu, Y., Kang, B.: The Simulation Method of the Galvanic Coupling Intrabody Communication With Different Signal Transmission Paths. IEEE Trans. Instrum. Meas. 60, 1257–1266 (2011)
10. Cho, N., Yoo, J., Song, S.-J., Lee, J., Jeon, S., Yoo, H.-J.: The Human Body Characteristics as a Signal Transmission Medium for Intrabody Communication. IEEE Trans. Microw. Theory Tech. 55, 1080–1086 (2007)
11. Wegmueller, M.S., Oberle, N., Kuster, N., Fichtner, W.: From Dielectrical Properties of Human Tissue to Intra-Body Communications. In: 2006 IFMBE Proceedings World Congress on Medical Physics and Biomedical Engineering, vol. 14, pp. 613–617 (2007)

12. Callejón, M.A., Roa, L.M., Reina, J., Naranjo, D.: Study of attenuation and dispersion through the skin in Intrabody Communication Systems. Submitted to IEEE Trans. Inf. Technol. Biomed. (April 2011) (under review)
13. Callejon, M.A., Castano, M.M., Roa, L.M., Reina, J.: Proposal of propagation model through skin for attenuation study in Intrabody Communication Systems. In: 27th Annual Conference Spanish Biomedical Engineering Society, CASEIB (2009) (in Spanish)
14. Tronstad, C., Johnsen, G.K., Grimnes, S., Martinsen, Ø.G.: A study on electrode gels for skin conductance measurements. Physiol. Meas. 31, 1395 (2010)
15. Gabriel, S., Lau, R.W., Gabriel, C.: The dielectric properties of biological tissues: III. Parametric models for the dielectric spectrum of tissues. Physics in Medicine and Biology 41, 2271 (1996)
16. Hachisuka, K., Takeda, T., Terauchi, Y., Sasaki, K., Hosaka, H., Itao, K.: Intrabody data transmission for the personal area network. Microsystem Technologies 11, 1020–1027 (2005)
17. Wegmueller, M.S., Huclova, S., Froehlich, J., Oberle, M., Felber, N., Kuster, N., Fichtner, W.: Galvanic Coupling Enabling Wireless Implant Communications. IEEE Trans. Instrum. Meas. 58, 2618–2625 (2009)
18. Ruiz, J.A., Shimamoto, S.: Experimental Evaluation of Body Channel Response and Digital Modulation Schemes for Intra-body Communications. In: IEEE International Conference on Communications, ICC 2006, vol. 1, pp. 349–354 (2006)

An Ultra-Low Power MAC Protocol
for In-body Medical Implant Networks

Ashutosh Ghildiyal[1], Balwant Godara[2], and Amara Amara[2]

[1] Parc d'affaires noveos, 4 avenue Réaumur, 92140 Clamart, France
[2] Institut Supérieur d'Electronique de Paris,
28 Notre Dame des Champs,76005 Paris, France
{ashutosh.ghildiyal}@sorin.com
{bgodara,amara.amara}@isep.fr

Abstract. We present an ultra low power MAC designed for battery-operated subcutaneous implants. Our MAC protocol addresses special communication needs of medical implants like latency, emergency messaging, priority etc., while maintaining an extremely low power-consumption profile. The paper presents the design choices made for a practical cardiac intra-body network and exploits the inherent asymmetries of the network to reduce power consumption. We present a new scheme for deriving analytically the power-optimised TDMA frame parameters like beacon interval and discuss a hardware solution to manage synchronisation overhead. Equations for deriving the duty-cycling efficiency are presented and the packet error rate is calculated for the in-body wireless channel. Our results and simulations show that our protocol is several times more efficient than the state of the art ultra low power protocols. Thus, we illustrate and validate our solution for a very real use case: cardiac networks. However, our new methodology can be applied for any Body Area Network. In this sense, our paper presents a 'universal' solution.

Keywords: Biomedical implants, Body area networks, IEEE 802.15.6, Media Access Layer, Power optimization.

1   Introduction

A body area network (BAN) is a network of sensor nodes which are either on-body or implanted inside the body or both. Since BANs operate in the vicinity of human body, and they have to last for several years in human body, they have a distinct set of requirements of their own which cannot be addressed by the wireless sensor networks (WSN) domain. In view of these requirements, the IEEE has setup a new workgroup (IEEE 802.15.6) which is currently in the process of drafting standards for BANs [1]. The draft intends to propose the medium access layer and physical layer standards for BAN. Since standardization is done for a generic set of requirements, it is not usually optimal for a particular network. In this paper we discuss one such implanted network, an in-body cardiac network and present the design of media access layer for the same. We present the first principles of such a design and justify the design

K.S. Nikita et al. (Eds.): MobiHealth 2011, LNICST 83, pp. 9–15, 2012.
© Institute for Computer Sciences, Social Informatics and Telecommunications Engineering 2012

choices thereof. We present a new approach of designing a TDMA-based protocol, for nodes which have different data rates, priorities and latencies. In doing so, we developed an analytical model to determine the trade-off between latency and power consumption, the key issues for medical devices. The paper is organized as follows. In section 2, we present the distinguishing features of an implanted network and discuss our cardiac network and its requirements. In section 3 we present the design of our MAC and its methodology and in section 4 we present the results of our MAC.

2   Requirements of Implanted BANs and Our Use Case

Inherent in the nature of medical implants are requirements and features which make them distinct from other networks. Since these implants deal with critical medical data and have some characteristic communication needs, it is imperative that we understand them before designing the network. Table 1 summarizes the characteristics of the implanted BAN. Having discussed the general requirements of all BANs, we now present the specific use case we use to illustrate our MAC design methodology: a cardiac network.   Our cardiac network consists of six different sensors, subcutaneously implanted inside the human body. These sensors are controlled by the pacemaker which functions as the master node placed 15-20 cm away from the nodes. The pacemaker is generally placed just underneath the chest skin. Table 2 specifies the corresponding priorities and data rates of these sensors.

Table 1. Requirements of an implanted BAN

Parameters	Choices
Topology	Star
Nature of traffic	Uplink (mostly)
Power	Ultra low power
Latency	Low and predictable
Priority	Needed
Asymmetry	Between master and slave

Table 2. Specifications of cardiac-BAN

Sensors	Data rates	Priority
PEA	10kbps	High
EGM	5kbps	High
G2D	2kbps	Low
BioZ	1.28 kbps	Low
MV	80bits/second	Low
Temperature	.2 bits/second	Low

3   Design of the New MAC for Cardiac BAN

Given the design choices, we choose TDMA as the access scheme since TDMA eliminate*s collisions, idle listening, overhearing* the major sources of power consumption [3]. Moreover TDMA is the only scheme that can *guarantee* a predictable QoS, a much needed feature for delay bound medical networks *[2]*. Despite the aforementioned advantages, TDMA suffers from the fact that the periodic *synchronization* phase must be performed every frame to keep the nodes *synchronized* according to their slots[3]. The other disadvantage of TDMA is that *since nodes* get to speak each frame, the latency of the frame is determined by the frame period. This could be of crucial importance to BANs.

The key parameter in designing a TDMA based MAC is the duty cycling interval (the interval between 2 transmission slots of a node). This interval not only controls latency, but also has a direct effect on the power consumption. In order to conserve power we would like to reduce the sleep period (ON time of radio) as much as possible and then transmit rapidly during our slot and sleep again. However, in BAN unlike the WSN, sensor events are periodic. Therefore the longer the node's sleep, the more data it has in the buffer. It would then have to wake up for a proportionately longer time to send the buffered data to the master. Hence duty cycling the node to reduce power consumption would have only a finite advantage. This implies that if we consider power consumption to be most crucial parameter of optimisation, *there would be a point beyond which sleeping leads to no advantage in power but leads to increase in latency*. We now proceed to determine this point analytically and then present the results for each sensor node of our use case.

Let us consider a duty-cycled system having a current consumption of I_{on} and I_{sl} for on and sleep times respectively. We define t_{sl}, t_{su} and t_{trx} as the radio sleep, start-up and transmission times respectively. Then, the average current drawn over the duty cycling period would be:

$$I_{avg} = [(t_{su}+t_{trx})*I_{on}+ I_{sl}*t_{sl}] / (t_{su}+t_{trx}+t_{sl}) \qquad (1)$$

We define 'R' to be the sampling data rate (in bits-per-second, bps) and 'DR' to be the data-rate over physical layer. So, after sleeping for t_{sl}, the amount of data to be sent and the time to send it are:

$$\text{Data to send} = (t_{su}+t_{sl})* R \text{ bits, } R \text{ is the data sampling rate;} \qquad (2)$$

$$\text{Time to send the data} = t_{trx} = (t_{su}+t_{sl})*(R/DR) \qquad (3)$$

Substituting (3) in (1) we get:

$$I_{avg} = \frac{[t_{su} + (t_{su} + t_{sl}) \times (R/DR)] \times I_{on} + I_{sl} \times t_{sl}}{(t_{su} + (t_{su} + t_{sl}) * (R/DR) + t_{sl})} \qquad (3)$$

We see that for high-rate sensors (high 'R'), the time taken to send the data increases as sleep time t_{sl} increases. Hence the duty cycling efficiency of the sensor node is related to the sampling rate of the inherent sensor. Figure 1 show the graphical description of change in I_{avg} with increase in duty cycle period (up to 5s). The different lines correspond to the sensors of our use-case. The values of t_{su}, R, DR, I_{on}, I_{sl} were taken from the data sheets of the TI-Chicpon CC2430, a popular low power radio [4]. Figure 1 shows clearly the finite effects of duty cycling. Table 3 presents the results for easier comprehension. We see that for high-rate sensors, sleeping beyond 5 or 10 seconds does not lead to any significant reduction in consumption. On the other hand, for low-rate sensors (MV and Temp), the energy minimum is achieved around 100s. Hence, we see that each sensor has a different energy minimum point.

In table 4, we have underlined the ideal duty cycle points for each sensor. For example, for the PEA sensor, going from a 5s beacon interval to 10s leads to reduction of only 3μA. If we can tolerate 3μA of additional consumption, we can reduce the latency from 10s to 5s. We could also duty-cycle the PEA at 1s intervals but the energy cost would be 30μA higher. This table therefore takes us back to the latency-versus-energy trade-off that we have just discussed. *Furthermore, the duty-cycle points naturally lead us to the ideal beacon intervals for each of these sensors in the TDMA scheme.* Now, how do we incorporate these different ideal beacon intervals of each sensor in the TDMA scheme? We propose to do so by choosing the beacon interval such that it meets the latency and power requirements for the higher- rate and priority sensors.

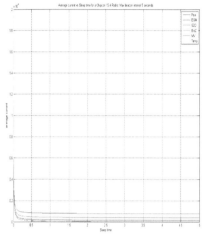

Table 3. Average current consumption(μA) of our sensors for different duty cycle intervals

Sensors	.1s	.5s	1s	5s	10s	100s
PEA	1147	847	808	<u>777.9</u>	774	770
EGM	777	471	432	<u>401</u>	<u>397</u>	393
G2D	548	238	198	170	<u>163</u>	160
BioZ	493	182	142	110.8	<u>106</u>	103
MV	399	87	47	15.39	11	<u>7.99</u>
Temp	393	80	40	8.997	4.99	<u>1.4</u>

Table 4. Cost of Synchronizing for different sensors

Snsors	Oringinal Iavg (μA)	Iavg with early start(μA)	Additional Cost(μA)
PEA:	777.9	793.5	15.6μA
EGM:	401	417.2	16.2μA
G2D:	167	184.3	17.3μA
BioZ:	110.8	127.6	16.8μA
MV:	7.99	9.4	1.41μA
Temp:	1.4	3.0	1.6μA

Fig. 1. Decrease in average current as duty cycle increases for all sensors

In our case, the high-rate sensors are also those that have higher priority: EGM and PEA. We thus choose a beacon interval according to these two sensors, and then make the other sensors wake up after every 'N' beacon intervals. We choose a beacon interval of 5s, the MV and Temp sensors, which have the ideal beacon interval of 100s, wake up after every 20 (100/5) beacons. Thus, not all nodes wake up with every beacon: Nodes which have more data to transmit (PEA/EGM) wake up every 5s; nodes with less data to send (G2D/BioZ) wake up after 10s; and so on. Therefore the TDMA scheme devised by us does not mandate that each sensor wake up with each beacon. Having decided the *beacon interval* we now discuss the crucial problem of *synchronisation* of sensors.

The more the nodes sleep, the more time drift they accumulate. So, nodes that wake up after 20 beacons (100s) would have much higher time uncertainty about their time slots and could instead transmit in their neighbouring slot. We propose to resolve this problem as follows: If we have a crystal of tolerance 'ε' ppm, the amount of timing accuracy 'δ' over the duration of sleep interval (same as the beacon interval BI) will be($\pm \varepsilon \times BI$). Since we analytically know ε, and the maximum timing error that we can encounter (δ), we can avoid missing the beacon by forcing the radio start-up at (BI-δ) instead of BI. We call this the '*cost of synchronising*' with the beacon. This cost will be higher for nodes which sleep longer. Table 4 shows the cost of synchronising with beacon in terms of current for different sensors for the minimum-energy points of table 4. Note that for low-rate sensors, the relative cost is insignificant (2μA). For the higher-rate sensors, it is low.

Slots Determination
Table 5 shows the time required by each sensor to do so. We have assumed conservative values of PER, overhead and retransmission to arrive at the worst-case scenario. We chose the slot interval to be 20ms, since 20ms was close to the least.

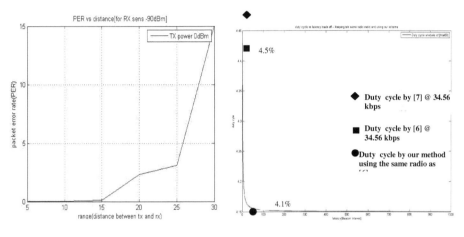

Fig. 2. PER versus distance simulation   **Fig. 3.** Gain in duty cycle from 4.51% [6] to 4.1%

Table 5. Slot size determiniation by considering various factors

Sensors	BI	Total Data @BI (BI*DR)	Time To Send (@250kbps)	+ Over -head (15%)	+ PER (@5%)	+ACK(20%)	#of slots (20 ms slots)
PEA	5s	50kb	200ms	230ms	242ms	290ms	15
EGM	5s	25kb	100ms	115ms	121ms	145ms	8
G2D	10s	20kb	80ms	92ms	97ms	116ms	6
BioZ	10s	12.8kb	51.2ms	59ms	62.5ms	75ms	4
MV	100s	8kb	32ms	37ms	39ms	47ms	3
Temp	100s	.02kb	.08ms	.108ms	.11ms	.13ms	1

4   Results

We simulated the network in the network simulation software OMNET++, a popular discrete event network simulator [5]. The MICS channel model as specified by the IEEE 802.15.6 was used to model physical layer behaviour [1]. We carried out a packet error rate (PER) analysis of our protocol inside a slot. For a receiver sensitivity of -90dBm, we tried to find out PER behaviour for various distances [4]. This gives us a good idea of the range of our network. Figure 6 shows the plot of PER as we vary the range for MICS band. We obtain an acceptable PER (3.1%) for distances upto 25cm inside body. We now compare our work with the two ultra-low power TDMA-based BAN protocols by Omeni [7] and Marinkovic [6]. Note that a power analysis depends on the underlying hardware, hence radios which are more power-efficient and have faster data rates tends to give better power consumptions. The radios used by [6] and [7] have much lower data rates (34.56kbps and 50kbps respectively) compared to our solution (250kbps). A direct analysis on the basis of duty cycle or power consumption would automatically favour our implementation. So, we try to provide two analyses by: a) Comparing our scheme "as it is" with these two protocols, b) Adapting our cycle scheme to use the physical radio characteristics of these protocols. Duty-cycle analysis is generally considered a good figure of merit for any

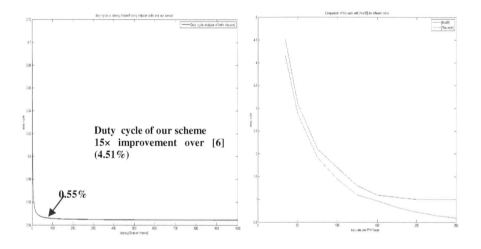

Fig. 4. Duty cycle analysis for a 2.5kbps sensor radio

Fig. 5. Comparison of our scheme using our scheme with [6] for different radios

TDMA protocol [6]. It measures how much time one's receiver is 'on'. The protocol published in [6] reports a duty cycle of of 4.51% for 1.25kbps sensor and 5.7% for 2.5kbps. We carried out this analysis for the protocol of [6] and [7] and found that using the same physical and MAC parameters and simply altering the beacon interval, we can reduce the duty cycle from 4.51% to 4.10%, as shown in figure 3. This is the reduction for each frame; the overall gains in the energy over the life-time of sensor would be significant. Using our radio and our MAC scheme, the same data-rate sensor can be duty-cycled to 0.55%, a factor of more than 9 times improvement over and 15.5 times over [6](figure 4) . Figure 5 shows the duty-cycle analysis as we change the data rate over physical layer (use faster radios). We see that our scheme shows greater gains as we move to faster and better radios. These results assume more importance because of the fact that most commercial low-power radios have higher data rates (>200kbps)[4],[1]. Furthermore, our protocol takes into account retransmission while other protocols permit some packet loss which could be critical for medical data.

References

1. https://mentor.ieee.org/802.15/documents?is_group=0006
2. Keshav: Engineering Approach to Computer Networking. An: ATM Networks, the Internet, and the Telephone Network (1996)
3. Ye, W., Heidemann, J., Estrin, D.: An energy-efficient MAC protocol for wireless sensor networks. In: Proc. IEEE 21st Ann. Joint Conf. IEEE Comput. Commun. Soc. (2002)
4. Texas instruments, Chipcon CC2430 data sheet
5. http://www.omentpp.org
6. Marinkovic, S.J., et al.: Energy-Efficient Low Duty Cycle MAC Protocol for Wireless Body Area Networks. IEEE Transs. on Information Technology in Biomedicine (2009)
7. Omeni, O., et al.: Energy Efficient Medium Access Protocol for wireless medical body area sensor networks. IEEE Trans. on Biomedical Circuits and Systems (December 2008)

On the Effectiveness of Relaxation Theory for Controlling High Traffic Volumes in Body Sensor Networks

Naimah Yaakob[1], Ibrahim Khalil[1], and Jiankun Hu[2]

[1] School of Computer Science and Information Technology,
RMIT University, Melbourne VIC 3000, Australia
[2] School of Engineering and Information Technology,
University of New South Wales, Northcott Drive, Canberra
{yaakob.naimah,ibrahim.khalil}@rmit.edu.au, J.Hu@adfa.edu.au

Abstract. Congestion related issues are major concerns in any network-ing system including Body Sensor Networks (BSN). This is due to the number of disastrous effects (e.g. high packet loss rate and service inter-ruption) it may cause on the system's performance. BSN, which normally involves with life-threatening measurements, are found to be very much affected by this problem. The incorporation of its real-time applications with life-death matters may likely put people at high risk during con-gestion. To address this challenge and alleviate congestion in BSN, we explore the feasibility of a new rate limiting technique known as Relax-ation Theory (RT). Uniquely distinctive from the typical rate limiting schemes, the novelty of our approach lies in the ability to 'relax' or post-pone the excessive incoming packets to a certain extent, and avoid con-gestion from occurring in the first place. An insight performance analysis on one of BSN applications in healthcare monitoring (Electrocardiogram - ECG) shows promising results.

Keywords: Congestion control, Relaxation Theory, Engineering Level.

1 Introduction

The limited buffer spaces in Body Sensor Networks have created substantial performance boundaries for the underlying applications [1]. With very limited capacities of bandwidth and storage, the buffer is highly likely to be overwhelmed by large volumes of sensors reading. This problem, which is known as *congestion*, has appeared to be even more challenging in BSN's domain applications that mainly involve with life-threatening measurements. Congestion may cause many other problems such as the increase in packet loss rate, high energy consumption and high service delay, all of which are detrimental to BSN's performance.

Fig. 1 shows BSN scenario in healthcare application and how congestion ini-tiates. Heavy congestion may occur when multiple leads have data to be sent to the gateway or intermediate nodes, and sensors from several patients are send-ing their readings simultaneously. This will create high reporting rates that may

K.S. Nikita et al. (Eds.): MobiHealth 2011, LNICST 83, pp. 16–23, 2012.

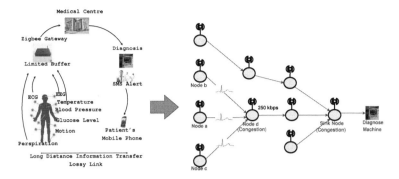

Fig. 1. BSN in Healthcare monitoring and congestion scenario

overload either the intermediate nodes or gateway's buffer and cause massive information loss. Ensuring reliable and timely packet delivery in healthcare applications are therefore of utmost importance to deliver accurate information. This might really help in avoiding false alarm that might lead to wrong diagnosis and affect the monitored patient. Late delivery and packet loss may result in obsolete data and thus inaccurate information. The loss or unreliable transmission of this kind of information might lead to patient's death, who would have survived should the packets arrived safely at the destination.

To overcome the above-mentioned problems, in this paper, we have introduced a novel scheme for managing and controlling high traffic volumes that can cause buffer overflow. This is done at an early stage of physiological data transmission at a sensor node using Relaxation Theory (RT) [2]. To the best of our knowledge, this is the first attempt to solve congestion issue directly associated with BSN. All the existing approaches [3–8] are not applicable for BSN due to its unique characteristics, colossal amount of traffics and different network topology. In addition, most of the existing techniques are mainly focusing on tackling the congestion after it has already occurred, while RT is designed to ensure that the system is steered clear from congestion in the first place. Our method avoids the formation of unfinished works in any nodes, hence ensures reliable packet delivery to the destination. Also, this method is able to eliminate the packet loss occurrence that have been major dilemma in BSN applications. The unfinished works, or sometimes termed as incomplete works, is defined as the number of excess packets still queued in the buffer at end of transmission. The term 'works' refers to packets and is used interchangeably in this paper.

By considering the healthcare monitoring system as shown in Fig. 1, our goal is to find an appropriate Engineering Level (EL), $L(f, \xi)$ [1] to ensure zero unfinished works and packet loss at the end of transmission. Note that the word

---

[1] The Engineering Level is defined as the number of packet slots in a combined pool of bandwidth, in order to be able to place the incoming packets, potentially after some buffering, without any additional loss or delay.

EL and $L(f, \xi)$ are used interchangeably to represent the Engineering Level. In particular, we aim at tackling the issue of losing important packets; either still queued in the buffer or lost due to congestion.

2   Related Research

The use of rate limiting in solving congestion issues has been widely explored in sensor networks as a fundamental approach to control high traffic volumes. While many of the proposed mechanisms [3–6] mainly focus on mitigating the congestion, we aim at preventing this problem at an early stage before its initiation.

The method found in [3] employed Additive Increase Multiplicative Decrease (AIMD) to adaptively control the sending rate based on the congestion notification from the child nodes. When the previous packet is successfully forwarded to destination, the intermediate node will increase the sending rate by a constant α. Otherwise, the sending rate is multiplied by the factor of β. In contrast, Event-to-Sink Reliable Transport (ESRT) [5] monitors the local buffer occupancies and notifies the source nodes to slow down the sending rates once buffer level exceeds certain threshold. Despite their success, these two methods did not considered the amount of incomplete works in the buffer at the end of transmission which is an important criteria for ensuring reliability in BSN. Another rate limiting technique known as Pump-Slowly, Fetch-Quickly (PSFQ) [6] also does not solve the packet loss problem due to congestion, hence is not feasible for BSN.

Our primary objective is basically similar to what is found in [7, 8]. However, in LACAS [7], the authors try to equate the packet arrival rate with packet service rate, the assumption that may not always hold for BSN. While this method can gracefully ensures congestion-free network in Wireless Sensor Networks (WSN), it might not be applicable to BSN. Huge excessive data over the service rate, might lead to high number of packets drop. On the other hand, CODA [8] also throttle nodes' transmission once congestion is detected. Though this technique can achieve energy efficiency, the reliability issue which is the main concern in healthcare application has not been addressed in this approach.

Different with those typical rate limiting approaches [3–8], our approach prevents congestion from occurring by deriving an appropriate $L(f, \xi)$ that constitutes together the service rate and buffer size of the system, so that the occurrence of packet loss due to congestion can be avoided in advance. Once the correct $L(f, \xi)$ is obtained, packets are either directly applied to the output pool, or kept in the specified buffer for at most ξ cell slots.

3   Relaxation Theory (RT): Concept and Overview

The main objective behind RT [2] is to find the EL value that can adhere and satisfy with the defined Quality of Service (QoS), so that the arriving packets can be 'relaxed' or 'postponed' to any value of ξ without additional loss or delay (still within the pre-defined period). Once the EL is properly defined, there will be a fairly allocated bandwidth to each packet without any contention. In reference to

Table 1. Mathematical Notations

No	Symbol	Explanation
1	ξ	Allowable unit delay
2	$L(f,\xi)$, h	Engineering Level
3	$f(t)$	Packets arrival function
4	$B = h\xi$	Buffer size
5	c_n	The point on x-axis where the average intersects with $f(t)$
6	$O(Xn)$	Accumulative unfinished works at specific time t
7	$O(b)$	Unfinished works at end of transmission time b

healthcare monitoring, avoiding contention among the sensed packets is crucial in ensuring packets loss from occurring and avoid misleading interpretation of the data. All the notations used can be listed in Table 1.

3.1 The Constraints

All the related constraints in obtaining the correct EL are defined as follows:

1. $O(b) = 0$
2. $Loss = 0$
3. $O(X_n) <= h\xi \ \ for \ all \ n$
4. $\xi < b - a$

The first and second constraints are important criteria for real-time healthcare monitoring system. In fact, BSN itself is very sensitive to packet loss [1] as this might trigger wrong diagnosis that may harm the critical ailing patients. The third constraint requires that the incomplete works at any time between a and b should be less than the buffer size $(h\xi)$. This is to avoid the formation of buffer overloading. The largest unfinished works should occur at point c_n in such a way that $O(c_n) = h\xi$. The last constraint implies that the unit delay ξ should be less than $b-a$, or otherwise, the system may require a very large buffer to gracefully discharged all the packets, which is very impractical in scarce resources BSN.

3.2 Engineering Level (EL), $L(f, \xi)$

There are four parameters that need to be known in advanced in order to achieve the required $L(f, \xi)$ which are: Packets arrival function $(f(t))$, start time (a), end time (b), and allowable unit delay (ξ). Based on these parameters, the correct $L(f, \xi)$ can be derived as follows:

$$h = L(f, \xi) = max[h1, h2.., hn] \tag{1}$$

where h_n are various EL candidates. The h_n value is derived from the average arrival rate (A) given as $A(f, b-a) = \dfrac{1}{b-a} \int_a^b f(t)\,dt = 0$. Then the $L(f, \xi)$ can be obtained as follows:

Table 2. Experimental Setup

No	Input Parameters	Setup
1	Bandwidth	250 kbps
2	Arrival Rates	1-12 leads
3	Packet Size	30 bytes
4	ξ	1, 2, 3
5	Sensor Leads	1-12

$$L(f,\xi) = max[max[\frac{1}{b-a}\int_a^b f(t)\,dt, \frac{1}{b-c_n}\int_{c_n}^b f(t)\,dt], \frac{1}{c_n - a + \xi}\int_a^{c_n} f(t)\,dt] \tag{2}$$

Value c_n is obtained from the intersection point between (A) and the x-axis. Note that $O(c_n)$ is the largest unfinished works. Finally, we need to check whether $O(c_n) = h\xi$ and $O(b) = 0$, which are the two significant validation tests to know that the obtained $L(f,\xi)$ is absolutely correct. In other words, the incomplete works that exceed the buffer size at any time in the interval, may violate the rules and thus may result in some packet loss.

3.3   Experimental Setup

We tested the potentiality of RT using some numerical analysis which have been conducted based on some configuration setup as specified in Table 2. It is suffice to mention that this experiment is conducted based on the assumption that the data are collected from only one patient at this point. We believed that controlling the incoming traffics and managing the buffer at this early stage are the key success of the congestion control in the entire system. If the buffer can be emptied at each sensor node including the intermediaries, the congestion can highly likely be alleviated from the whole network system. For the purpose of this experiment, we generate random number of packets using the Random Number Generator (RNG). This has been carried out in C language in order to create different arrival rates at different time unit. Once the incoming packets have entered the input trunk, RT is immediately applied and the system will check for the available slots of transmission based on the calculated $(L(f,\xi))$.

4   Results and Analysis

This section demonstrates some parts of the results obtained through the numerical analysis. The performance have been evaluated using two performance metrics: Number of unfinished works at time b, $(O(b))$ and Number of Packet Loss.

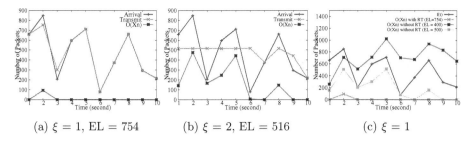

(a) $\xi = 1$, EL $= 754$ (b) $\xi = 2$, EL $= 516$ (c) $\xi = 1$

Fig. 2. The resulting $O(Xn)$ with the calculated EL for different ξ values (a) $EL = 754$ (b) $EL = 516$ (c) Performance comparison with and without RT

Fig. 2a shows the resulting number of $O(Xn)$ for $\xi = 1$. From the figure, we noticed that there is no unfinished works at all time except once while $t = 2$. The calculated EL for this case is 754. Thus, if $f(t)$ at any time t are less than the EL, they will be transmitted as per arrival. At $t = 2$, there is excessive arrival of 95 packets beyond the available EL slots. These extra packets are therefore stored in the buffer. Since the buffer is still empty and not yet being occupied, at $t = 2$, there are enough spaces to keep the packets until the next transmission. In this case, the buffered packets can be delayed to the future until one unit of time.

At $t = 3$, there are only 206 packets arrived, thus there are ample of empty spaces to accommodate together those stored in the buffer. That is explained by the transmission line that is slightly above the packets arrival in Fig. 2a at that particular time. As such, there is no unfinished works left ($O(b) = 0$) which validates the finding of EL. On top of that, the absence of packets loss at any time t, further verifies the obtained results.

We extended the value of ξ to 2 and 3 and the results obtained are as depicted in Fig. 2b. To our surprise, the results for $\xi = 2$ and 3 are absolutely the same. This is because, the shape of the graph is determined by the obtained EL, which is the same for both the cases. We have a strong belief that this scenario is created when the obtained value of $h2$ is always higher than $h1$. Hence, no matter what ξ is, we will always get the same EL value.

The EL for this case is 516. Since EL value is far below almost half of the arrival function $f(t)$, there are buffered packets as early as $t = 1$ until $t = 6$. All the packets stored in the buffer, are then being transmitted together with the current arrival of packets, and sent at the maximum of $(EL - f(t))$. At $t = 6$, there are still 443 $O(Xn)$ in the buffer. However, the sum of those in the buffer and the new arrival exceed the EL for about 7 slots. Hence, these extra packets will be kept for transmission in the next round. The processes are repeated until end of time b, whereby all of the works have been fully transmitted. Since the EL obtained in both cases satisfy all the defined constraints, no packets are lost or dropped as can be seen in the graphs. Therefore, the ELs have been properly selected.

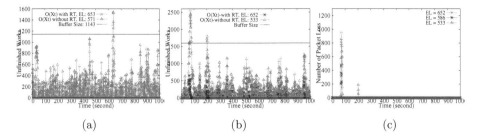

Fig. 3. (a) Performance comparison between proper and improper selection of EL. $O(Xn)$ above the buffer size is considered lost. (b) Another performance comparisons for different sets of data (c) The resulting packet loss for (b) in different EL.

Performance comparison of $O(Xn)$ with and without RT is shown in Fig. 2c. Our method outperforms the one without RT all the times, and the resulting $O(b)$ is also 0 as evident in Fig. 2a and Fig. 2b. Therefore, the implementation of RT in this experiment has substantially improved the performance and eliminated the occurrence of congestion and packet loss in advance.

The significant contributions of zero unfinished works and packet loss presented in this paper have eliminated the issue of incomplete transmission of packets at the end of transmission, while gracefully solving the information loss problems. These contributions are crucially needed in providing and ensuring a reliable transmission in healthcare monitoring. The losses of vital signals in such real-time environment may affect the ongoing diagnosis and lead to false interpretation of the end result that may harm the monitored patients. Since the diagnosis may involve with life-death matters, careful handling of the collected readings is of the major concern.

As we extended the experiment further to $b = 1000$, the number of $O(Xn)$ happened to appear more frequently since it is an accumulative incomplete works. However, with the correct derivation of EL, these $O(Xn)$ tend to be vanished as the time approaches the final point b. This is depicted in Fig. 3a.

In comparison with that, the value lower than the calculated EL (in this case is 571) produced high number of $O(Xn)$ to the extent that violate the RT rules at some points (points that exceed the buffer size) and results in the number of packet loss due to improper selection of EL. Another RT performance comparison is shown in Fig. 3b. This new data set possess a different EL value (652). Again, two different values (586 and 533) of lower rates have been chosen for comparison purposes. As expected, the correct EL gives outstanding performance compared to the other two. As discussed earlier, BSN performance decreases without the use of RT. This can be obviously noticed in the graph. In fact, some of the packets have surpassed the limited capacity of the buffer, and thus lost as depicted in Fig. 3c. These information loss may be detrimental to BSN operation as it might contain useful knowledge. On the other hand, no packet loss has occurred when the correct EL value was used.

All the results obtained in this experiment show promising features in assisting good BSN services. The outcomes have brought significant improvement to BSN since it is very sensitive to any information loss. The significant contribution of zero unfinished works presented in this paper has eliminated the issue incomplete transmission of packets (in the buffer) at the end of transmission, while gracefully solving the packet loss problems.

5 Conclusion and Future Works

In this paper, we have introduced a new rate limiting technique using an approach known as Relaxation Theory. Based on the preliminary analysis, we are able to demonstrate that the properly selected EL is capable of improving BSN performance and avoid congestion in advance. We have also highlighted several constraints that need to be followed in order to obtain the desired EL. The preliminary results demonstrate promising features for our future experimentations in solving many congestion-related issues in BSN. As part of our future works, we would also be interested to study the RT performance in larger network size and investigate the performance trade-off in terms of delay and buffer size.

References

1. Yang, G.Z.: Body sensor networks, research challenges and applications
2. Minoli, D.: Broadband Network Analysis and Design. Artech House, Inc., Norwood (1993)
3. Woo, A., Culler, D.E.: A transmission control scheme for media access in sensor networks. In: Proceedings of the 7th Annual International Conference on Mobile Computing and Networking, MobiCom 2001, pp. 221–235. ACM, New York (2001), http://doi.acm.org/10.1145/381677.381699
4. Stann, F., Heidemann, J.: Rmst: reliable data transport in sensor networks. In: Proceedings of the First IEEE International Workshop on Sensor Network Protocols and Applications, pp. 102–112 (May 2003)
5. Akan, O., Akyildiz, I.: Event-to-sink reliable transport in wireless sensor networks. IEEE/ACM Transactions on Networking 13(5), 1003–1016 (2005)
6. Wan, C.-Y., Campbell, A., Krishnamurthy, L.: Pump-slowly, fetch-quickly (psfq): a reliable transport protocol for sensor networks. IEEE Journal on Selected Areas in Communications 23(4), 862–872 (2005)
7. Misra, S., Tiwari, V., Obaidat, M.: Lacas: learning automata-based congestion avoidance scheme for healthcare wireless sensor networks. IEEE Journal on Selected Areas in Communications 27(4), 466–479 (2009)
8. Wan, C.-Y., Eisenman, S.B., Campbell, A.T.: Coda: congestion detection and avoidance in sensor networks. In: Proceedings of the 1st International Conference on Embedded Networked Sensor Systems, SenSys 2003, pp. 266–279. ACM, New York (2003), http://doi.acm.org/10.1145/958491.958523

Design and Validation of a Secure Communication Platform for Mobile Health

Beatriz Martín de Juan[1], Miguel Ángel Valero Duboy[1], Diana Soler[2], José Manuel Azorín[3], and Rafael Conde[1]

[1] T>SIC, Information & Knowledge Society Technologies,
Technical University of Madrid EUIT Telecomunicación,
Ca. Valencia, km. 7. 28031, Madrid, Spain
{bmdjuan,mavalero,rconde}@diatel.upm.es
[2] Modelos de Atención Gestionada
C. Bruc, 35, 5°, 08010 Barcelona, Spain
diana@mag.es
[3] Vodafone España
C. Grañón (Pau Las Tablas), 22, 28050 Madrid, Spain
jose-manuel.azorin@vodafone.com

Abstract. This paper focuses on the design and validation experience of an eHealth system that provides end users with web and mobile access to their personal health records (PHR). The system has been tested with different mobile devices to ensure full compliance with privacy and confidentiality requirements. As a result, both patients and medical doctors can share sensible medical data in a secure and efficient way. The pilot site has been evaluated under the cooperation of 375 volunteers from a global mobile operator, a regional medical company, a SME applications developer and a state university with deep background on eHealth. Users´ feedback has been quite satisfactory and promising.

Keywords: Mobile health, medical confidentiality, secure communications.

1 Introduction

Mobile health (mHealth) [1] requires suitable network, software and hardware technologies to provide patients with mobile health services in a flexible, realistic and usable way at the point of need [2]. Telemedicine services are meant to provide at this point not only medical care but services provision depending on the availability of reliable networks to make medical information fully accessible for clinical purposes. However, mobility goes further beyond this aim since patients' location is not restricted and they can manage clinical information.

Telemedicine and eHealth must be secure not only in terms of safety but also regarding data confidentiality and integrity. This research paper focuses on the design and validation of a secure platform to provide mobile patients with integral PHR. The experience has been trailed in Spain thanks to the participation of 375 volunteers from a global mobile operator, a regional level medical company, a SME applications developer and the research work of an estate university with wide background on eHealth and telemedicine.

K.S. Nikita et al. (Eds.): MobiHealth 2011, LNICST 83, pp. 24–31, 2012.

1.1   Background

European health care policies do always face new challenges to manage huge amounts of medical data due to the increasing penetration of PHRs. The combination of mobile and web technologies paves the way for efficient solutions to provide citizens with private access to their medical data. Since health care services are highly fragmented, patients and medical doctors have multiple difficulties to easily consult requested clinical information. The enhancement of care quality provision is directly related to the availability of ad hoc medical data. Computer and communication technology advances, mobile networks and affordable devices have made data information more manageable for patients that can now easily access to their personal information distributed in compliance with privacy and confidentiality requirements.

Although mHealth has developed a framework to allow more and more patients to benefit from ubiquitous transmission of massive medical information in a quick way, it sets up new inevitable issues such as open information exposition to malicious intruders [3]. The transmission of patients' information over wireless networks must grant confidentiality, integrity, availability and privacy as HL7/ISO standards do specify.

Several initiatives have already tackled various key problems when integrating secure PHR management with usable and efficient mobile access. However, while wireless applications for healthcare can be divided in: monitoring applications and patient communication and supporting applications [4] [5] [6], the innovative side of this proposal is the secure data exchange. Patients records security is not only based on complex mechanisms to support encryption or authentication to servers or network access, but the integration of solutions in a widespread and highly demanded device such as mobile phone from which information is potentially transferable to the social environment and widely used by citizens.

1.2   Platform Requirements

The actual system consists of a web platform that can be consulted directly or by means of a mobile application. The main goal is to promote proximity and universal access to the patient's PHR; it supports access from anywhere, at any time using standard clinical information [7] and enables interfacing communication with other systems and providers by transferring the actual contents to the patient's mobile phone. The system allows automatic updates, either by further interventions and updating their data, or by means of upgrading health advisories and alerts generated automatically by the system. It also guarantees the secure data storage and transmission ensuring the confidentiality and security of data throughout the whole process [8], in compliance with current regulations [9].

mHealth data needs to be protected against unauthorized access. The issues concerning mHealth information are mainly related to: information recollection, retention, distribution and use. In the first instance, the provider must be authorized by the user before the information is given [2]. In order to protect the access to the medical information, the patient will have the right to know if such information exists and when necessary if it has been consulted by another person. Personal identifiable health information about individuals should also have the fully informed consent of patients [10].

Secure Systems of Information should incorporate politics, standards and procedures that can assure data confidentiality, integrity and availability. The current legislation must be obeyed and be aware of potential digital danger to face up to. Furthermore, information needs to be shared between different systems, so it will be a must to guarantee and enable secure communication and protection by means of secure communication channels and integration platforms between systems [11]. A mHealth application must also ensure that data remain confidential, integrity is always guaranteed throughout the whole process and only authorized personnel can have access to it [12]. There are mainly four requirements that such a system has to fulfil:

I.   Anonymity – in this kind of service anonymity is essential, since users' personal data cannot be linked with their identity while are being transmitted. The system must protect participants' anonymity throughout the whole process. Occasionally, it could be relevant to request users' personal data, as sex, age, occupation, etc so it is essential that they can assure their confidential information is absolutely anonymous and there isn't any way to verify their truthfulness. Consequently, if users think their data may have been incorrectly treated, they could always decline to include their true information.

II.   Confidentiality - Along with anonymity, the application must guarantee that there is no way to link health information with the actual source as well as data confidentiality in transmission. An unauthorized source should never access to the content.

III. Authentication – Health information needs to be authenticated at both the user and application source level in response to the need to ensure anonymity. This service must guarantee that mobile application installed in the mobile is only received by the patient.

IV.   Integrity – It's essential the information has not been manipulated in their way from the provider server to the patient's mobile.

2   Design and Implementation

The deployed architecture consists of a mobile device in which a mHealth application is installed; a Proxy Server carries out the security function; a portal web; and a Provider's Server for data management. The web platform offers a management tool for the patient's medical history and active processes. Further, it enables proximity and trustworthiness throughout the interaction with the health workers and providers involved in the process. It also includes a subsystem to be adapted to different mobile phone models in order to achieve a fair presentation of the record in the mobile phone. Due to the use of a transparent proxy security server the functionality is not affected by the incorporation of the proposed security mechanisms to the web platform.

Data have to be protected in each and every phase of the process: storage, updating, search and retrieval. The system has to assure that data are accurate, correct, and valid and that they have not been modified in the whole process. The mechanism developed to carry out this function is the digital signature that is also used to validate the identity of the user. Cryptography is meant to ensure the network traffic integrity and confidentiality so as the communication between the user and the service provider. The provider-side is validated by means of certificates and SSL protocol guaranteeing that authenticity is always verified [13].

The mHealth application has been developed using J2 Micro Edition (J2ME) technology. This technology is mainly supported by Symbian and BlackBerry which own a high percentage of the market. This is the main reason to develop the security library in this language. Since BlackBerry is not fully compatible with all the security libraries like J2ME, it was required to avoid obfuscation and others methods like socket push-registry that are not supported. Figure 3 points out the Global statics of the different mobile OSs in the last 2 years. These data reveal that handsets with J2ME, Symbiam OS, BB OS, Sony Ericsson and some Samsung devices cover more than 50% of the market [14], although Android devices are gaining ground with a 16,23% in the last year. Data are based on mobile web usage and not on physical handsets. Once the operation in J2ME compatible devices was validated, the Android version was fully developed. This platform has been also verified with testing applications and it is actually being integrated within the mHealth service.

The library of security has been implemented using the BouncyCastle lightweight API that works as with J2ME and JDK 1.6. Android has its own libraries of security used to the implementation. Android is a software stack for mobile devices that includes an operating system, middleware and key applications. The Android SDK provides the essential tools and APIs to begin developing applications on the Android platform, using the Java programming language. Each Android application has its own security sandbox once it's been installed on a device.

2.1   Security Mechanisms

The security mechanisms carried out by the security library are the keys generation and storage, the ciphered and signed of the information. Encryption is the main tool used to prevent from reading confidential information. The keys used are RSA. The generation time is a key factor to be considered for usability reasons since a user could become impatient if the application spends too much time in this process. Table 1 shows the measured results of these tests in a common smartphone.

Table 1. Time keys' generation

Size	Time
4096 bits	>5 minutes
2048 bits	approx. 4 min and 30 sec.
1024 bits	approx. 40 sec.
512 bits	approx. 15 sec.

Both J2ME computer simulators and real market devices allowed the generation of keys up to 4096 bits, although their generation took too much processing time. Consequently, we opted for using a size of 1024 bits since the keys have an adequate size, compared with other security services and the waiting time for the patient isn't excessive. The size of the server's key is 2048 bits. The keys are stored as in Base64 as Byte arrays in a database with the purpose of being recovered later. Every time the application is open, the system checks those keys that have been stored which means the application has been correctly installed. If the keys weren't stored should be generated and stored in the device database.

2.1.1 Configuration and Information Storage

The configuration of the library is made based on a file sent to the mobile along with the installation file. This file contains the address of the Security Server ciphered with an AES key that is well-known by the mobile application.

There is a register of the different communication that occurs between the server and application. Each register consists of a nickname of the communication initiator, the date and part of the information that has been exchanged. This information is stored inside the mobile device after being ciphered with the same AES key that was used to decipher the configuration file.

2.1.2 Keys Generation and Storage

The library uses two different couples of keys to sign, cipher, and decipher the data and to verify the signature during the user's authentication process in front of the Security Server.

Access to data stored in the Record Management System of the application is granted only if the authorization mode of the RecordStore allows access by the current MIDlet suite. In our case, the access is limited through the parameter "authmode" which is private. The information is also stored ciphered using AES. When some of the information is sent, it will be ciphered and signed in this case with the set of RSA keys and the server public's key so a malware won't be able to interpret it.

The application stores the public key's modules and exponent while the recovering of the private keys is necessary to store modulus, public exponent, private exponent, prime1, prime2, exponent1, exponent2 and coefficient. Each parameter is ciphered with the AES key before being stored.

2.1.3 Cipher and Digital Signature

The data sent from the application to the Security Server are ciphered using the server's public key to ensure that only the Security Servers can access to the information. The confidentiality requirement is fulfilled since the information can only be deciphered by the Security Server. The cipher is made by using the algorithm RSA, in mode Electronic Code Book (ECB) and the padding PKCS1Padding. The signature is made up of the SHAwithRSA algorithm that has been implemented by the BouncyCastle libraries. The digital signature ensures the authentication and integrity requirements.

2.1.4 Security Server

Secure information submission by the application requires a proxy server whose functions are signing/verification and encryption/decryption of information. This server is also in charge of transferring information requests to the Provider's Server, receiving and ensuring the security of the data given as a response.

The requested tasks of the Security Server are identified subsequently:

I. To send public key contains in its certificate.

The server sends data needed to the application in order to rebuild the server's public key that will be used to the ciphered of the information. Data sent are the module and the exponent of the RSA key of 2048 bits.

II. To transfer the information to the destiny:

Once the information is received, the application has to verify the digital signature and, if it is valid, it will later decipher the information since the mobile application sends its public key to the server. All data involved in the process have to be signed; as a result it will be necessary to implement a specific method capable of verifying such a signature and another in charge of carrying out the digital signature and sending it to the mobile application. The Security Server contains also one method to cipher the data sent to the mobile application and another to decipher the information received from it.

Once the Security Server has verified and recovered the information, it's the right time to send the information to the destiny. The destiny can be either the mobile application or the Provider's Server depending on which one was the emitter. The address of the Provider's Server is part of the cipher information that the Security Server receives from the mobile application. The address of the Security Server is included in the configuration file that is then sent to the mobile along with the installation file.

3   Validation

Reliable tests were performed within 40 mobile devices of 5 different brands: Nokia, BlackBerry, Samsung, Sony Ericsson and LG, with a percentage of compatibility of 85%. A device is compatible providing it passes the following tests: a WAP Push SMS reception; the installation file download; the full installation; the initial launch; the keys generation and initial data load; the manual launch; the test of navigability and visualization; the update starting; the data updating and the automatic close. Nowadays, the test application is also compatible with tactile devices. However, some devices have failures in the visualization because of the presentation of base64 characters that are incorrectly showed. This is another source of incompatibility in devices as KU990 Viewty or Samsung M1.

The pilot site of the eHealth platform is being carried out amongst 375 registered volunteers of Vodafone Spain. The pilot will have 9 month of duration. After two months of pilot over 90% of the volunteers have already introduced their data in the web platform and the 34% of them have downloaded the mobile application. The application's installation success rate is 90% and the failures 10% that were caused by non-compatible devices like iPhone or Android.

A sociological study is also taking place with a double methodology, quantitative, the surveys, and qualitative, the face interview and profiles. An initial survey was carried out amongst volunteers from the age of 17 up to 55, before the platform was tested. This survey has identified different profiles between the registered users and an exhaustive evaluation of the usage reasons will be also conducted within a short period of time. The reasons are mainly: hardly go to the doctor; the possible impact of the application over their health; the time factor the future problems prevention and pregnant.

A face to face interview, whose main goal is to find out volunteers' prospects regarding this kind of systems, will be carried out. At last, a use case survey is carried out each three months to improve the application's features based on volunteers' opinion. This specific survey has three parts related to the web application, mobile application and fulfilled prospects.

The results of the first use case survey shows the rate of acceptance by the users. The Table 2 also reveals the answer of the users to the question: Is it worth having your medical information in the mobile phone? More than 60% of the volunteers agree that the mobile phone application is worth versus the 6% who disagree.

Table 2. Is it worth having your medical information in the mobile phone?

Fully disagree	Disagree	Indifferent	Agree	Fully agree	Total
2,73%	2,73%	28,18%	38,18%	28,18%	100,00%

4 Conclusions and Future Research

The developed architecture provides usable security mechanism in the exchanging process of sensible data between devices and external entities. Both the device used, a smartphone, and the technology chosen for the developed platform allow to integrate mHealth application with other existing network technologies. It is portable and operational in a timely appropriate period and also compatible with a high rate of available devices of the market regardless of the operating system: RIM or Symbian that covers most of mobile devices.

For the time being, the conclusions drawn, considering the analysis of data and statistical association tests (Chi2, C Pearson and V Kramer), state that: gender and age don't influence the inserted data; having a chronic pain may raise the prospects, but the number of data inserted: maternity/paternity can also gradually/highly increase the interest towards the platform, even as a service for children.

Work is in progress on the Android version mobile application integration within the eHealth platform and on the development of new features to improve the system interactivity.

Acknowledgements. This research has been supported by the Ministerio de Industría, Turismo y Comercio under project TSI-020302-2009-85 and the Ministerio de Ciencia e Innovación of Spain under project TIN2010-20510-C04-01. The authors would like to acknowledge the contribution of the volunteers from Vodafone Spain.

References

1. Tessier, C.: Management and Security of Health Information on Mobile Devices. American Health Information Management Association (AHIMA), Chicago (2010)
2. Huang, X., Jiang, Y., Liu, Z., Kanter, T., Zhang, T.: Privacy for mHealth presence. International Journal of Next-Generation Networks (IJNGN) 2(4), 33–44 (2010)
3. Ren, Y., Werner, R., Boukerche, A.: Monitoring Patients via a Secure and Mobile Healthcare System. Wireless Technologies for EHealthcare. IEEE Wireless Communications, 59–65 (2010)
4. Alasaarela, E., Nemana, R., DeMello, S.: Drivers and challenges of wireless solutions in future healthcare. In: 2009 IEEE International Conference on eHealth, Telemedicine, and Social Medicine, pp. 19–21 (2009)
5. Curioso, W.H., Karras, B.T., Campos, P.E., Buendía, C., Holmes, K.K., Kimball, A.M.: Design and Implementation of Cell-PREVEN: A Real-Time Surveillance System for Adverse Events Using Cell Phones in Peru. In: AMIA Annu. Symp. Proc. 2005, pp. 176–180 (2005)
6. Kumar, A., Chen, J., Paik, M., Subramanian, L.: ELMR: Efficient Lightweight Mobile Records. In: MobiHeld 2009, Barcelona, Spain, pp. 69–70 (2009)
7. Health Level Seven International, http://www.hl7.org
8. GSMWorld: GSMA Mobile Privacy Initiative Discussion Document: Privacy Design Guidelines For Mobile Application Development (2011)
9. Ministerio del Interior. Gobierno de España: Ley Orgánica 15/1999 de Protección de Datos de Carácter Personal (1999),
   http://www.mir.es/SGACAVT/derecho/lo/lo15-1999.html
10. Papadopoulos, H., Pappa, D., Gortzis, L.: Legal & Clinical Risk Assessment Guidelines in Emerging m-Health Systems. In: Itab 2006, Congress Center Du Lac, Ioannina (2006)
11. Rubin, A.D.: Security Considerations for remote electronic voting. Communications of the ACM 45(12), 39–40 (2002)
12. Report of the WHO Global Observatory for eHealth, World Health Organization (2006)
13. Smith, M., Buchanan, W., Thuemmer, C., Hazelhoff Roelfzema, N.: Analysis of Information governance and patient data protection within primary health care. International Journal for Quality in Health Care, 1353–4505 (2010)
14. StatCounter Global Stats. Worldwide Mobile OS, http://gs.statcounter.com/

Activity Recognition Using Smartphones and Wearable Wireless Body Sensor Networks

Ioannis Kouris and Dimitris Koutsouris

National Technical University of Athens,
9, Heroon Polytechniou str., 15773 Zografou, Athens, Greece
{ikouris,dkoutsou}@biomed.ntua.gr

Abstract. This paper explores automated activity recognition using a WIreless Sensor nEtwork (WISE) connected via Bluetooth to a smartphone. Automated activity recognition enables patients, such as diabetics, to keep more accurate logs of their activities (intensity and duration of the activity) and so to prevent short-terms complication, such as hypoglycaemias. We developed a platform records motion using two wearable sensory devices equipped with 3-axis accelerometers, worn on the waist and the shank, and a wireless heart rate monitor. Data are transmitted via Bluetooth to a smartphone, annotated and analyzed to recognize user activity. WISE platform architecture is described along with recognition accuracy performed by multiple classifiers.

Keywords: activity recognition, body sensor network, classification, chronic diseases.

1 Introduction

Technical advances in sensors miniaturization has extend their usage in various daily and medical applications, while most of the sensors tend to be wirelessly interconnected. Their small size enabled the development of wearable devices that can be used to create Body Sensor Networks (BSNs) to transmit medical data in real time. Our implementation targets to provide patients and health professionals with an innovative mechanism to easily and accurately recognize daily physical activities, in terms of type, duration and intensity. A smart and non invasive activity recognition mechanism provides objective quantification of user's locomotion, helping health professionals to have a better overview of their patient's daily habits and allowing them to modify their medication or educate them more effectively. Smartphones have been used to promote patients' self-management [1], although their usage will expand further as they are interconnected with wearable wireless sensors to transmit user's vital signals. Automated activity recognition enables patients suffering from chronic diseases (Diabetes, Parkinson's Disease, etc) to keep more accurate logs on their daily activities and help them maintain more stable blood glucose levels (diabetics), or monitor and classify objectively the severity of their disease induced movements (parkinsonians) [2].

K.S. Nikita et al. (Eds.): MobiHealth 2011, LNICST 83, pp. 32–37, 2012.

2   Related Work

A number of relate studies has been performed using five biaxial accelerometers placed on the right ankle, the left thigh, the waist, the left upper arm and right wrist, trying to distinguish whole body movements and activities involving partial body movement (standing still, folding laundry, brushing teeth, watching TV or reading) [3]. Lester used one board embedded with eight sensors on the shoulder to classify physical activities such as sitting, standing, walking etc [4]. Ravi *et al* proposed the usage of an accelerometer placed near the pelvic region to detect activities, demonstrating the ability of distinguishing daily physical activities with a single accelerometer [5]. Parkka *et al* performed measurements in free living conditions investigating various activities recognition, although, the devices used were quite bulky and heavy [8].

In contrast to the related work, our platform (sensors and smartphone application) targets to provide a versatile, compact, lightweight, low cost and power efficient platform for activity monitoring of patients and healthy individuals using mainly open-source hardware and software, with high accuracy.

3   Platform Description

Our target on the development of the WISE BSN was to provide both health professionals and patients with a comfortable, non invasive platform to monitor vital parameters and motion. In that sense, we developed two similar sensors to acquire the data and transmit them wirelessly to a smartphone via Bluetooth protocol.

Two versions of the WISE sensors were developed, a basic and an extended version. Both are equipped with 3-axis, low noise and low power, accelerometers (ADXL335) to record motion with a sampling rate of 20 samples per second. Sampling rate proved adequate to identify daily activities described below, while keeping power consumption at low levels. The extended sensor version is equipped additionally with Polar Heart Rate Module (RMCM01) to receive and parse data transmitted by a Polar, chest worn, heart rate monitor. The recorded data are transmitted via Bluetooth protocol to a smartphone running Android 2.2 operating system. Data are stored locally for further analysis.

The sensors are equipped with Li-Pol rechargeable batteries, power consumption of the devices is relatively low, and during our measurements we succeeded to acquire data for more than 8 hours in a row without recharging, proving that it can be used for daily monitoring purposes. This, along with the fact that the devices are equipped with a USB port to recharge the battery, seems to make the sensors adequate for medical applications where data should be collected for long periods of time. The size of the sensors is relatively small. Packaging dimensions are 6.8 x 3.0 x 1.0 cm, and the weight is 50 gram (battery included), making them fairly portable and lightweight for everyday usage, as they can be worn under the clothes. The packaging of the sensors consists of a belt-clip to attach the sensor around the waist and a stretching strap that allows the sensor to be attached on the shank. Both sensors are equipped with LEDs to provide their status (power, connected to the mobile, heart rate received etc).

Fig. 1. Graphical representation of the *WISE Body Sensor Network*. One sensor is placed around the shank and the other one is placed around the waist. The heart rate monitor transmits the heart beat to the waist sensor and they both transmit the measurements via Bluetooth to the smartphone. GPS signal is also acquired to determine accurately the speed of the user.

4   Feature Extraction and Classification

4.1   Methodology

During our tests, healthy individuals, male and female, participated in data collection to train the classifiers. The average age of the participants was 28 ± 8 years old. We used both versions of WISE sensors, one place on the waist and one on the shank. Data collection performed following a predefined protocol, which included stair ascending and descending, standing, walking, jogging and running. Average recording time of all the individuals was almost 3 hours of data. The smartphone application developed and used for data annotation allowed the storage of speed and elevation data, acquired by the GPS receiver, embedded in the mobile phone. This information used to automatically discriminate activity intensity.

Data annotation performed by the users, using the prototype data collection mobile application. Upon data collection, raw accelerometer signal and heart rate were divided into epochs. We used half overlapping sliding window to classify the data. Window size selected to be 1 second long, allowing data identification of both rapid and slow movements. Each window includes 40 frames of data (20 samples per second per WISE sensor). The features we used to classify the recordings were mean, variance, energy and heart rate (19 features). Energy calculated as the sum of the absolute values of the FFT components. Data analysis performed using WEKA toolbox, an open source data analysis software developed by the University of Waikato [6]. The analysis performed using all the extracted features, but the primary results showed that by reducing the number of the features from 13 to 7, the results are better. Then, speed data provided by the GPS signal included as an extra feature, resulting to 14 features. The 10-folder cross-validation used to evaluate classification accuracy. All the test cases were put in one dataset and then randomly divided it into 10 equal-sized folders. Each time one folder used the test dataset and the rest the training dataset.

4.2   Classification and Results

The activities identified were: walking (various at speeds), jogging, running, ascending and descending stairs, standing or sitting and driving. Classification performed using Naive Bayes, Ensembles of Nested Dichotomies [10], Multilayer Perceptron with back-propagation (one hidden layer with 12 hidden nodes, learning rate 0.3 and momentum 0.2, 500 epochs sigmoid for activation), Decision Trees implementing C4.5 pruned algorithm, Random Forest of 10 trees considering 4 random features classifiers and Functional Trees [7] [9]. The results of the classification executed on the collected datasets are summarized in Tables 1 and 2.

4.3   Result Analysis

Examining the results shown in Tables 1 and 2, it can be assumed that WISE platform provides relatively accurate results, regarding activity recognition, especially when Functional Trees (FT) are used for data classification. Examining the confusion matrix of the FT classifier (Table 2), it is clear that most of the activities can be discriminated from the other and that the activities that cause misclassification errors are stairs versus walking. This can be explained due to the fact that annotation "walk" is given when there is absence of GPS signal (no speed data), something that also happens upon ascending/descending stairs because most of the times happens inside a building.

Table 1. Classification results using 14 features and various classifiers

Classifiers	14 Features			
	Correctly Classified	Precision	Recall	F-Measure
Naive Bayes	84.52%	85.30%	84.50%	83.70%
END	95.93%	96.10%	95.90%	95.90%
Multilayer Perceptron	95.99%	95.80%	96.00%	95.80%
Functional Trees	**97.28%**	97.20%	97.30%	99.10%
Random Forest	93.89%	94.30%	93.90%	98.90%
Decision Tree C4.5	94.84%	94.90%	94.80%	94.80%

Table 2. Confusion matrix of Functional Trees classifier

	Activity	a	b	c	d	e	f	g	h	i	j
a	Walk	2917	0	1	0	0	0	13	13	98	0
b	Walk ≤ 3 km/h	0	5502	2	0	0	0	3	0	16	0
c	Walk 3 ~ 5 km/h	2	3	5640	0	0	0	6	0	3	1
d	Walk 5 ~ 7 km/h	0	0	0	427	0	0	0	0	0	0
e	Walk 7 ~ 9 km/h	0	0	0	0	40	0	1	0	0	0
f	Running	1	2	2	0	0	536	77	10	1	0
g	Jogging	7	6	20	2	2	60	1318	0	11	0
h	Standing	6	0	0	0	0	6	3	3947	6	0
i	Stairs	152	29	4	0	0	0	35	22	747	0
j	Driving	0	0	1	0	1	0	0	3	1	1561

5   Conclusions – Future Work

Our first field tests of WISE platform prototype provided highly accurate activity recognition rates. The next steps of our research shall focus on testing the platform with a wider range of healthy individuals and people suffering from chronic diseases, so that to examine activity recognition accuracy in miscellaneous groups. We also work on extending the range of the activities recognized, covering a wider range of daily activities, while performing extended analysis on system accuracy under various circumstances (absence of GPS signal, no heart rate monitoring, loss of connection of one of the sensors etc).

References

1. Kouris, I., Mougiakakou, S., Scarnato, L., Iliopoulou, D., Diem, P., Vazeou, A., Koutsouris, D.: Mobile phone technologies and advanced data analysis towards the enhancement of diabetes self-management. Int. J. Electron Healthc. 5, 386–402 (2010)
2. Kirk, A., Mutrie, N., MacIntyre, P., Fisher, M.: Increasing Physical Activity in People With Type 2 Diabetes. Diabetes Care 26, 1186–1192 (2003)
3. Bao, L., Intille, S.S.: Activity Recognition from User-Annotated Acceleration Data. In: Ferscha, A., Mattern, F. (eds.) PERVASIVE 2004. LNCS, vol. 3001, pp. 1–17. Springer, Heidelberg (2004)
4. Lester, J., Choudhury, T., Borriello, G.: A Practical Approach to Recognizing Physical Activities. In: Fishkin, K.P., Schiele, B., Nixon, P., Quigley, A. (eds.) PERVASIVE 2006. LNCS, vol. 3968, pp. 1–16. Springer, Heidelberg (2006)
5. Nishkam, R., Nikhil, D., Preetham, M., Littman, M.L.: Activity Recognition from Accelerometer Data. In: Proceedings of the Seventeenth Conference on Innovative Applications of Artificial Intelligence, pp. 1541–1546 (2005)
6. Cunningham, S.J., Denize, C.P.: A Tool for Model Generation and Knowledge Acquisition. In: Proceedings International Workshop on Artificial Intelligence and Statistics, pp. 213–222 (2003)

7. Bauer, E., Kohavi, R.: An Empirical Comparison of Voting Classification Algorithms: Bagging, Boosting, and Variants. Machine Learning 36, 105–139 (1999)

8. Parkka, J., Ermes, M., Korpipaa, P., Mantyjarvi, J., Peltola, J., Korhonen, I.: Activity classification using realistic data from wearable sensors. IEEE Transactions on Information Technology in Biomedicine 10, 119–128 (2006)

9. Gama, J.: Functional Trees. Machine Learning 55, 219–250 (2004)

10. Dong, L., Frank, E., Kramer, S.: Ensembles of Balanced Nested Dichotomies for Multi-class Problems. In: Jorge, A.M., Torgo, L., Brazdil, P., Camacho, R., Gama, J. (eds.) PKDD 2005. LNCS (LNAI), vol. 3721, pp. 84–95. Springer, Heidelberg (2005)

Decision Support for the Remote Management
of Chronic Patients

Sara Colantonio[1], Giuseppe De Pietro[2], Massimo Esposito[2],
Alberto Machì[2], Massimo Martinelli[1], and Ovidio Salvetti[1]

[1] Institute of Information Science and Tecnologies, ISTI-CNR
Via G. Moruzzi, 1-56124, Pisa, Italy
{sara.colantonio,massimo.martinelli,ovidio.salvetti}@isti.cnr.it
[2] Institute for High Performance Computing and Networking, ICAR-CNR
Via P. Castellino, 111-80131, Napoli, Italy
{giuseppe.depietro,massimo.esposito,alberto.machi}@icar.cnr.it

Abstract. Chronic diseases may cause major limitations in patients' daily living due to acute or deterioration events, which can happen more or less frequently and, often, cannot be totally relieved, causing a worsening of patients' conditions. In the last years, a strong effort is being spent in the development of intelligent ICT applications for patients' telemonitoring, aimed at maximizing the quality of life of chronic patients by means of a regular collection of information about their status and actions in a long-stay setting. In this paper, a knowledge based decision support system is presented, which is aimed at aiding clinical professionals in managing chronic patients on a daily basis, by assessing their current status, helping face their worsening conditions, and preventing their exacerbation events.

Keywords: Decision Support, Knowledge Formalization, Ontology.

1   Introduction

Chronic diseases are generally characterized by complex and difficult prognosis and treatment, and experience, more or less frequently, exacerbation events that require patients' hospitalization. A key role in the management of chronic diseases is recently being assumed by smart systems specifically devised to constantly monitor patients for perceiving changes of their status and anticipating or detecting the occurrence of acute events to be treated. In particular, an effective and profitable solution for supporting the decisional processes cannot be limited to the gathering of patients' data via dedicated sensors, but it should also offer advanced facilities for analyzing such data in order to present only the most relevant information to clinicians without overwhelming their clinical activity.

According to these considerations, this paper proposes a knowledge-based *Clinical Decision Support System* (CDSS) for managing chronic patients by interpreting data acquired through a sensor infrastructure deployed in patients' normal life environment. This system combines acquired data with patients' clinical information, issues possible alarms and supplies motivated suggestions to clinicians. The application scenario was

K.S. Nikita et al. (Eds.): MobiHealth 2011, LNICST 83, pp. 38–45, 2012.
© Institute for Computer Sciences, Social Informatics and Telecommunications Engineering 2012

represented by two specific diseases, namely Chronic Obstructive Pulmonary Disease (COPD) and Chronic Kidney Disease (CKD), even if the approach can be applied, more in general, to any chronic disease.

The system was developed within the EU IST Project CHRONIOUS which is aimed at defining a generic platform schema for health status monitoring, addressed to and specialized for people suffering from chronic diseases [1]. In the following, the adopted strategy is detailed, presenting the CDSS design and describing implementation details and results achieved.

2   Clinical Decision Support System Design

The CDSS was designed to support the remote management of chronic patients by interpreting data collected via a sensing infrastructure, deployed to acquire a number of records about patients' disease signs, behavior, activity, and contextual situations. Such records are meant to assess, on a daily basis, patients' situation and, hence, follow up their response to therapy on a long period, verify that their life style is mostly correct, and identify the onset of possible disease exacerbations.

More precisely, a platform of services was designed to acquire, store and interpret these sensor data and alert clinicians whenever a worrying situation is detected. The platform comprises a sensing infrastructure which collects the following patient's data:

- disease signs and symptoms, collected by a sensorized vest able to record patients' (i) electrocardiographic and (ii) respiratory activities, (ii) arterial oxygen saturation, (iii) skin temperature, (iv) cough and snoring, (v) motion activity and fall. Commercially available devices are also employed for measuring (vi) body weight, (vii) blood pressure and (viii) blood glucose;
- contextual data, collected by an environmental device installed in patients' living room for acquiring information about (i) ambient light, the presence of (ii) carbon monoxide, (iii) volatile organic compound and (iv) air particle;
- patient's inputted data, collected through questionnaires proposed on a touch-screen workstation for acquiring information pertinent to (i) patients' lifestyle, (ii) food and (iii) drug intake, (iv) psychological conditions.

All these data are collected regularly and stored into the platform repository which is also used to stock patient's data collected during clinical visits (through external communications with clinical sites, according to HL7 medical standard).

For the interpretation of these data, in order to achieve a notable level of reliability and flexibility, two levels of intelligence were conceived: a first level of intelligence deployed on a *Personal Device Assistant* (PDA) is in charge of detecting changes of the patients' status by applying simple rules, only working on the acquired parameters. Whenever a significant alterations is detected, clinical staff is alerted and, contextually, a second level of intelligence, i.e. the CDSS, is invoked, since able to correlate a wider set of information pertaining the patient and provide more reliable and accurate answers.

Figure 1 summarized the main components of the whole monitoring platform (for more details about the entire platform, please refer to [2]).

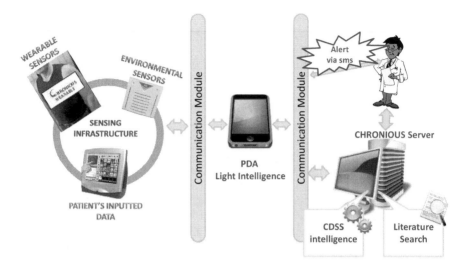

Fig. 1. The remote monitoring infrastructure: the Sensing Infrastructure with the different types of collected data and the two levels of intelligence for interpreting patient's collected data

As it often happens in the development of clinical support systems, a knowledge based approach was followed for developing the CDSS. However, differently from other clinical problems, the remote management of chronic patients is an emerging procedure, not yet well assessed. The clinical guidelines currently available provide clinicians with general instructions for disease diagnosis and treatment within the practice of clinical environment (see, e.g., COPD guidelines GOLD [3]). They do not contain any information about which parameters should be acquired for monitoring patients in their daily life, how these should be interpreted and how an exacerbation event could be identified.

Since some experiments of patients' telemonitoring are being carried out [4], the most viable solution appeared the elicitation of the pertinent knowledge directly from clinicians involved in such experimental activities. The CDSS Knowledge Base (KB) was then built by formalizing the knowledge brought forth by clinicians with specific expertise in telemedicine programs from the Fondazione Salvatore Maugeri for COPD, and from the University of Milan and the San Carlo Hospital in Milan for CKD.

The knowledge elicitation and formalization process is detailed in the next section; to make this process easier, a scenario based approach was adopted so as to better identify the CDSS interventions and three main scenarios were identified for invoking the data interpretation:

- an *Alarm Checking* scenario, for the assessment of patients' status after an alarm issued by the PDA due to a possible exacerbation detected;
- a *Home Monitoring* scenario, for a periodic assessment of patients' status (once a day), even without any alerting exacerbation;
- a *Clinical Assessment* scenario, for the evaluation of patients' status after a clinical visits.

In all these cases, according to the knowledge modeled into its KB, the CDSS first analyzes the received data and correlates them with historical patient's data and, then, alerts clinicians when an acute event happens, also providing suggestions about actions to be performed.

2.1   Knowledge Formalization

For the knowledge elicitation process, several meetings were carried out with clinicians for agreeing how all the information gathered inside the monitoring platform are correlated and can be interpreted to identify worrying conditions. A set of evidence-based statements resulted from this process; each of them relates a condition about the possible values of the acquired sensor data with a conclusion that can be drawn on patient's status. Knowledge contained into the available clinical guidelines, among them the GOLD [3] for COPD and the NKF-KDOQI-02 [5] for CKD, was also considered, especially for interpreting and relating data acquired during clinical visits.

The formalism selected for encoding all relevant knowledge was selected considering:

- the "*condition-conclusion*" form of the knowledge elicited from clinicians;
- the characteristic of such elicited knowledge to resemble not standard clinical procedures, but more a cross-analysis of patients' vital signs parameters aimed at assessing or predicting the occurrence of an acute event;
- the peculiarity of existing methods for modeling clinical guidelines which are, so far, only modeling methodologies, without a really working engine for running the encoded guidelines;
- the possibility to develop an application for allowing clinicians to upgrade the encoded knowledge.

These considerations fostered the selection of an encoding formalism based on *production rules*, i.e. a set of conditional statements expressed in form of "*if antecedents **then** consequent*".

In order to make the KB interoperable and define a well assessed terminology for the KB, the production rules were defined on the top of an ontology. This contains all the information pertaining a chronic patient (i.e. medical history, patient general information, laboratory assays, patient monitoring measurements or environmental measurements gathered at the patient's home, questionnaires about mental problems or symptoms) and also the suggestions generated by the CDSS to be reported to the clinicians. The combination of ontology and rules consists in expressing rule antecedents and consequents by using concepts, properties and relations of the ontology. In particular, for making the rule writing process simpler, all the relevant parameters to be used were defined as concept properties (i.e., the *datatype* properties of the *Ontology Web Language* – OWL [6]). In deed, the use of concept relations (i.e., the *object* properties of OWL) would have required the specification of complex and scarcely intuitive chains of concepts and relations in the rule antecedents and consequents.

The ontology concepts were characterized by a well-specified semantics through axioms and restrictions, whereas properties have been defined in terms of domain, i.e. the admissible subject concept, and range, i.e. the admissible object concept or a value type. In more detail, concepts and properties was organized in two groups, named *Patient Situation* and *CDSS Suggestion*. This kept the ontology maintenance simpler and assured ontology easily reusability and extensibility. Both these groups of concepts were specialized by sub-concepts and/or properties.

In particular, the group named *Patient Situation* models a sort of synthetic clinical summary of chronic patient and includes all the information required to define the clinical situation of the patient, while being constantly and remotely monitored at home. Such information was further categorized under the main concept *PatientSituation* and placed into five sub-groups, named respectively *Clinical Information*, *Medical History*, *Monitoring Information*, *Patient General Information* and *Therapy Prescription*. These were further specialized as shown in Figure 2.a.

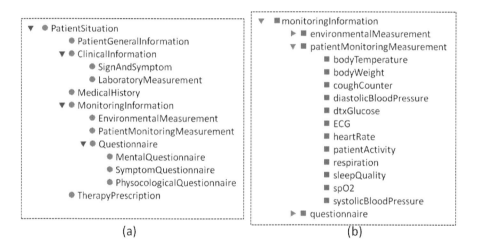

Fig. 2. (a) the taxonomy of concepts; (b) a fragment of the taxonomy of properties, in particular the properties of the *PatientMonitoringMeasurement* (when having the same name, concepts and properties are differentiated by uppercase and lowercase initial letter)

This group of concepts were specified by defining a number of properties, which correspond to each piece of information (i.e., patient's monitoring data, clinical visit data, patient's medical history, and so on) relevant for the CDSS KB. As an example, Figure 2.b reports all the properties associated to the concept *MonitoringInformation*.

The group *CDSS Suggestion* specifies the results of the inferences generated by the CDSS in terms of suggestions to be reported to the clinicians. Such suggestions were expressed in terms of alerts, i.e. messages with a different severity, varying in accordance with the current patient condition, that can require or not the attention of a clinical operator. In more detail, the group contains the concept *CDSS_Suggestion* and a set of properties which describe the CDSS outputs. More precisely, the defined properties are:

- *alertSeverity*, which indicates the severity the suggestion produced by CDSS has to be reported with, in order to appropriately alert a specific clinical operator, e.g. a nurse, a clinician or the emergency department of an hospital. Such a severity was formalized in terms of a color which can assume values in this set *{white, green, yellow, red}*;
- *patientStage*, which specifies the severity inferred by CDSS for each chronic disease (i.e. COPD and CKD). Such a severity is expressed in terms of stages, assuming values between 1 and 4 for COPD and 1 and 5 for CKD, in accordance with the guidelines for the COPD and CKD severity evaluation [3, 5];
- *patientCondition*, which indicates a variation of the health status of the patient with respect to the morbidity he/she is affected by. For example, it indicates a *"renal function worsening"* when the morbidity stage in a CKD patient is changed from 3 to 4, or an *"invariant renal function"* whenever the morbidity stage in a CKD patient remains stable.
- *guideline*, which describes in natural language the specific clinical guideline applied by CDSS to infer the corresponding suggestion;
- *suggestedAction*, which expresses the action which has to be performed in response to a determined suggestion generated by CDSS. Such an action should be reported either to the patient, when his/her condition is not critical, or to a clinical operator, whenever the situation is very critical for the patient.

Based on this ontology, production rules were devised for the three reported scenarios, representing, this way, actions and suggestions to be generated (i.e., the so called *procedural knowledge*).

Each rule is composed of one or more antecedents, expressed in terms of ontology properties concatenated by logical conjunctive operators, which can be evaluated to be either true or false. Disjunction was not supported. Moreover, each rule has exactly one conclusion, which can be an assignment to some parameters or a CDSS suggestion. Some simple rules, organized according to the reported structure, are reported in Table 1.

Table 1. Three examples of elicited rules

Antecedent	Consequent		
Description	Suggested Action	Patient's Condition	Alert Severity
systolicBloodPressure >140mmHg and spO2<95%	Alert medical doctor	Fluid overload	Red
diastolicBloodPressure >90mmHg and spO2<95%	Alert medical doctor	Fluid overload	Red
systolicBloodPressure<100 and heartRate>115 and nauseaOrVomiting='true'	Doublecheck the alarm and alert immediately emergency medical service	Hypo-Volemia	Red

3   Implementation and Results

The CDSS was modularly conceived and realized in order to be straightforwardly connected to any platform by using a service oriented approach. It was indeed structured as a set of decisional services which are called on demand when specific events occur, in accordance with the set of scenarios introduced in the previous section. These were implemented as two Web Services, named COPD_Decisional_Service and CKD_Decisional_Service, realized respectively for COPD and CKD diseases. Each service was delineated from a functional perspective in terms of its operations, where each operation is coarse-grained and models how the CDSS works for one of the identified scenarios.

For what concerns the KB, the ontology was developed using OWL [6] and currently, consists of 28 concepts and 860 properties. The rules were formalized in the Jena rule language [7], structured in scenarios and divided between the two pathologies: totally the base of rules contains 435 rules for CKD and 273 rules for COPD. Results provided by the CDSS consist in an advice about the status of the patient and a suggestion about the action to be undertaken for managing the situation. As an example, Figure 3 shows the results generated by CKD_Decisional_Service in terms of suggestions and explanations reported in response to abnormal values due to a condition of HypoVolemia. This is suitably displayed on a clinicians' graphical user interface developed into the platform of services, not pertaining only the work carried out for the CDSS implementation [2].

The system was tested for evaluating the performance of the Web Services; in particular both functional and load tests were performed by using SOAP-UI tool [7]. The system has been released and a validation phase is planned to start at the end of year.

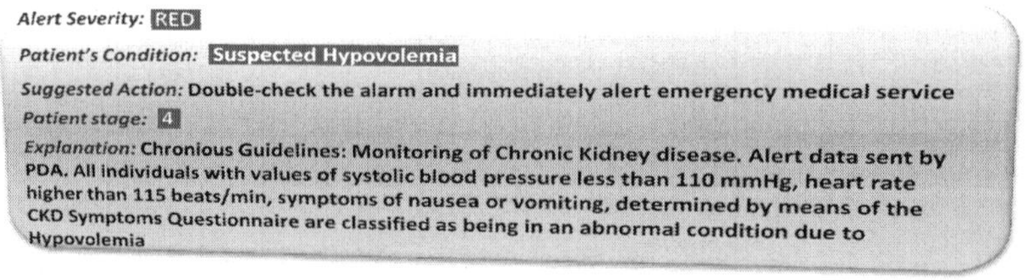

Alert Severity: RED

Patient's Condition: Suspected Hypovolemia

Suggested Action: Double-check the alarm and immediately alert emergency medical service

Patient stage: 4

Explanation: Chronious Guidelines: Monitoring of Chronic Kidney disease. Alert data sent by PDA. All individuals with values of systolic blood pressure less than 110 mmHg, heart rate higher than 115 beats/min, symptoms of nausea or vomiting, determined by means of the CKD Symptoms Questionnaire are classified as being in an abnormal condition due to Hypovolemia

Fig. 3. Example of information suggested by the Clinical Decision Support System

4   Conclusions

A key role in the management of chronic diseases is covered by smart systems specifically devised to constantly monitor patients for perceiving changes of their status and anticipating or detecting the occurring of acute events to be treated. A great added value to these systems is the development of intelligent applications able to automatically interpret the acquired data.

In this frame, a Clinical Decision Support System has been presented in this paper, which is aimed at aiding clinical professionals in managing chronic patients on a daily basis, by analyzing sensor data for assessing patients' current status and presenting a more personalized advice and feedback to clinicians. The CDSS was developed by encoding the pertinent knowledge elicited from clinicians with a specific experience in patients' telemonitoring. A formalism based on one ontology and a base of rules was selected since the most suitable in terms of the form of elicited knowledge and the purposes of the system. The system was developed according to a scenario-based approach, implementing each identified scenario as a Web Service operation. This assures the interoperability of the system and the possibility to plug it into any monitoring platform of the same kind.

Currently, the CDSS has successfully passed the functional and load tests applied to the developed web services. A clinical testing process is planned to start at the end of the year within the EU project CHRONIOUS.

Acknowledgments. This work has been partly funded by the EC IST Project FP7-ICT-2007–1–216461 CHRONIOUS: http://www.chronious.eu/. The authors wish to acknowledge their gratitude and appreciation to all the project partners for the development of the ideas and concepts presented in this paper.

References

1. Rosso, R., Munaro, G., Salvetti, O., Colantonio, S., Ciancitto, F.: CHRONIOUS: an open, ubiquitous and adaptive chronic disease management platform for Chronic Obstructive Pulmonary Disease (COPD), Chronic Kidney Disease (CKD) and Renal Insufficiency. In: Armentano, R.L., Monzon, J.E., Hudson, D., Patton, J.L. (eds.) Proceedings of Annual Int. Conf. of the IEEE, EMBC 2010 - Engineering in Medicine and Biology, Buenos Aires, Argentina, August 31-September 4, pp. 6850–6853. IEEE (2010)
2. Lawo, M., Papadopoulos, A., Ciancitto, F., Dellaca, R.L., Munaro, G., Rosso, R.: An open, ubiquitous and adaptive chronic disease management platform. In: 2009 6th International Workshop on Wearable Micro and Nano Technologies for Personalized Health (pHealth), June 24-26, pp. 69–71 (2009)
3. Gold 2008. Global initiative for chronic obstructive lung disease. A collaborative project of the National Heart, Lung, and Blood Institute, National Institutes of Health, and the World Health Organization. Bethesda (2008), http://www.goldcopd.org
4. Vitacca, M., Bianchi, L., Guerra, A., Fracchia, C., Spanevello, A., Balbi, B., Scalvini, S.: Tele-assistance in chronic respiratory failure patients: a randomised clinical trial. Eur. Respir. J. 33, 411–418 (2009)
5. NKF-KDOQI-02 Clinical Practice Guidelines for Chronic Kidney Disease: Evaluation, Classification and Stratification (2002),
   http://www.kidney.org/professionals/kdoqi/
   guidelines_ckd/toc.htm
6. OWL (2009), http://www.w3.org/TR/owl-features/
7. Carroll, J.J., Dickinson, I., Dollin, C., Reynolds, D., Seaborne, A., Wilkinson, K.: Jena: implementing the semantic web recommendations. In: Proceedings of the 13th Int.l World Wide Web Conference on Alternate Track, pp. 74–83. ACM, New York (2004)
8. (2011), http://www.soapui.org/

Risk Assessment Models for Diabetes Complications: A Survey of Available Online Tools

Lefteris Koumakis[1], Franco Chiarugi[1], Vincenzo Lagani[1],
Angelina Kouroubali[1], and Ioannis Tsamardinos[1,2]

[1] Institute of Computer Science, Foundation for Research and Technology – Hellas (FORTH),
P.O. Box 1385, 100 N. Plastira St, Heraklion, 70013, Crete, Greece
[2] Department of Computer Science, University of Crete, 71409 Heraklion, Greece
{koumakis,chiarugi,vlagani,kouroub,tsamard}@ics.forth.gr

Abstract. Predictions, risk assessment and risk profiling are among the various decision support techniques that medical professionals increasingly rely on to provide early diagnose in patients with elevated risks and to slow down the rapid increase in prevalence of chronic diseases. The introduction of risk assessment tools and applications for chronic diseases in large scale longitudinal clinical studies, presents many challenges due to the nature of the data (studies last around a decade) and the complexity of the models. In this paper, we give an overview of research work on risk assessment tools and applications for diabetes complications. We also introduce the REACTION project and its vision in the field of risk assessment for diabetes complications.

1 Introduction

Risk factors for diabetes complications have been intensively studied during the last decades, and these studies greatly improved the current scientific knowledge about the biological processes underlying diabetes. Risk factors have been commonly used in risk assessment models for the prediction of diabetes complications. Risk assessment models are the backbone of risk assessment tools used in the clinical practice. These tools as parts of clinical/medical applications are able to stratify diabetes patients according to their probability of developing complications or experiencing adverse events.

A risk assessment tool is based on one or more models which could be any type of algorithm or mathematical formula (e.g., a set of rules, a decision tree or a weighted sum) for assessing the overall statistical probability of certain situations to occur in the future. Medical risk assessment may provide probabilistic statements as the likelihood that certain complications may occur given the present and historic health status.

Several risk assessment models for diabetes complications have been proposed in the literature. In the overall clinical management of people with diabetes special attention has to be dedicated to the prevention of short-term as well as long-term complications. Even though "short" and "long term" are commonly used terms in the context of diabetes complications, there is not a clear and universally accepted

K.S. Nikita et al. (Eds.): MobiHealth 2011, LNICST 83, pp. 46–53, 2012.
© Institute for Computer Sciences, Social Informatics and Telecommunications Engineering 2012

distinction between the two; here thereafter, we indicate as short term complication any pathological process or event related to diabetes that is expected to arise within weeks or few months, while long term complications may arise even after several years. In the context of the REACTION project (see Section 5), our research group work exclusively on long term risk assessment models; thus, we consider short term models (especially insulin management tools) out of the scope of this paper.

The most common predictive risk assessment models for diabetes complications are not able to deal with all the major complications, but are mainly focused on cardiovascular diseases, coronary heart disease and diabetic retinopathy (long-term complications).

The paper is organized as follows: the major clinical studies for diabetes and its complications have been reviewed in Section 2, while in Section 3 the risk assessment tools and applications for diabetes complications have been examined. Advantages and limits of present tools are discussed in Section 4, while Section 5 presents the EU funded project REACTION, an ICT based initiative that will develop and integrate new risk assessment models for diabetes complications.

2   Major Clinical Studies for Diabetes

Long term risk assessment tools and applications are usually built upon data collected during large scale, longitudinal clinical studies. Such type of studies typically last around a decade, involve thousands of patients in numerous health centres, and measure different aspects of patient's clinical/medical profile. Thus, it is not surprising that the data collected in each study can be employed for deriving multiple risk assessment models, differing from each other for predicted outcomes, involved parameters or analytical techniques.

Some of the well known clinical studies related to diabetes complications which will be discussed here, are DCCT/EDIC [1], Qrisk [2] and UKPDS [3].

2.1   DCCT/EDIC Study

A study of long term risk assessment related to diabetes and complications is the Diabetes Control and Complications Trial (DCCT). DCCT [1] is a landmark medical study conducted by the United States National Institute of Diabetes and Digestive and Kidney Diseases (NIDDK). The DCCT involved 1,441 volunteers, ages 13 to 39, with type 1 diabetes and 29 medical centres in the United States and Canada. DCCT is a multicenter, randomized clinical trial designed to compare intensive with conventional diabetes therapy with regard to their effects on the development and progression of the early vascular and neurologic complications. Volunteers had to have had diabetes for at least 1 year but no longer than 15 years. The study compared the effects of standard control of blood glucose versus intensive control on the complications of diabetes. Intensive control meant keeping glycated haemoglobin (HbA1c) levels as close as possible to the normal value of 6 percent or less.

A new study started after the DCCT, called Epidemiology of Diabetes Interventions and Complications (EDIC). EDIC [4] is a follow up study on 90% of the participants from DCCT that looked into cardiovascular disease and the effects of intensive control on quality of life and cost effectiveness.

2.2   Qrisk Study

Another large study in the field is the QRisk [2]. Qrisk aims to develop a cardiovascular disease risk algorithm which will provide accurate estimates of cardiovascular risk in patients from different ethnic groups in England and Wales and to compare its performance with the modified version of Framingham score recommended by the National Institute for Health and Clinical Excellence (NICE). QRisk study is based on a cohort of 2.3 million patients aged 35-74 with 140,000 cardiovascular events with and without diabetes. Overall population (derivation and validation cohorts) comprised 2.22 million people who were white or whose ethnic group was not recorded, 22,013 south Asian, 11,595 black African, 10,402 black Caribbean, and 19,792 from Chinese, other Asian or other ethnic groups.

Although Qrisk is not a study focused on diabetes (the initial cohort consisted of 0.61 million patients who were free of diabetes and existing cardiovascular disease) the consequently developed risk assessment Qrisk2 calculator takes as risk factor the presence of diabetes.

2.3   UKPDS Study

The United Kingdom Prospective Diabetes Study [3] (UKPDS) is a landmark randomized controlled trial which showed that both intensive treatment of blood glucose and of blood pressure in diabetes can lower the risk of diabetes-related complications in individuals newly diagnosed with Type 2 diabetes. The UKPDS cohort consists of 5,102 patients, followed for a median of 10.7 years. Between 1977 and 1991, general practitioners in the catchment areas of 23 participating UKPDS hospitals were asked to refer all patients aged 25 to 65 years presenting with newly diagnosed diabetes. Patients in the UKPDS had biochemical measurements, including HbA1c, blood pressures, and lipid and lipoprotein fractions, recorded at entry to the study, at randomization in the study after a three-month period of dietary therapy, and each year subsequently.

3   Risk Assessment Applications for Diabetes Complications

To our knowledge, almost all of the risk assessment applications for diabetes complications provide risk profiles in real time. Most of these tools compute complications related to cardiovascular diseases, but it is also possible to find calculators for kidney failure, eye problems and foot problems. For short term diabetes complications there is no tool or software available. The risk assessment applications for diabetes complications resulting from this research and discussed in this section are: Diabetes PHD, Qrisk2, Framingham, Risk Score calculator, UKPDS Risk Engine and HARP.

3.1   Diabetes PHD

Diabetes PHD [5] is an online tool that calculates the risk for diabetes and complications associated with it. Diabetes PHD is based on a math model called Archimedes [6] that represents the anatomy, physiology and pathology related to diabetes and its complications. In the virtual world of Archimedes, every element corresponds to the same elements in the real world, one-to-one. The model is populated with thousands of simulated people, all of whom are living simulated lives, sometimes developing simulated diabetes and/or its complications.

The user has to set basic information (age, weight, gender, etc.), basic family health history, blood pressure, cholesterol levels, HbA1c, Fasting Plasma Glucose, health history and current medications related to diabetes.

The Diabetes PHD calculator gives percentages of risk for diabetes (if the person doesn't have already) and 5 complications of diabetes (type 1 and 2). Probabilities are for the next 30 years and are displayed as graphs. The report also provides an accurate picture of how the individual can alter those risks with lifestyle modification actions such as losing weight, quitting smoking, or reducing blood pressure or cholesterol levels.

3.2   Qrisk2 Cardiovascular Risk Score

Qrisk2 [7] is a cardiovascular disease risk calculator which is designed to identify people at high risk of developing CVD and need to be recalled and assessed in more detail in order to reduce their risk of developing CVD. The Qrisk2 score estimates the risk of a person developing CVD over the next 10 years.

Qrisk is using a cohort of 1.28 million patients without evidence of diabetes mellitus or cardiovascular disease. Patients were followed up for >5 years, looking for the first development of cardiovascular disease as an endpoint. Qrisk2 has been specifically developed by doctors and academics for use in the United Kingdom.

The Qrisk2 calculator uses the following parameters (if known - missing values are calculated by a complex averaging procedure called multiple imputations): patient age (35-74), patient gender, current smoker, family history, existing treatment with blood pressure agent, postcode related Townsend score (an area measure of deprivation), body mass index, systolic blood pressure, total and HDL cholesterol, self-assigned ethnicity, rheumatoid arthritis, chronic kidney disease, atrial fibrillation.

The output of the calculator is a score (percentage) of probability to have cardiovascular disease the next 10 year. The calculator also gives the score of a typical person (good reference) with the same age, sex, and ethnicity and a relative risk (patient's risk divided by the typical risk).

3.3   Framingham Heart Study

The Framingham Heart Study [8] is a risk predictor for the risk of various cardiovascular disease outcomes in different time horizons and it is available as score sheets or direct risk functions. In the Framingham Heart study there are six groups of participants: Original Cohort, Offspring Cohort, Third Generation Cohort, New Offspring Spouse Cohort, Omni Generation 1 Cohort and Omni Generation 2 Cohort.

The inputs for a risk prediction about stroke were defined as follows: systolic blood pressure, diabetes, history of diabetes, smoking and if yes how many cigarettes per day, CVD, history of myocardial infarction, angina pectoris, coronary insufficiency, intermittent claudication or congestive heart failure, atrial fibrillation, history of atrial fibrillation, left ventricular hypertrophy on electrocardiogram.

The output of the model is probability of stroke within 10 years and heart disease for 10 or 30 years.

3.4   Risk Score Calculator

The risk score calculator [9] is a score which is derived from data on 47,088 men and women who participated in eight randomized controlled trials of drug treatment for high blood pressure in Europe and North America. Average follow-up was over 5 years and 1,639 patients died of cardiovascular disease (1,031 coronary heart disease, 371 stroke and 237 other). Two of these trials were on 21,750 British subjects and their data have been used to give a country-specific probability of cardiovascular death linked to a person's risk score.

Inputs for the risk score calculator are: Age, Sex, Current cigarette smoker, Systolic blood pressure, Total cholesterol, Creatine, Height, Has diabetes, Has left ventricular hypertrophy, Had myocardial infarction and Had stroke.

The output consists in (a) the risk of death due to a cardiovascular cause in the next 5 years, (b) a graph with the risk of cardiovascular mortality, and (c) a graph with the distribution of risk scores for the specific sex.

3.5   UKPDS Risk Engine

The UKPDS Risk Engine [10] is a type 2 diabetes specific risk calculator based on 53,000 patients' years of data from the UK Prospective Diabetes Study.

The UKPDS risk engine calculates complications of type 2 diabetes based on age, sex, ethnicity, smoking status, presence or absence of atrial fibrillation, levels of HbA1c, systolic blood pressure, total cholesterol and HDL cholesterol.

The UKPDS Risk Engine provides risk estimates and 95% confidence intervals, in individuals with type 2 diabetes not known to have heart disease, for:

- Non-fatal and fatal coronary heart disease
- Non-fatal and fatal stroke

The UKPDS Risk Engine is intended primarily for use by health care professionals to assist in the management of people with type 2 diabetes.

3.6   HARP Risk Calculator

HARP risk calculator [11] is based on the extensive work done with the Westbay diabetes project. The calculator aims to determine the risk of people with chronic or complex care needs presenting to hospital for treatment in the following 12 months.. The risk screen is based on presenting clinical symptoms, service access profile, self-management, and psycho-social issues. This screening categorizes a person into one of four risk categories: low, medium, high and urgent.

HARP risk calculator developed to measure predictable level of risk for diabetes related complications in the next 12 months.

4   Discussion

During the last decades diabetes and its complications have been intensively studied and clinical trials have been designed in order to understand and handle better the therapy of diabetes and its complications. Table 1 gives a summary of the available risk assessment tools and applications for diabetes complications with their major features. There are several remarks that have to be done related to the use of these risk assessment tools. As we can see from table 1 all the risk assessment applications except diabetes PHD are based on big studies such as UKPDS, Qrisk or DCCT. These studies measure and store a large number of risk factors for a long period for every patient. DCCT and EDIC studies were conducted from 1983 to 1993. UKPDS study ran from 1977 to 1991 in 23 UK clinical sites and Qrisk study started in 1995. Having in mind the evolution in clinical sensors and in electronic health care records someone could claim that the data of these studies is obsolete. Measurements for specific risk factors that could not be measured twenty years ago now are available with specialized sensors and sensor accuracy and reliability has been significantly increased. Furthermore, the normal values for various risk factors, such as blood pressure, have been revised during the years.

Table 1. A summary of the available risk assessment applications (web based or standalone) for diabetes complications

	Diabetes PHD	Qrisk 2	Framingham Heart Study	Risk Score calculator	UKPDS Risk Engine	HARP Risk Calculator
Study/Training of model	Archimedes simulator	2.3 million patients	Framingham Study cohort	47,088 patients from Europe(UK) and North America.	UK Prospective Diabetes Study	Westbay diabetes project
Diabetes Type	Type I & II	with & without diabetes	with & without diabetes		Type II	
Region	No	UK	USA	UK & USA	UK	No
Type of application	web page	web page	Online calculators and Excel spreadsheets.	web page	Standalone application	Forms with a final score
Num. of attributes needed	more than 20	16	depends on the disease	11	10	40
Prediction	Diabetes, Heart attack, stroke, kidney failure eye problems, foot problems	diabetes, heart disease, stroke for the next 10 years	stroke, heart diseases.	country-specific probability of cardiovascular death	heart disease, coronary heart disease, stroke	risk for diabetes related complications in the next 12 months.

Another limitation is the close relation of diabetes and its complications with the geographical location of the patient. Environmental characteristics such as special diet and lifestyle characteristics such as socioeconomic status of the patients' population are strongly associated with diabetes mellitus [12]. For that reason Qrisk2 and UKPDS risk engine are risk calculators focused on the United Kingdom population.

Diabetes PHD seem to overcome theses limitations in the sense that it is based on Archimedes model which is a large-scale simulation model of human physiology and health care systems. Of course the use of virtual data as cohort is a limitation itself.

5   The REACTION Project

The REACTION project aims to research and develop an intelligent service platform that can provide professional, remote monitoring and therapy management to diabetes patients in different healthcare regimes across Europe. A range of REACTION applications will be developed mainly targeting insulin-dependent type 1 patients.

Part of this effort consists in reviewing, developing and implementing tools able to provide long term risk assessment evaluations based on patient's current health state and history. The tools will allow for integration between instantaneously measured data from sensors, historic data from EPR, statistical data from stratification studies, statistical database and evidence based case management repositories.

A continuous update and calibration with independent off-line measurements of relevant biomarkers and diagnostic tools is possible. For the developed tools, emphasis will be put on usability and clearness of personalized feedback. This will allow easy and readily available risk assessment tool to both clinicians and patients, which can be customized according to user needs and preferences.

Last but not least, REACTION decision support services will undergo a preliminary validation phase within the clinical sites involved into the project. The relatively limited project's duration does not allow an exact evaluation of the accuracy of the predictive models: assessing the validity of predictions related to events that can arise within a time horizon of thirty years requires ad hoc longitudinal studies, that are out of the scope of the REACTION project. However, our preliminary validation phase will provide precious information about the impact that decision support tools can have on the daily clinical management of diabetes patients.

6   Conclusion

The present work introduced and discussed the main clinical studies for diabetes risk factors. Moreover, the risk assessment tools developed on the basis of the data collected during these studies were reviewed as well. Finally, we described a new EU project, REACTION, that will develop and integrate a series of risk assessment models for diabetes complications.

Acknowledgment. This work is supported by the European Commission's Seventh Framework Program in the area of Personal Health Systems under Grant Agreement no. 248590 (REACTION FP7-IP-No 248590).

References

1. The Diabetes Control and Complications Trial Research Group: The effect of intensive treatment of diabetes on the development and progression of long-term complications in insulin-dependent diabetes mellitus. N. Engl. J. Med. 329(14), 977–986 (1993)
2. Hippisley-Cox, J., Coupland, C., Vinogradova, Y., et al.: Derivation and validation of QRISK, a new cardiovascular disease risk score for the United Kingdom: prospective open cohort study. BMJ, June 27 (2007)
3. UKPDS Group: UK Prospective Diabetes Study VIII: Study design, progress and performance. Diabetologia 34, 877–890 (1991)
4. Epidemiology of Diabetes Interventions and Complications (EDIC): design, implementation, and preliminary results of a long-term follow-up of the Diabetes Control and Complications Trial cohort. Diabetes Care 22, 99–111 (1999)
5. Eddy, D.M., Schlessinger, L.: Validation of the Archimedes Diabetes Model. Diabetes Care 26, 3102–3110 (2003)
6. Schlessinger, L., Eddy, D.M.: Archimedes: a new model for simulating health care systems - the mathematical formulation. Journal of Biomedical Informatics 35, 37–50 (2002)
7. Hippisley-Cox, J., Coupland, C., Vinogradova, Y., Robson, J., Minhas, R., Sheikh, A., Brindle, P.: Predicting cardiovascular risk in England and Wales: prospective derivation and validation of QRISK2. BMJ 336, 1475 (2008) doi: 10.1136/bmj.39609.449676.25
8. D'Agostino, R.B., Wolf, P.A., Belanger, A.J., Kannel, W.B.: Stroke Risk Profile: Adjustment for Antihypertensive Medication, Stroke (1994)
9. Pocock, S., McCormack, V., Gueyffier, F., Boutitie, F., Fagard, R., Boissel, J.-P.: A Score for Predicting Risk of Cardiovascular Death in Adults with Elevated Blood Pressure. British Medical Journal 323(7304), 75–81 (2001)
10. The Mount Hood 4 Modeling Group: Computer modeling of diabetes and its complications. The Mount Hood modeling group, Diabetes Care 30, 638–1646 (2007)
11. http://clearinghouse.adma.org.au/browse-resources/assessment-tool/2.html
12. Green, C., Hoppa, R.D., Young, T.K., Blanchard, J.F.: Geographic analysis of diabetes prevalence in an urban area. Soc. Sci. Med. 57(3), 551–560 (2003)

Clinical Effectiveness of the "Healthwear" Wearable System in the Reduction of COPD Patients' Hospitalization

Alexis Milsis, Theodoros Katsaras, Nicolaos Saoulis,
Evita Varoutaki, and Aggelos Vontetsianos

e-Health Unit, 'Sotiria' Hospital,
Athens, Greece
{amilsis,tkatsa,agelvonte}@gmail.com,
{nsaoulis,evitavaroutaki}@hotmail.com

Abstract. Patients with Chronic Obstructive Pulmonary Disease (COPD) experience frequent exacerbations and hospital readmissions. Early hospital discharge schemes have been proved effective and safe approaches for suitable patients. Forty-eight (n=48) COPD patients were included in a randomized control trial (ratio 1:1), after their hospital admission due to an acute exacerbation. The aim of the study was to evaluate whether they could be early discharged and successfully continue their treatment at home, assisted by the use of wearable systems. Study group patients were discharged early (3rd-5th day) and monitored at home through the wearable "Healthwear" system while control group patients underwent conventional care. Intervention patients intensive home monitoring included ECG, heart and respiratory rate, oxygen saturation, activity and body position, combined with 3G mobile video sessions. The results indicated a significant reduction of in-hospital days, outpatient clinic and emergency room visits, as well as in readmission rates. Wearable systems that allow continuous wireless monitoring of bio-signals, can play a significant role in early hospital discharge.

Keywords: e-Health, Remote monitoring, Smart clothes, Smart textiles, wearable, COPD, RCT, Early hospital discharge.

1   Introduction

Chronic obstructive pulmonary disease is a leading cause of chronic morbidity and mortality. Hospital admissions due to exacerbations, especially in the winter period, constitute a major problem in the management of the disease, due to their negative impact on health-related quality of life, prognosis and costs [1,2].

Early hospital discharge (ED) schemes is one of the most promising approaches for efficient healthcare system intervention, in order to control cost and provide high quality services at patients' premises [3,4,5,6].

K.S. Nikita et al. (Eds.): MobiHealth 2011, LNICST 83, pp. 54–60, 2012.
© Institute for Computer Sciences, Social Informatics and Telecommunications Engineering 2012

New technologies can play a major role towards this transition. Moreover innovative, non-invasive wearable systems that allow continuous wireless monitoring of patients' status, can play a significant role to ED schemes.

The aim of the present study was to evaluate the clinical effectiveness of the "Healthwear" wearable system in the reduction of in-hospital length of stay (LoS), for COPD patients who were admitted to clinical wards of "Sotiria" Hospital, after an acute exacerbation.

2   Materials and Methods

The wearable solution used is based on the "Wealthy" prototype [7,8]. The "Healthwear" system consisted of a wearable garment (easy to wear, washable, available in variable sizes, yet suitable only for male patients at the time of the study) with biosensors embedded into the textiles, coupled with a small, lightweight (145gr) electronic device, called Portable Patient Unit (PPU) (Fig. 1). The PPU is easy to use, with two LEDs and a buzzer for user-warning purposes and a button to let a manual trigger of an alarm. Data transmission is done over a General Packet Radio Service (GPRS) link. The device is powered by a Li-Ion battery, autonomous up to 4 hours with real time streaming of all signals. The PPU collects and transmits the signals from the knitted sensors, as well as from other external medical devices connected via an available RS-232 port (in our case an oximeter was used). The bio-signals (6 lead ECG, pulse rate, respiratory rate, oxygen saturation, skin temperature, and body position), were transmitted via a GPRS mobile connection to a central server.

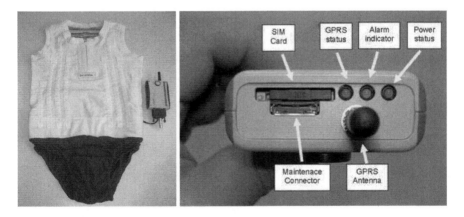

Fig. 1. Healthwear garment and PPU

Complementary, regular 3G cell phones could be used to perform videoconferencing sessions with the patients (Fig. 2).

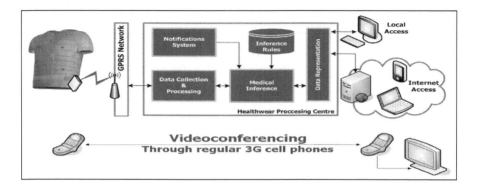

Fig. 2. Architecture of the "Healthwear" System

Medical measurements were stored into patient's electronic health record (EHR) and accessed via a secure TCP/IP connection by the attending physician or other authorized healthcare professionals, in near real-time or off-line mode, using a specifically designed software application with a suitable graphical representation (Fig. 3).

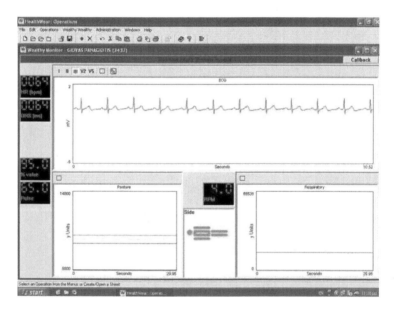

Fig. 3. Biosignals obtained by "Healthwear" system

Eligible patients for the trial were COPD exacerbated patients, admitted as emergency cases to the pulmonary wards of "Sotiria" Chest Diseases Hospital of Athens. Patients requiring inpatient imperative management or investigation for other medical problems and co morbidities were excluded. Additionally, patients who were not residents of Athens, homeless, living under extremely poor social conditions, or unable to give informed consent, were not considered eligible.

A total number of forty eight (n=48) male patients, were included and randomized (1:1 ratio) to intervention (early discharge) and control group (conventional inpatient care) by a third party staff member. The necessary equipment (garment, PPU, external oximeter) was provided to each patient of the intervention group, while they were hospitalized. During their in-hospital stay, the system functionality was validated and the patients were trained on its usage (Fig. 4). The patient learned how to wear the garment and set-up the system in order to facilitate the bio-signal collection and transmission whenever he was instructed to.

Fig. 4. Patient being monitored during his in-hospital training and while performing outdoor activities, using the "Healthwear" system

A multidisciplinary team (specialists, nurses, social workers and physiotherapists), in close cooperation with the physician in charge of the patient, designed a personalized care plan [9] to be followed at home, after the patient's discharge. Between the 3rd and 5th day of hospitalization, the patient was early discharged by the attending physician and continued his treatment at home. The patient was given unlimited telephone access to a respiratory nurse (case manager) of the hospital's home care team, who was responsible for the implementation of the prescribed care plan. The patient, using the PPU, was also capable to alert the case manager in case of deterioration, or initiate a session of bio-signal transmissions, according to the given instructions.

During the follow up period of the patient, a number of remote sessions (e-visits) were performed, consisting of bio-signal transmissions and simultaneous videoconferencing with the case manager or the attending physician (Fig. 5). The frequency and the duration of the sessions were adjusted as required, according to the patients' health status and findings. Usual sessions were scheduled while the patients were at their home premises or during outdoor activities, while performing prescribed physical exercises. Scheduled or patient initiated sessions of vital sign transmissions (unattended monitoring) could also be performed and be reviewed later on by the healthcare professionals (off-line mode).

During the first week after ED, patients were monitored more intensively with about three scheduled sessions per day, while one scheduled session per day was approximately performed, in the following three weeks.

Fig. 5. e-Visit (Biosignal transmission combined with video conference)

At the end of the two months follow-up period, length of stay (LoS), readmission and mortality rate, as well as visits to the emergency room or outpatient clinic, were assessed for both groups [5].

The control group patients were discharged according to the current criteria and practices followed in traditional in-hospital care.

3   Results

The average length of in-hospital stay for the intervention group patients was 3.6 days, versus 6.8 days of the control group. From the control group, three patients were readmitted, compared to one patient of the intervention group. For the study group two emergency room (ER) visits and two outpatient clinic visits were recorded, while for the control group eight and thirty two visits were observed respectively. No mortality incidents were reported for either group, during the studied period (Fig. 6).

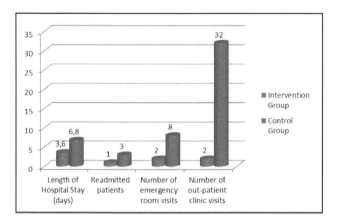

Fig. 6. Length of Stay, Readmission Rate and health services usage

The mean number of biosignal monitoring sessions was 50.3 per patient, while the number of videoconferencing sessions was 7.3 for the follow-up period. From a total of 1388 transmission sessions, 1207 (87%) were successful (175 of them combined with videoconference) (Fig. 7).

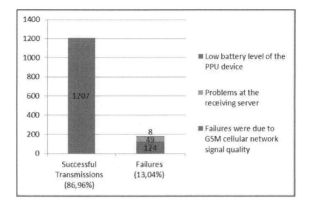

Fig. 7. Biosignal transmission statistics

Twenty out of twenty four intervention group patients considered the garment as acceptable, while remaining four patients reported various reasons for inconvenience.

4   Discussion

Our trial's results presented notable reduction of in-hospital length of stay, which was the main expectation of the trial. Reductions in using hospital emergency services, such as unnecessary ER visits and readmission rate were also observed.

From a technical point of view, the system supported successfully bio-signal transmissions, as most of the failures were not directly related to "Healthwear" (poor GSM/GPRS signal).

In the past, patients with acute exacerbations of COPD were successfully early discharged from the hospital, with the support of visiting respiratory nurses [3,5,9]. According to our findings, the "Healthwear" system supported efficiently the ED intervention, allowing patients' remotely home monitoring, decreasing the need of multiple nurse home visits. The combined use of videoconference sessions contributed substantially to the patients' accessibility, convenience and feeling of safety.

Technical limitations related to the system's capabilities (e.g. garment suitable only for male patients or reduced autonomy of the PPU's battery), are expected to be resolved in upcoming designs. Additionally other services or capabilities, widely available today, are anticipated to be integrated in future implementations (i.e. smart phones / PPU integration, UMTS etc.).

Wearable solutions are new era's tools, facilitating successfully the implementation of early hospital discharge schemes and the reduction of nurse home visits. Extensive research is expected to prove the value of wearable devices used in the treatment of other chronic diseases such as Chronic Heart Failure or diabetes. It's also expected to demonstrate the utilization of these systems in other services supporting chronic patients' care management and personalized care, such as telemonitoring during exercise and home based rehabilitation.

Acknowledgements. The trial was implemented as part of the EU / e-TEN funded project "Healthwear".

References

1. Pauwels, R.A., et al.: GOLD Scientific Committee. Global strategy for the diagnosis, management, and prevention of chronic obstructive pulmonary disease. NHLBI/WHO Global Initiative for Chronic Obstructive Lung Disease (GOLD) Workshop summary, Am. J. Respir. Crit. Care Med. 163, 1256–1276 (2001)
2. Garcia-Aymerich, J., et al.: Risk factors for hospitalization for a chronic obstructive pulmonary disease exacerbation. EFRAM study. Am. J. Respir. Crit. Care Med. 164, 1002–1007 (2001)
3. Ram, F., Wedzicha, J., Wright, J., Greenstone, M.: Hospital at home for patients with acute exacerbations of chronic obstructive pulmonary disease: systematic review of evidence. BMJ (2004)
4. Puig-Junoy, J., Casas, A., Font-Planells, J., Esxarrabill, J., Hernandez, C., Alonso, J., Farrero, E., Vilagut, G., Roca, J.: The Impact of Home Hospitalization on Healthcare Costs of Exacerbations in COPD Patients. European Journal Health Economy (2007)
5. Taylor, S., Eldridge, S., Chang, Y.-M., Sohanpal, R., Clarke, A.: Evaluating hospital at home and early discharge schemes for patients with an acute exacerbation of COPD. Chronic Respiratory Disease (2007)
6. Casas, A., et al.: Integrated care prevents hospitalisations for exacerbations in COPD patients. Eur. Respir. J. 28, 123–130 (2006)
7. Paradiso, R., Alonso, A., Cianflone, D., Milsis, A., Vavouras, T., Malliopoulos, C.: Remote health monitoring with wearable non-invasive mobile system: The Healthwear project. In: IEEE Eng. Med. Biol. Soc. Conf. Proc., vol. 1, pp. 1699–1702 (2008)
8. Malliopoulos, C., Milsis, A., Vavouras, T., Paradiso, R., Alonso, A., Cianflone, D.: Continuous mobile services for healthcare: The Healthwear Project. In: ICMCC Conf. Proc. (2008)
9. Hernandez, C., Casas, A., Escarrabill, J., Alonso, J., Puig-Junoy, J., Farrero, E., Vilagut, G., Collvinent, B., Rodriquez-Roisin, R., Roca, J.: Home Hospitalisation of Exacerbated Chronic Obstructive Pulmonary Disease Patients. European Respiratory Journal (2003)

Towards an Accessible Personal Health Record

Ioannis Basdekis[1], Vangelis Sakkalis[1], and Constantine Stephanidis[1, 2]

[1] Institute of Computer Science, Foundation for Research and Technology – Hellas
[2] Department of Computer Science, University of Crete
{johnbas,sakkalis,cs}@ics.forth.gr

Abstract. Patient empowerment frameworks, including personal health records (PHR), actively engage technology empowered citizens in their healthcare. Particularly today, with the current increase of chronic diseases, the high growth rate of the elderly and disabled populations and at the same time the much higher cross-border patient mobility, such systems may prove to be lifesaving, cost effective and time saving. Currently, there are many different online applications promoted as being functional, user-friendly and detailed enough to provide a complete and accurate summary of an individual's medical history. However, it seems that most of the Web services available do not fully adhere to well known accessibility standards, such as those promoted by the W3C, thus turning them away from people with disability and elderly people, who most probably need them most. Additionally, support for mobile devices introduces additional obstacles to users with disability when trying to operate such services. This paper presents fundamental (*design for all*) guidelines for the successful implementation of an accessible ePHR service that can be operated by any patient including people with disabilities irrespective of the device they use to access this service.

Keywords: E-Accessibility, WCAG, Disabled people, Personal health record (PHR), Accessible electronic PHR (ePHR).

1 Introduction

Personal health record (PHR) systems are widely used to maintain a dynamic and up-to-date health profile, including a variety of different data that are not necessarily limited to medical family history, medications, laboratory tests, diagnostic studies and vaccination, but may also contain lifestyle information, medication compliance data, emotions, physical activity, etc. These records are intended to provide a complete and accurate summary of an individual's medical history in order to be useful, as well as (re)usable for clinicians and healthcare professionals to correctly evaluate the condition of a patient, without the need for time consuming and costly examinations. Thus, there exists significant value in making this information accessible online for all citizens, while complying with patient data privacy and security ethics.

In western countries, e-accessibility of public information and e-services, including ePHR, provided by governmental agencies (e.g., health insurance organizations, hospitals, etc.), is mandatory by law. For instance, in the U.S.A., "The Americans with Disabilities Act of 1990" [1], applies to all goods and services provided by the government and requires that all public facilities, not just those receiving federal

K.S. Nikita et al. (Eds.): MobiHealth 2011, LNICST 83, pp. 61–68, 2012.
© Institute for Computer Sciences, Social Informatics and Telecommunications Engineering 2012

funding, be accessible to the disabled population. More specifically, websites and e-services are required to comply with the technical provisions of the U.S. Rehabilitation Act of 1973 (Sections 504 and 508 – a subset of WCAG 1.0 with a few additions) [21], [22]. In the European Union, besides on-going legislation in some Member States, latest policy developments include the eHealth action plan to facilitate a more harmonious and complementary European approach to eHealth, with specific references promoting the accessibility of eHealth services, particularly for elderly or disabled persons [8]. With regards to technical specifications, the W3C's Web Content Accessibility Guidelines (WCAG) 2.0 has been adopted as the de facto accessibility standard (adopted also in Australia, Canada, France, Germany, Greece, Hong-Kong, Ireland, Italy, New Zealand, UK and elsewhere [10]). Besides those specific policy cases and technical specifications, it is also worth mentioning Article 25 of the UN Convention on the Rights of Persons with Disabilities, which states that "States Parties shall take all appropriate measures to ensure access for persons with disabilities to health services that are gender sensitive, including health-related rehabilitation" [20]. However, despite the worldwide recognized importance of e-accessibility, several studies indicate that many available e-services, based on visual concepts, are largely inaccessible to the elderly and to people with disability [5], [6]. More specifically, concerns indicate poor or no integration of specific technical accessibility requirements, while usability barriers are recorded on PHRs usage by elderly, disabled, and immigrant patients [7], [12], [13], [19], [29]. Therefore the need for indentifying Universal Access design challenges is more prominent than ever. Thus, appropriate design processes and methods must be applied to existing or to the newborn eHealth care platforms with smart surroundings.

According to the HIMSS (Healthcare Information and Management Systems Society) an ePHR is supposed to be "*a universally accessible, layperson comprehensible, lifelong tool for managing relevant health information, promoting health maintenance and assisting with chronic disease management via an interactive, common data set of electronic health information and e-health tools*". It should therefore be operated by the patient himself, aiming at the provision of access to such services for anyone, anywhere and at anytime, through any kind of devices. Such an approach implies an explicit design focus to address diversity, as opposed to reactive or ad hoc approaches, and additional consideration towards redefining the concept of *Design for All* in the context of Human Computer Interaction [15]. In that context, as an ePHR is a health record that is handled by an individual user himself, it is necessary to make this information accessible online to anyone who has the necessary electronic credentials to view the information.

In addition to functional limitations, someone has to also take into account that users increasingly demand more freedom to choose their preferred hardware-software combination (i.e., iphone or android mobile devices) for accessing all kinds of e-services through the browser of their choice. Following this trend, new and existing e-services are being (re-)designed in order to be accessed through mobile devices, as well as traditional PCs. However, as recent studies indicate, e-services which are designed basically for visual interaction are largely inaccessible to people with disability, raising as a consequence barriers to mobile device users as well [23].

As with a typical e-service, the development of a fully accessible and interoperable ePHR introduces new challenges to the accessibility provisions that have to be adopted from the early design stages [2], otherwise development costs rises [3]. Due to the importance of the ePHR in comparison to other e-services, the design process is even more demanding compared to a typical interoperable e-service [4], [9], as the considerations mentioned previously have to be carefully addressed. Toward this end, This paper evaluates some of the most widely used ePHRs (section 2) and presents specific design characteristics addressing accessibility and usability considerations that should be taken into account aiming to the development of a fully accessible ePHR, available through mobile devices as well as traditional desktop PCs equipped with assistive technology. We argue that an electronically accessible PHR web based service (ePHR) must be offered directly to individuals, so that information can be inserted, at a later (or earlier) stage during a medical/clinical act, accurately via online web based forms or other kind of online software tools linked directly with their personal record. Our contribution, in this paper, is to identify the main challenges and propose specific (experience - based) design guidelines that web developers must follow in order to comply with WCAG 2.0 [24], as well as with the *Mobile Web Best Practices* version 1.0 [25].

2   E-Accessibility Support of ePHRs

Recent advances of the Web 2.0 and wireless network communication have altered the traditional way people, including those with disability and the elderly, use computers and e-services. Now people engage in social networks, perform various everyday activities and are willing, to a certain extent, to share personal health data. Microsoft HealthVault (http://www.microsoft.com/en-us/healthvault), Google Health (http://www.google.com/intl/en-US/health/about), Patientslikeme (http://www.patientslikeme.com), PatientSite (https://www.patientsite.org), WebMD Health Record (http://www.webmd.com/phr), MyPHR (http://www.myphr.com/), My Revolution of RevolutionHealth (http://www.revolutionhealth.com/my-revolution/promo) and NoMoreClipboard.com (http://www.nomoreclipboard.com) are only some of the well-known available Web-based ePHRs, mainly based in U.S.A., that enable the patient to manage health data such as medical family history, medications, laboratory tests, diagnostic studies, surgeries, vaccination, and allergies. In addition to this basic functionality, some PHRs provide extra services such as drug interaction checking or messaging between patients and medical providers. One question that arises is whether disabled or elderly people could utilise this functionality, or if these services can be operated effectively with the use of assistive technology solutions.

In order to determine the e-accessibility level of these representative ePHRs, they were evaluated against WCAG 2.0 conformance level AA. The evaluation was carried out during the period October 2010 to June 2011 and the test sample included at least 5 different web interface screens from each ePHR (e.g., submission forms and view pages). The tools used for the evaluation were the TAW [16] and Total Validator [18], supported by manual testing provided by experts to ensure the accuracy of the automated assessment (in cases of manual checks). The manual testing included rendering without style sheets, scripting on–off, alternatives to JavaScript, use of placeholder images without alternative text, accuracy of alternative text description of content images, markup validity pseudo errors, presence of frames, disturbing animation, image-maps, pop-ups, utilization of keyboard, deprecated techniques for text alignment, etc. The results of the automatic testing conducted across this sample, supported by manual testing by experts to ensure the accuracy of the results of the automated assessment, found poor e-accessibility conformance results. As Table 1 indicates, none of the aforementioned ePHRs achieved Level AA conformance of WCAG 2.0 that ensures good accessibility level for several categories of disabled and elderly individuals.

Table 1. E-accessibility evaluation of eight selected ePHRs mainly against WCAG 2.0

Service Name	WCAG 2.0 level AA Conformance	Markup Validity	Mobile version (MWBP 1.0)
Microsoft HealthVault	Fail	Fail	N/A
Google Health (discontinued)	Fail	Fail	Iphone App N/A
Patientslikeme	Fail	Fail	As Desktop – N/A
PatientSite	Fail	Fail	As Desktop – N/A
WebMD Health Record	Fail	Fail	As Desktop – N/A
MyPHR	Fail	Fail	As Desktop – N/A
My Revolution	Fail	Fail	As Desktop – N/A
NoMoreClipboard.com	Fail	Fail	As Desktop – N/A

Such trend is not surprising. Similar findings related to inadequate e-accessibility levels have been also reported diachronically for e-services in general [5], [6], [14]. Up to early 2000, Web (mainly static) content was comprised mostly of text with images and interactive Web forms. These types of components could easily be identified by assistive technologies (e.g., Braille display, screen reader, enhanced keyboards, switches, etc.). However, the newly introduced Web technologies utilize new features that can cause problems to disabled Web users, especially those using screen readers. Such problems include:

- inaccessibility of built-in refreshable scripting technologies that triggers the browsers XMLHttpRequest object and cannot be handled by current versions of screen readers (although WAI-ARIA [26] is making progress in this specific area)
- lack of non-scripting alternatives or media without captioning
- in general the use of authoring practices based on a WYSIWYG metaphor
- dynamic behavior in selection of segments of content, that utilises the drag and drop interaction metaphor with a pointing device (without providing keyboard equivalent behavior)
- lack of liquid designs (for text only enlargement)
- lack of semantics in the content that provide non visual cues of information structure,
- use of embedded applications which do not provide accessibility features (e.g., Flash objects, Active-X controls, embedded video players).

Therefore, with all these technology advancements, the question arises as to whether it is possible for disabled and elderly patients to become end-users of ePHRs, since they present not only accessibility but readability [17] problems as well. Currently, the role of the disadvantaged or excluded groups, including the unskilled, the disabled and the elderly, is limited, since traditionally the delivery of these e-services has been biased towards: a) the "typical" or "average" able-bodied user, familiar with the notion of the "desktop" and the typical input and output peripherals and b) WYSIWYG notions supported by authoring tools or more sophisticated platforms and eServices, which generate final code without considering accessibility issues or the inability to use a pointing device.

It is argued that e-accessibility and device independence can be achieved only if design standards are applied from day one of the design process. In the case of an interoperable and accessible ePHR, the designer should comply with even more strict constraints than those targeted only to desktop solutions, since the screen size of the mobile device or the interaction style may be totally different compared to the desktop environment. To this end, design and usability guidelines for mobile design can contribute significantly towards ensuring that the final outcome addresses functional limitations such as visual disabilities, hearing impairments, motor disabilities, speech disabilities and some types of cognitive disabilities. From a usability point of view, applicable principles can be derived from guidelines improving mobile web usability [9]. For example, excellent usability experiments demonstrate that the most effective navigation hierarchy for use with mobile devices is one with only four to eight items on each level [2].

In order to develop multiplatform and fully accessible ePHRs, specific technical guidelines can be derived from similar e-services. The proposed design approach is built upon the flexible authoring methodology [4], [11], which has been successfully used in the implementation of the following e-services: a) the interoperable accessible portal of the Hellenic General Secretariat for Research and Technology [27] and b) the www.Ameanet.gr portal, developed in the context of the National funded project "Universally Accessible eServices for Disabled People" [28]. These guidelines imply designing according to this larger set of rules, performing tests and at the end re-evaluate and re-visit the designs, prior to any implementation. Once the design space has been documented, the resulting designs need to be encapsulated into reusable and extensible design components.

3   Proposed Guidelines

Commonly in clinical practice taking history data by a clinician is not an easy task mostly due to time restrictions or missing/ lost information available only in paper. Medical personnel must interview the patient, prior to any medical action invoked, and complete such records as accurately as possible. In order to be able to provide such a system, for all possible actuators (i.e. patients, clinicians, etc.) and all possible access devices the necessary design guidelines that will enable the interaction of end-users with an ePHR system must be defined, allowing them to be able to share their personal health data (independently of storage restrictions, utilizing experience, environment of use, time limitations and information requested). The aim is to enable disabled, elderly, low vision and blind, keyboard / ear or other groups of users via assistive technology solutions to use these ePHR services which are currently designed only for optimal visual presentation by "able-bodied" individuals. The practical experience acquired during the design process for a number of accessible and interoperable e-services such as the ones mentioned above, resulted in the consolidation of the following fundamental steps:

1. *Identify device-specific constraints or capabilities.* In this phase the different limitations or features of the computing devices should be identified. The identified characteristics can be organized according to their type. Thus, a typical classification should contain: a) Output interaction capabilities (such as the screen size of the device, screen resolution, number of colors, speech synthesizer, etc.), and   b) supported input interaction modes, such as physical or virtual keyboard, size of keys, touch screen, stylus, speech recognition, etc.). As a result, different presentation elements (implemented with the use of CSS versions) and adaptation logic (e.g., forms with more than 5 elements can be divided in more than one steps) should be used.

2. *Identify the context of use for each device and provide meaningful (sub-) sets of functionalities.* This phase comprises the analysis of the contexts of use for each device. In most cases, the devices are neither used in the same context nor interchangeably.

3. *Select the 'worst case' device for each function.* The computing device that appears to have the highest number of important limitations against all the diverse contexts of use should be selected in this phase. In most cases a mobile device is the most suitable candidate.

4. *Design the first user interface prototype according to the device-specific limitations.* Using well-established prototyping techniques, such as paper and pencil, mock ups, etc., proceed with the development of the first prototype for the selected device.

5. *Infer a generic set of requirements based on the first UI design.* Specific design requirements can emerge from the first prototype regarding, e.g., navigation, content structure, presentation, accessibility, etc.

6. *Design the user interface prototypes for the other devices applying the set of generic requirements.* Proceed with the user interface prototype development for the remaining devices taking into consideration the design requirements elaborated in the previous step. Additional design specific requirements may emerge for the alternative devices. These design artifacts can be incorporated and extend the set of the generic requirements.

7. *Decide which user interface components can be automatically transformed between the diverse computing devices.*

8. *Utilize e-accessibility standards for each interface component:* for desktop only functionality adhere with WCAG 2.0 level AA (including subjective 14.1 whenever possible), with the use of valid XHTML, while in case of mobile make use of most of MWBP 1.0 possible, and make use of valid XHTML Basic 1.1. For all those templates test against web accessibility with evaluation tools (e.g., TAW, Firefox Web development toolbar, W3C's mobileOK Checker, TAW mobileOK Basic Checker, etc.). In addition perform manual checks (e.g., rendering without style sheets, test the accuracy of alternative text descriptions, etc).

9. *Evaluate the user interface prototypes for all the different devices.* An appropriate usability evaluation methodology should be selected to identify

potential usability problems in the user interface prototypes. The selection of the evaluation method depends upon several factors such as available resources, evaluators with expertise, time to complete the project, etc.

10. *Revisit the set of requirements and the prototypes according to the findings.* This stage requires an analytical review of the design requirements based on the evaluation findings, as well as a review of the user interface prototypes in order to amend potential usability problem or inconsistencies between the diverse computing devices.

4   Conclusion

This paper proposes the adoption of specific guidelines in the context of delivering accessible and interoperable ePHRs, to be used by disabled and elderly people with the same success rate as with the "able-bodied" end-users. From the results of the accessibility evaluation presented, it can be derived that well known ePHRs do not consider accessibility standards, thus present barriers to those mostly in need of this kind of services. By following a strict procedure from the beginning of the design process, it is possible to deliver fully accessible and usable e-services that can be utilized by assistive technology solutions, altering the present status quo of well known ePHRs and largely improving worldwide acceptability of such a service. This is the reason that this set of guidelines is applicable not only to a general purpose web-based application but also to any modern ePHR systems that can use this design framework in their early design stages.

Acknowledgment. Present work was partially supported by the community initiative program INTERREG III, project "ΥΠΕΡΘΕΝ", financed by the EC through the European Regional Development Fund (ERDF) and by national funds of Hellas and Cyprus.

References

1. Americans with Disabilities Act (ADA): ADA standards for accessible design, http://www.usdoj.gov/crt/ada/stdspdf.htm
2. Geven, A., Sefelin, R., Tscheligi, M.: Depth and breadth away from the desktop: the optimal information hierarchy for mobile use. In: Mobile HCI 2006, pp. 157–164 (2006)
3. Basdekis, I., Alexandraki, C., Mourouzis, A., Stephanidis, C.: Incorporating Accessibility in Web-Based Work Environments: Two Alternative Approaches and Issues Involved. In: Proceedings of the 11th HCI International 2005, Las Vegas, Nevada, USA (2005)
4. Basdekis, I., Karampelas, P., Doulgeraki, V., Stephanidis, C.: Designing Universally Accessible Networking Services for a Mobile Personal Assistant. In: Stephanidis, C. (ed.) UAHCI 2009, Part II. LNCS, vol. 5615, pp. 279–288. Springer, Heidelberg (2009)
5. Basdekis, I., Klironomos, I., Metaxas, I., Stephanidis, C.: An overview of web accessibility in Greece: a comparative study 2004-2008. Universal Access in the Information Society 9(2), 185–190 (2010)
6. Cabinet Office: eAccessibility of public sector services in the European Union (2005), http://webarchive.nationalarchives.gov.uk/20060925031332/cabinetoffice.gov.uk/e-government/resources/eaccessibility/
7. DeJong, G., Palsbo, S.E., Beatty, P.W., Jones, G.C., Knoll, T., Neri, M.T.: The organization and financing of health services for persons with disabilities. Milbank Quarterly 80, 261–301 (2002)

8. E.U. Communication COM/2004/0356 final, The eHealth action plan, `http://eur-lex.europa.eu/LexUriServ/LexUriServ.do?uri=CELEX:52004DC0356:EN:HTML`

9. Buchanan, G., Farrant, S., Jones, M., Thimbleby, H., Marsden, G., Pazzani, M.: Improving mobile internet usability. In: Proceedings of the 10th International Conference on World Wide Web, Hong Kong, May 1-5, pp. 673–680 (2001)

10. Government Accessibility Standards and WCAG 2.0, `http://blog.powermapper.com/blog/post/Government-Accessibility-Standards.aspx`

11. Karampelas, P., Basdekis, I., Stephanidis, C.: Web User Interface Design Strategy: Designing for Device Independence. In: Stephanidis, C. (ed.) UAHCI 2009, Part I. LNCS, vol. 5614, pp. 515–524. Springer, Heidelberg (2009)

12. Lober, W., Zierler, B., Herbaugh, A., Shinstrom, S., Stolyar, A., Kim, E., Kim, Y.: Barriers to the use of a Personal Health Record by an Elderly Population. In: AMIA Annu. Symp. Proc., pp. 514–518 (2006)

13. Neri, M.T., Kroll, T.: Understanding the Consequences of Access Barriers to Health Care: Experiences of Adults with Disabilities. Disability and Rehabilitation 25(2), 85–96 (2003)

14. Nomensa: United Nations global audit of web accessibility (2006), `http://www.un.org/esa/socdev/enable/documents/fnomensarep.pdf`

15. Stephanidis, C.: User Interfaces for All: New perspectives into Human-Computer Interaction. In: Stephanidis, C. (ed.) User Interfaces for All - Concepts, Methods, and Tools, pp. 3–17, 760 pages. Lawrence Erlbaum Associates, Mahwah (2001) ISBN 0-8058-2967-9

16. TAW tool, `http://www.tawdis.net/taw3/cms/en`

17. Taylor, D., Hoenig, H.: Access to health care services for the disabled elderly. Health Serv. Res. 41 (Pt 1), 743–758 (2006)

18. Total Validator, `http://www.totalvalidator.com/`

19. Tsiknakis, M., Spanakis, M.: Adoption of innovative eHealth services in prehospital emergency management: a case study. In: 10th IEEE International Conference on Information Technology and Applications in Biomedicine (ITAB), Corfu (November 2010)

20. UN - Convention on the Rights of Persons with Disabilities, `http://www.un.org/disabilities/convention/conventionfull.shtml`

21. U.S.Rehabilitation Act (1973). Section504 (1973), `http://www.dol.gov/oasam/regs/statutes/sec504.htm`

22. U.S.Rehabilitation Act (1973). Section 508 (1973), `http://www.section508.gov/index.cfm?fuseAction=stdsdoc`

23. W3C-WAI, Shared Web Experiences: Barriers Common to Mobile Device Users and People with Disabilities (2007), `http://www.w3.org/WAI/mobile/experiences`

24. W3C-WAI, Web Content Accessibility Guidelines 2.0, `http://www.w3.org/TR/WCAG20/`

25. W3C-WAI, Mobile Web Best Practices 1.0, `http://www.w3.org/TR/mobile-bp/`

26. W3C-ARIA Overview, `http://www.w3.org/WAI/intro/aria`

27. Web Portal of the Hellenic General Secretariat for Research and Technology, Ministry of Education and Lifelong Learning, `http://www.gsrt.gr`

28. Web portal Universally Accessible eServices for Disabled People, `http://www.ameanet.gr`

29. West, D.M., Miller, E.A.: The digital divide in public e-health: Barriers to accessibility and privacy in state health department websites. Journal of Health Care for the Poor and Underserved 17, 652–667 (2006)

Developing Advanced Technology Services for Diabetes Management: User Preferences in Europe

Angelina Kouroubali and Franco Chiarugi

Computational Medicine Laboratory,
Institute of Computer Science, Foundation for Research & Technology-Hellas
100 N. Plastira St, Heraklion, 70013, Crete, Greece
{kouroub,chiarugi}@ics.forth.gr

Abstract. This paper analyzes the methodology and preliminary results of four focus groups conducted for the diabetes project Reaction, a four year European project that aims to develop an intelligent service platform for remote monitoring of glucose levels with diabetes management and therapy to patients in different healthcare regimes across Europe. The focus groups aimed to investigate opinions, concerns, issues and ideas about management of disease, technology services, privacy and confidentiality of people involved with diabetes. The focus groups were conducted in four different European countries to assist in the user centric design of the services to reach beyond the boundaries of the project.

Keywords: Focus groups, diabetes management, user preferences.

1 Introduction

Diabetes mellitus has reached worrying proportions in western countries making, diabetes one of the fastest growing chronic conditions in the world. Diabetes has been estimated to affect 60 million Europeans. Given the increasing trends towards sedentary lifestyles and obesity related problems, this number is expected to increase in the coming years.

Diabetes can cause many complications if the disease is not adequately controlled. Adequate treatment of diabetes, as well as increased emphasis on blood pressure control and lifestyle factors, may reduce the risk of long-term complications. Self-management of diabetes is an area that offers exceptionally good prospects, both in clinical and business terms [1]. Information and communication technologies (ICT) may offer useful capabilities to improve illness prevention and safety of care. ICT may facilitate active participation of patients and enable the personalisation of care, allowing new opportunities in health and disease management [2].

This paper analyzes the methodology and preliminary results of four focus groups conducted for the diabetes project REACTION[1]. REACTION is a four year European project that aims to develop an intelligent service platform for remote monitoring of glucose levels with diabetes management and therapy to patients in different healthcare regimes across Europe. The REACTION platform plans to execute various

---

[1] http://www.reactionproject.eu/news.php

K.S. Nikita et al. (Eds.): MobiHealth 2011, LNICST 83, pp. 69–74, 2012.
© Institute for Computer Sciences, Social Informatics and Telecommunications Engineering 2012

clinical applications, such as monitoring of vital signs, feedback provision to the point of care, integrative risk assessment, event and alarm handling as well as integration with clinical and organizational workflows and external Health Information Systems focusing on improvement of continuous blood glucose monitoring and tight/safe glycaemic control. Diabetic outpatients will be able to better control their disease, with prompt feedback from formal carers and medical systems and appropriate risk assessment services that can be deployed in any healthcare system in Europe. On the other hand, REACTION is expected to have an impact on formal carers in hospital wards by improving glycaemic control of admitted patients with diabetes using continuous blood glucose monitoring and therapy feedback.

The organization of the focus groups aims to be an exploratory study of user preferences beyond the boundaries of the project in four different European countries. It is a novel work in that it explores user preferences of technology features involved in diabetes management while the technologies are being implemented outside of the boundaries of the project. Rather than testing an already developed service, the study explores preferences about the REACTION services using end users who are not involved in the project. Also, the study offers an expanded view of technology management of diabetes that involves individualized care and cultural differences.

The focus groups examines what diabetic patients, nurses, doctors as well as healthcare professionals and informal carers expect from technology, in addition to identifying values, beliefs, hopes, concerns and needs related to the use of tele-monitoring services. Focus groups also highlight how the use of information technology could potentially change the experience of living with diabetes. Understanding societal factors is a core prerequisite for addressing ethical and social issues at the design stage of technology development.

2   Methodology

A series of four focus groups have been organized in different European countries between November 2010 and May 2011. The results of the focus groups are collected, analysed, and formulated in terms of requirements that give direction to the iterative design process. The main objective of the focus groups is to understand the relevant personal, social and cultural factors related to diabetes management and the REACTION services. In order to explore the potential impact of Reaction services beyond the boundaries of the project, focus groups were organized in European countries that do not participate in the clinical trials foreseen in the project. They took place in Greece, Italy, Cyprus and France with a range of participants including doctors, nurses, social scientists, technical personnel, patients, carers, nutritionists and lawyers. Each focus group included 6-8 participants. The questions that guided the discussions focused on several topics including information and risk management of diabetes, security privacy and confidentiality issues, quality of living, monitoring and alert systems, device and sensors design, technical skills, daily activities, concerns and suggestions. Discussions were informal, encouraging all participants to express their opinion and relate their experiences with diabetes care. Table 1 presents the focus group list.

Table 1. Focus group details

Focus Group	City & Country	Participants
1	Thessaloniki, Greece	2 doctors, 2 patients, 2 nurses, 1 carer
2	Florence, Italy	3 doctors, 2 patients, 1 dietician
3	Nicosia, Cyprus	3 doctors, 2 patients, 1 nurse, 2 carers
4	Paris, France	2 doctors, 2 patients, 1 nurse, 1 social scientist

The moderators of the focus groups guided the discussions and kept extensive notes as participants were expressing their opinions. Recording devices were not used and answers have not been attributed to specific participants in order to maintain confidentiality and privacy [3]. Data analysis of the focus groups includes translation of interviews conducted in Greek, to English, and identification of themes across European countries. Based on the analysis, suggestions and requirements will feed the iterative design process of REACTION services.

3 Results

Preliminary analysis of the focus group discussions identified several topics involved in diabetes and new technologies. Focus group participants discussed issues about autonomy and self-management, privacy and confidentiality, person – care provider relationship, and health management. The following paragraphs present some of the views and opinions of participants about these topics. Table 2 provides a summary of the results.

Table 2. Themes and participant opinions

Theme	Participant opinions
Autonomy & self-management	Technology as personal assistant & facilitator of autonomy Expanded medical view to include social & psychological management of diabetes Technology at affordable cost
Privacy & confidentiality	Complete disclosure of data to personal physician Need for encrypted data exchange Anonymized use of data for research purposes
Diabetic person – care provider relationship	Balanced exchange of information to avoid overload Technology as facilitator and not replacement of personal contact New models of care to include technology advancements
Health management	Perceived benefit will facilitate data entry Need for continuous motivational and psychological support to ensure compliance

Autonomy and self-management: A person who is afflicted with diabetes runs short and long term risks such as hypoglycaemia, ketoacidosis, retinopathy, nephropathy, cardiovascular disease. Technology can assist users to regain a measure of autonomy in managing their condition and preventing long term risks. Focus group participants also viewed technology as a potential personal assistant. An application on a mobile phone could be designed to provide personalized estimation of insulin needs, messages for motivation and support as well as alerts. Based on the opinion of a doctor involved in remote management of diabetes with a mobile application, continuous monitoring may not be essential for improved management of outpatient diabetes and could be seen as information overload without clear benefits. Participants expressed their view that diabetes is not just a physical condition, but also a social and psychological one. Participants felt that advances in technology design and functionality may improve not only the physical management of the disease but also the social and psychological one. However, they also believe that widespread availability of technology at an affordable cost and integration into the national health systems remain open issues.

Privacy and confidentiality: One of the objectives of data protection is to avoid the use of data concerning health for purposes different than disease management and, in particular, by someone who may abuse such information. At the same time, redirecting health data can be a successful part of the treatment of disease using remote monitoring systems. Focus group participants expressed their willingness to disclose all information relevant to their condition to their physician, without particular attention to how this information is transmitted. However, if they had the option they would like all information transmitted to be encrypted. Participants would like to be informed about who is seeing their data and are interested to be able to discern whether their information is shared with trustworthy persons and with someone who may abuse their data. However, participants were willing to allow their data to be used anonymously for research purposes. Participants with diabetes appeared less concerned about trust in the use of an internet platform in conveying personal medical information. However, they were concerned about downtime of devices, inaccuracy of advice and loss of personal information.

Diabetic person-care provider relationship: All participants expressed that diabetes management benefits from a close relationship between the person with diabetes and the health care provider team including the physician, nurses, nutritionist, psychology and others. Participants felt that technology may allow for a more accurate, faster response to crisis, as well as better overall management and prevention of complications. However, some participants felt that a careful balance between information and communication is important to avoid information overload and excess. Technology for diabetes management may bring about profound changes in self-care and empowerment of people with diabetes, as well as, in depth communication with the health providers. During the focus groups, it has been acknowledged that a model of care with the inclusion of technologies will be significantly different to the usual model of patient physician relationship of face to face interactions. Convenience, frequent monitoring, and better management are some perceived advantages offering increased autonomy, attention and guidance. While, less frequent face to face visits, potential downtime of services and impersonal care were perceived as possible disadvantages.

Health management: Management of diabetes requires specific information in predefined time intervals that coupled with personalized algorithms can assist in insulin dosage decisions. A technology platform for diabetes management would require frequent measurements and regular entries of data. Participant willingness for data entry depended on the perceived benefit of the technology services and the time that data entry would require. However, people with diabetes felt that parameters that affect management of their condition, such as emotional and psychological stress, menses, physical exercise, and variability of daily life, are more difficult to define, harder to monitor, predict and account for. A technology service that could take into consideration some of these parameters would be at a relative advantage compared to a service that uses only glucose measurements. In addition, participants felt that the amount of information required to be recorded and the ease of use of the technology services will play an important role in the acceptance and use of the services.

Further analysis is being conducted to investigate these themes for similarities and differences across countries.

4 Discussion

Information and communication technologies have the potential to change health care as well as life style for people with chronic conditions [4]. These changes are occurring concurrently and need to be considered for appropriate design of sensor-enhanced health information systems. Basic functionalities of technology services include emergency detection and alarm, disease management, health status feedback and advice [4] as well as social and psychological support. These functionalities are combined differently based on individual conditions as well as individual needs of chronic patients and their informal carers. Focus groups were conducted in four different countries to identify, needs, opinions and concerns of people directly involved with diabetes about technology services of diabetes care.

Common themes as well as differences are being investigated. Preliminary analysis reveals that management of diabetes is similar across countries. However, significant differences exist in the national health system support of diabetic patients. Differences also occur based on the cultural background of people with diabetes. Cultural background is involved in whether people are open about communicating their disease, asking for help and support, while others prefer secrecy, autonomy and self management with minimum social contact. Cultural differences also occurred in the organization of health care, care teams and patient doctor relationships. Further analysis is being conducted to explore different and common themes across countries in topics including perception of autonomy and the balancing of personal goals, protection of personal data, patient-care provider relationship, and health management.

Participants in general welcomed technology services for supporting diabetes self management and thought it would improve their quality of life, resulting in fewer complications. These findings are in accordance with those of other studies about technology management of chronic conditions [5, 6]. The introduction of new technologies into diabetes care would need to involve appropriate education and adjustment periods to ensure the motivational and psychological support of users. These practices are fundamental in enhancing use of technology services.

5   Conclusions

This paper has presented the methodology and preliminary results of four focus groups conducted in different European countries. The focus groups investigate management of diabetes across countries. Participants expressed their opinions, ideas, issues and concerns about monitoring and personalized feedback functions of technology services including device design, information management, privacy and confidentiality. Preliminary analysis shows that people with diabetes and their formal and informal carers are open to use new technologies for improving self-management of the condition, avoiding complications and enhancing the quality of their lives. The focus groups provide a way to reach beyond the boundaries of the REACTION project. Ideas and opinions expressed during the focus groups will be presented to the consortium to explore ways to incorporate them in the design of the REACTION services. Reaching out to the community during the development of a service provides is necessary in order to inform design and explore user preferences.

Further work is needed to explore formulate the themes addressed in the focus groups into requirements that could be implemented in the REACTION platform. Future work could explore opinions and ideas of a larger number of participants in other parts of Europe not covered in the current research. In addition, analysis of national health system diabetes management across Europe would provide an important foundation for exploitation of REACTION services.

Acknowledgment. This work was performed in the framework of FP7 Integrated Project Reaction (Remote Accessibility to Diabetes Management and Therapy in Operational Healthcare Networks) partially funded by the European Commission under Grant Agreement 248590.

References

1. Farmer, A., Gibson, O., Tarassenko, L., Neil, A.: A systematic review of telemedicine interventions to support blood glucose self-monitoring in diabetes. Diabet. Med. (10), 1372–1378 (2005)
2. Fursse, J., Clarke, M., Jones, R., Khemka, S., Findlay, G.: Early experience in using telemonitoring for the management of chronic disease in primary care. J. Telemed. Telecare 14(3), 122–124 (2008)
3. Miles, M., Huberman, A.: Qualitative data analysis. Sage, London (1994)
4. Haux, R., Howe, J., Marschollek, M., Plischke, M., Wolf, K.-H.: Health-enabling technologies for pervasive health care: on services and ICT architecture paradigms. Informatics for Health and Social Care 33(2), 77–89 (2008)
5. Bostocka, Y., Hanleya, J., McGowna, D., Pinnocka, H., Padfielda, P., McKinstrya, B.: The acceptability to patients and professionals of remote blood pressure monitoring using mobile phones. Primary Health Care Research & Development 10(4), 299–308 (2009)
6. Verhoeven, F., van Gemert-Pijnen, L., Dijkstra, K., Nijland, N., Seydel, E., Steehouder, M.: The Contribution of Teleconsultation and Videoconferencing to Diabetes Care: A Systematic Literature Review. Journal of Medical Internet Research 9(5), 37 (2007)

Wireless Microrobotic Oxygen Sensing for Retinal Hypoxia Monitoring*

Olgaç Ergeneman[1],[**], George Chatzipirpiridis[1], Salvador Pané[1], Georgios A. Sotiriou[2], Christos Bergeles[1], and Bradley J. Nelson[1]

[1] Institute of Robotics and Intelligent Systems, ETH Zurich, 8092, CH
{oergeneman,chgeorge,vidalp,cbergeles,bnelson}@ethz.ch
[2] Particle Technology Laboratory, ETH Zurich, 8092, CH
sotiriou@ptl.mavt.ethz.ch

Abstract. This paper presents a luminescence oxygen sensor for retinal-hypoxia monitoring. The sensor coats a wirelessly controlled magnetic microrobot that will operate in the human eye. The coating embodies Pt(II) octaethylporphine (PtOEP) dyes as the luminescence material and polystyrene as a supporting matrix. It is deposited on the microrobot as a thin film and this film is experimentally evaluated using a custom optical setup. Due to the intrinsic nature of luminescence lifetimes, oxygen concentration was determined using a frequency-domain lifetime measurement approach.

Keywords: wireless, microrobot, oxygen sensing, ophthalmology.

1 Introduction

Retinal hypoxia (i.e., inadequate oxygen supply at the retina) is related to age-related macular degeneration, retinal-vein occlusion, and glaucoma, diseases that are responsible for the most cases of legal blindness [1,2]. For better understanding and monitoring of the progress of these diseases, *in vivo* oxygen sensing is essential.

Microrobots are proposed for targeted drug delivery and wireless sensing in the human body [3], but their actuation and control remains a challenge. Recently, [4] introduced an electromagnetic control system capable of accurately controlling magnetic microdevices with five degrees-of-freedom. This system will be employed to magnetically guide microrobots operating in the human eye. The microrobots are inserted in the human eye through a small incision at the sclera and are wirelessly controlled to the locations of interest using position information from conventional ophthalmoscopic systems [5].

Implantable MEMS-based oxygen sensors for intraocular measurements have been proposed [6]. However, implantable sensors do not possess mobility, and are, thus, limited to oxygen measuring at a fixed position.

---

* This work was supported by the NCCR Co-Me and the Nr. 200020-126694 grant of the Swiss National Science Foundation.
** Corresponding author.

K.S. Nikita et al. (Eds.): MobiHealth 2011, LNICST 83, pp. 75–79, 2012.
© Institute for Computer Sciences, Social Informatics and Telecommunications Engineering 2012

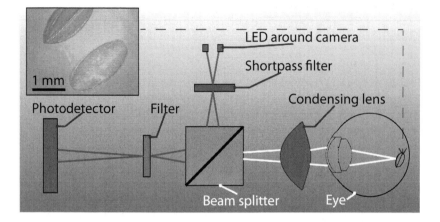

Fig. 1. Setup for oxygen detection. The oxygen-sensing microrobots can be used in the posterior eye segment. The inlet photo shows a microrobot coated with PS film with PtOEP dyes (top), and a microrobot coated with gold (bottom). A condensing lens is required for intraocular observation.

In this paper, a method to functionalize mobile microrobots for intraocular oxygen sensing is presented. The research builds on [7] and proposes luminescence-based oxygen sensing using thin-films. The sensors used in [7] are further miniaturized and the readout setup is improved to utilize smaller microrobot sensors. The coated microrobots are used in an experimental setup that acts as a human-eye phantom (Fig. 1). The experimental results show the feasibility of oxygen sensing using microrobots.

2 Luminescence Oxygen Sensor

Optical luminescence oxygen sensors work based on quenching of luminescence due to oxygen. A number of devices using this principle have been demonstrated, and the basic principles of different methods can be found in [8]. Luminescence-based oxygen sensors are attractive because they provide wireless readout, fast response, high accuracy, and they do not consume oxygen. They can be disposable and do not require reference electrodes or stirring. Additionally, they do not interfere with magnetic fields.

During sensing, the luminescent sensor is excited with a known input signal, and the intensity and lifetime of emission decrease due to quenching. Hence, the output signal can be correlated with oxygen concentration. The output signal intensity is difficult to control in an intraocular application. For example, it depends on the sensor's intraocular location, the input signal's incidence angle, etc. In this work, a lifetime measurement approach was chosen. The lifetime is an intrinsic property of the sensor material and is more robust to environmental conditions.

Lifetime is measured in the frequency domain. The sample is excited with a periodic signal that consequently causes a modulated luminescence emission at the identical frequency. Because of the lifetime of emission, the emission signal has a phase shift (i.e., time delay) with respect to the excitation signal. Measuring this phase shift provides the lifetime.

3   Experiments

3.1   Preparation of the Film Sensor

Figure 1 shows two assembled CoNi microrobots. For biocompatibility and surface functionalization they are first coated with a thin layer of gold by electroless deposition. To prepare the luminescence film, 3 mg of PtOEP (Frontier Scientific, UT, USA) and 197 mg of polystyrene (PS) were dissolved in 2 ml of chloroform by stirring. The microrobots are dip-coated and stored 2 hours, allowing evaporation of chloroform. Gold-coated silicon chips ($10 \, mm^2$) were also spin-coated with the prepared solution for characterization.

3.2   Characterization Setup

A Cary Eclipse fluorescence spectrophotometer (Varian Inc.) was used to characterize the luminescence of the sensors. Excitation scan, emission scan, kinetics, and lifetime measurements were performed. A custom flow cell was built and used in all experiments. Oxygen and nitrogen were mixed at different ratios using two gas flow controllers (Bronkhorst High-Tech B.V.) and applied to the cell. A total gas flow of 500 ml/min was maintained in all gas measurements.

Dissolved oxygen (DO) measurements were also performed using the same chamber with circulating water instead of gas. The oxygen concentration in water was changed by bubbling nitrogen or oxygen in a container. A commercial DO sensor (Oxi 340i, WTW Gmbh) was used to monitor the DO concentration in a second chamber in order to avoid possible interference by gas bubbles. The water was circulated using a pump, and the fluid flow rate was kept constant.

3.3   Intraocular Sensing Setup

A custom setup was built for wireless oxygen concentration measurements considering the anatomy of the eye and the control system described in [4]. A UV LED and a shortpass filter were used as the excitation source, and a Si photodetector (PD) (PD-100A, Thorlabs Gmbh) with a longpass filter were used for the read out. Using a beamsplitter (Edmund Optics) two separate optical paths were generated, one for the detecting system and the other one for the excitation system and tracking camera. Figure 1 shows an illustration of the measurement setup. A lock-in amplifier (HF2LI, Zurich Instruments) was used for the detection of phase change as a function of oxygen concentration. Its internal signal generator modulated the excitation circuit of the LED and acted as the reference signal for the detection of the PD signal. By this method, effective noise cancellation was obtained.

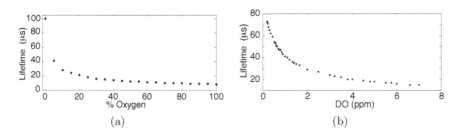

Fig. 2. (a) The lifetime of emission of a coated microrobot in response to different ratios of oxygen to nitrogen. The total flow rate was 500 ml/min, (b) The lifetime of emission of the coated microrobot at different dissolved oxygen concentrations in water. The oxygen concentration was obtained by the commercial oxygen sensor.

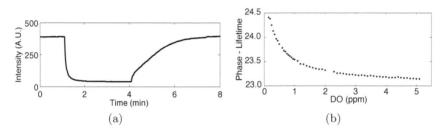

Fig. 3. (a) Response time of the film sensor. The high intensity state is 100 % nitrogen and the low intensity state is 100 % oxygen flow. (b) The response of the microrobotic sensor under different oxygen concentrations using the custom ophthalmic setup with the lock-in detection.

4   Results

4.1   Sensor Film Characterization

The excitation and emission characteristics were first obtained for the sensor film containing PtOEP. The peak emission wavelength was found to be 645 nm. Excitation wavelengths between 300 nm to 400 nm produced high emission intensity. Next, using the flow cell described, the oxygen sensitivity of the sensor was measured in gas and in water. Figure 2(a) shows the lifetime of emission of a coated microrobot with respect to different oxygen-to-nitrogen ratios under a constant flow. In Fig. 2(b), lifetime of emission of the same microrobot is shown in water with a flow rate of 3.15 l/min at different dissolved oxygen concentrations observed by the commercial oxygen sensor. An unquenched emission lifetime of 100 μs was observed. Finally, the response time of the sensor was obtained going from 100 % nitrogen gas to 100 % oxygen gas and back to 100 % nitrogen gas. The decay time of the sensor was determined to be approximately 30 seconds and the rise time approximately three minutes, as seen in Fig. 3(a).

4.2   Measurements Using Lock-in Amplifier

The film coated microrobot sensor was used in the custom-built setup. To mimic the optical properties of the eye an eye model was used. Using the commercial sensor the oxygen concentration was observed and the phase change was acquired from the lock-in amplifier. Figure 3(b) shows the response of the sensor under different dissolved oxygen concentrations. A curve similar to that obtained from the spectrophotometer was obtained, indicating that DO concentration can be successfully measured with the custom setup and the microrobot.

5   Conclusions and Future Work

Microrobotic oxygen sensors were developed using PS film with PtOEP dye. A custom setup for excitation and readout was implemented, and oxygen sensing was demonstrated. The sensors can be precisely controlled in the ocular cavity by applying magnetic fields as described in [4]. Future work will focus on using the readout system together with the control and tracking systems to create oxygen maps.

References

1. Congdon, N., O'Colmain, B., Klaver, C.C., Klein, R., Munoz, B., Friedman, D.S., Kempen, J., Taylor, H.R., Mitchell, P.: Causes and prevalence of visual impairment among adults in the United States. Archives of Ophthalmology 122(4), 477–485 (2004)
2. Galloway, N.R., Amoaku, W.M.K., Galloway, P.H., Browning, A.C.: Common eye diseases and their management, 3rd edn. Springer, Heidelberg (2005)
3. Nelson, B.J., Kaliakatsos, I.K., Abbott, J.J.: Microrobots for minimally invasive medicine. Annual Review of Biomedical Engineering 12, 55–85 (2010)
4. Kummer, M.P., Abbott, J.J., Kratochvil, B.E., Borer, R., Sengul, A., Nelson, B.J.: OctoMag: An electromagnetic system for 5-DOF wireless micromanipulation. IEEE Trans. Robotics 26(6), 1006–1017 (2010)
5. Bergeles, C., Kratochvil, B.E., Nelson, B.J.: Model-based localization of intraocular microrobots for wireless electromagnetic control. In: IEEE Int. Conf. Robotics and Automation, pp. 2617–2622 (2011)
6. Chen, P.J., Rodger, D., Saati, S., Humayun, M., Tai, Y.C.: Implantable parylene-based wireless intraocular pressure sensor. In: IEEE Int. Conf. Micro Electro Mechanical Systems, pp. 58–61 (2008)
7. Ergeneman, O., Dogangil, G., Kummer, M.P., Abbott, J.J., Nazeeruddin, M.K., Nelson, B.J.: A magnetically controlled wireless optical oxygen sensor for intraocular measurements. IEEE Sensors 8(1-2), 29–37 (2008)
8. Lakowicz, J.R.: Principles of fluorescence spectroscopy, 2nd edn. Kluwer Academic/Plenum Publishers (1999)

Meandered versus Spiral Novel Miniature PIFAs Implanted in the Human Head: Tuning and Performance

Asimina Kiourti and Konstantina S. Nikita

National Technical University of Athens, School of Electrical and Computer Engineering
akiourti@biosim.ntua.gr, knikita@ece.ntua.gr

Abstract. A meandered and a spiral stacked circular planar inverted–F antennas (PIFAs) are proposed for integration into head–implantable biomedical devices and wireless biotelemetry in the 402–405 MHz Medical Implant Communications Service (MICS) band. Designs only differ in the patch shape, feed, and shorting pin positions, while emphasis is given on miniaturization and biocompatibility. Phantom–related resonance detuning issues are addressed, and the PIFAs' radiation performance (radiation pattern, specific absorption rate (SAR) conformance with international guidelines, SAR distribution, and quality of up–link communication with exterior monitoring equipment) is evaluated and compared. Finite Difference Time Domain (FDTD) numerical simulations are performed.

Keywords: Biocompatibility, implanted biomedical devices, meandered antenna, medical implant communications service (MICS) band, spiral antenna.

1 Introduction

Wireless biomedical telemetry between implantable and exterior medical devices is an area of growing scientific interest [1]. In the most common scenario, the signals are transmitted wirelessly by means of antennas operating in the 402–405 MHz Medical Implant Communications Service (MICS) band, which is allocated for ultra–low–power active medical implants [2]. Miniaturization, biocompatibility, patient safety and quality of communication issues make the design of implantable antennas highly challenging.

In this study, we propose a meandered and a spiral MICS stacked planar inverted–F antennas (PIFAs) intended for integration into head–implantable biomedical devices (e.g. intra–cranial pressure sensors used in neurosurgery and neurology, brain–edema monitors for the paralyzed, position trackers for people with Alzheimer's disease etc). Microstrip designs are chosen because of their flexibility in design, conformability and shape, while shorting pins, high dielectric constant materials and surface optimization (meandering/spiraling) techniques are applied to increase the apparent size and reduce the physical dimensions of the structures. The proposed antennas are round–shaped, biocompatible, and identically–sized, achieving significant miniaturization compared to previously related works (e.g. [1], [3]–[7]). Tuning refinement is performed inside the skin tissue of a 15–tissue anatomical head

K.S. Nikita et al. (Eds.): MobiHealth 2011, LNICST 83, pp. 80–87, 2012.
© Institute for Computer Sciences, Social Informatics and Telecommunications Engineering 2012

model, and the radiation performance of the antennas (radiation pattern, specific absorption rate (SAR) conformance with international guidelines, SAR distribution, and quality of up–link communication with exterior monitoring equipment) is evaluated and compared. Finite Difference Time Domain (FDTD) numerical simulations are conducted in the XFDTD electromagnetic solver [8].

The paper is organized as follows. Section 2 presents the simulation set–ups and geometries of the proposed PIFAs. In Section 3, FDTD numerical results are presented regarding the performance of the proposed PIFAs implanted inside a 15–tissue anatomical head model. The paper concludes in Section 4.

2   Antenna Design

Since antennas are intended for skin tissue implantation, they are initially designed while in the center of a skin tissue simulating cube (Fig. 1(a)) (*initial PIFAs*) in order to speed–up the design process. Implantation and design refinement inside the skin tissue of a 15–tissue anatomical head model (Fig. 1(b)) are subsequently performed (*refined PIFAs*) [9], [10]. Tissue dielectric properties at 402 MHz are considered [11], while tissue mass densities are obtained from [9]. FDTD cells of $0.1 \times 0.1 \times 0.05$ mm^3, $2.5 \times 2.5 \times 2.5$ mm^3, $1.25 \times 1.25 \times 1.25$ mm^3 and $10 \times 10 \times 10$ mm^3 are used for the antenna, skin box, head model and free–space, respectively, in order to provide a high degree of spatial resolution in the simulations. Meshing is adaptive to avoid abrupt transitions. Free–space surrounds both simulation setups by 200 mm ($\lambda_0/4 \approx$ 186.5 mm, where λ_0 is the free–space wavelength at 402 MHz), and Liao absorbing conditions are considered in the boundaries to extend radiation infinitely far into space.

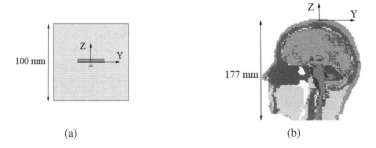

(a)   (b)

Fig. 1. Simulation set–ups: (a) skin tissue simulating cube, and (b) 15–tissue anatomical head

The proposed PIFA configurations are shown in Fig. 2. Antennas consist of a 4 mm–radius circular ground plane, and two identical stacked patches, relatively rotated by 180°. Both patches are fed by a 50 Ohm–coaxial cable (F) and radiate. Meandering/spiraling, patch stacking and shorting of the ground plane with the lower patch through a 0.2 mm–radius pin (S) lengthen the effective current flow path and miniaturize the PIFAs. Patches are printed on 0.25 mm–thick biocompatible alumina ($\varepsilon_r = 9.4$, tan$\delta = 0.006$) substrates, while a 0.15 mm–thick alumina superstrate covers the structures to preserve their biocompatibility and robustness. Throughout this study, the origin of the coordinate system is assumed to be at the center of the PIFAs' ground plane.

Designs of the proposed antennas were numerically studied and iterative simulation tests were performed to achieve the optimum resonance characteristics. The initial meandered PIFA has been presented by the authors in [9]. To obtain the initial spiral PIFA, patches are replaced by archimedean, single–arm, 1.9–turn spirals (inner radius of 1.1 mm), and the feed (F) and shorting pin (S) are re-positioned at (0 mm, 0 mm) and (3 mm, –1.5 mm), respectively. Tuning refinement is performed by suitably modifying the patch shapes while keeping all other PIFA design parameters constant. The refined meandered PIFA has meanders 1 and 5 (Fig. 2(b), (c)) lengthened by 0.5 mm, while patches are replaced by 1.95–turn archimedean spirals (inner radius of 1.05 mm) to obtain the refined spiral PIFA. The proposed designs achieve significant miniaturization compared to previously reported implantable PIFAs operating in the MICS band (e.g. [1], [3]–[7]), as indicated in Table 1.

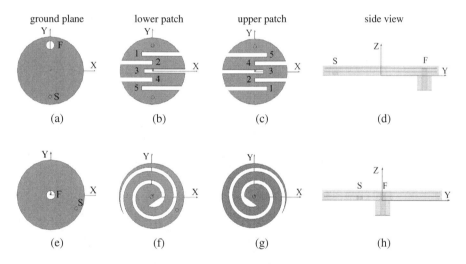

Fig. 2. Configurations of the proposed (a)–(d) meandered, and (e)–(h) spiral PIFAs

Table 1. Size comparison between the proposed and previously reported implantable PIFAs operating in the MICS band

PIFA	Volume occupied [mm^3]
[1]	$24 \times 32 \times 4 = 3072$
[3]	$22.5 \times 18.5 \times 1.9 = 790.9$
[4]	$23 \times 23 \times 1.25 \approx 661.3$
[5]	$\pi \times 7.5^2 \times 1.9 \approx 335.76$
[6]	$\pi \times 5^2 \times 1.815 \approx 142.6$
[7]	$8 \times 8 \times 1.9 = 121.6$
proposed	$\pi \times 4^2 \times 0.65 \approx 32.7$

3   Results and Discussion

3.1   Resonance Characteristics

Reflection coefficient ($|S_{11}|$) frequency responses of the proposed refined PIFAs implanted inside the anatomical head model are shown in Fig. 3. Even though the initial PIFAs are designed to resonate at 402 MHz while in the skin tissue simulating box, head implantation results in resonance frequency detunings by 11 and 5 MHz for the meandered and spiral PIFAs, respectively (solid). Detuning is attributed to the inherent dielectric loading of the surrounding tissues and exterior air on the antennas [9], [10], and is rectified by adequately modifying the patch shapes. Refined PIFAs resonate at 402 MHz when implanted inside the anatomical head model with broad 10 dB–bandwidths of 37 and 33 MHz, respectively (dotted).

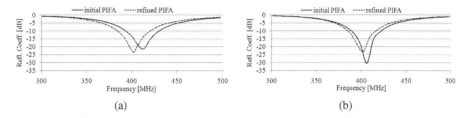

(a)   (b)

Fig. 3. Reflection coefficient ($|S_{11}|$) frequency responses of the (a) meandered, and (b) spiral PIFAs inside the anatomical head model

3.2   Far–Field Radiation Pattern

The 3D far–field gain radiation patterns of the refined PIFAs implanted inside the anatomical head model are shown in Fig. 4. Since the antennas are electrically very small, nearly omni–directional radiation is achieved. Comparable maximum gain values of –42.4 dB (in the $(\theta, \varphi) = (115°, 95°)$ direction, where θ and φ are the zenith and azimuth angles, respectively) and –42.9 dB (in the $(\theta, \varphi) = (95°, 270°)$ direction) are recorded for the meandered and spiral PIFAs, respectively. Low gain values are attributed to the miniaturized PIFA dimensions and human tissue absorption.

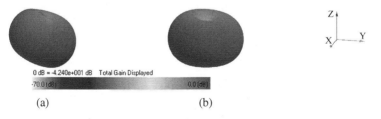

(a)   (b)

Fig. 4. Far–field gain radiation patterns of the refined (a) meandered, and (b) spiral PIFAs inside the anatomical head model

3.3   Specific Absorption Rate (SAR) Restrictions and Distribution

In order to ensure patient safety, the IEEE C95.1–1999 standard restricts the maximum SAR averaged over 1 g of tissue to less than 1.6 W/kg (1 g–avg SAR < 1.6 W/kg) [12], while the IEEE C95.1–2005 standard restricts the maximum SAR averaged over 10 g of tissue to less than 2 W/kg (10 g–avg SAR < 2 W/kg) [13]. Assuming a net–input power of 1 W incident to the proposed refined PIFAs, the maximum 1 g–avg and 10 g–avg SAR values induced in the anatomical head model are recorded (Table 2). Mass averaging procedures recommended by IEEE are applied [14]. The maximum allowed net–input power levels (P_{max}) which guarantee conformance with both IEEE standards are also given, indicating that the IEEE C95.1–1999 standard is much stricter.

Table 2. Maximum SAR values and power restrictions (P_{max}) for the refined PIFAs inside the anatomical head model

	Meandered PIFA	Spiral PIFA
1–g avg SAR	666.13 W/kg	676.01 W/kg
10–g avg SAR	80.21 W/kg	81.37 W/kg
P_{max} (IEEE C95.1–1999) [12]	2.402 mW	2.367 mW
P_{max} (IEEE C95.1–2005) [13]	24.93 mW	24.58 mW

Similar local SAR distributions are induced by the meandered and spiral PIFAs in the surrounding tissues, as indicated in Fig. 5 (the slices where maximum local SAR values have been calculated are depicted). To satisfy the strictest SAR limitations set by the IEEE guidelines, the net input power has been set to 2.402 mW and 2.367 mW for the meandered and spiral PIFAs, respectively (Table 2). For comparison reasons, all local SAR results have been normalized to 2 W/kg.

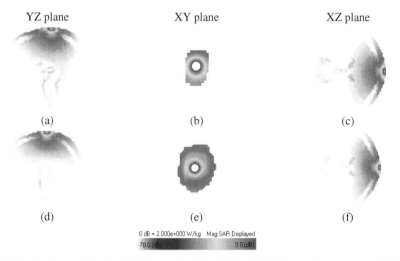

YZ plane   XY plane   XZ plane

(a)   (b)   (c)

(d)   (e)   (f)

0 dB = 2.000e+000 W/kg   Mag SAR Displayed
-70.0 [dB]   0.0 [dB]

Fig. 5. Distribution of the local SAR on the YZ, XY and XZ slices of the anatomical head model where maximum local SAR values have been calculated for the (a)–(c) meandered (net input power = 2.402 mW) and (d)–(f) spiral (net input power = 2.367 mW) PIFAs

3.4   Quality of Up–Link Communication

The quality of the wireless communication up–link which is formed between the proposed refined PIFAs implanted inside the anatomical head model (transmitter) and exterior monitoring equipment (receiver) is assessed. A simplified half–wavelength ($\lambda_0/2$, $f_0 = 402$ MHz) dipole antenna, placed vertically and symmetrically around the y–axis (i.e. centered at $(\theta, \varphi) = (90°, 270°)$), is considered to account for the exterior receiving antenna (Fig. 6(a)).

The transmission performance of the meandered and spiral PIFAs is shown in Fig. 6(b) in terms of the simulated transmission coefficient ($|S_{21}|$) versus distance between the antennas (d). The transmission coefficient, $|S_{21}|$, quantifies power transmission in the wireless link, so that $|S_{21}|^2 = P_r/P_t$, where P_t is the available power at the transmitter, and P_r is the power delivered to a 50 Ohm–load terminating the receiver antenna [15]. Higher $|S_{21}|$ values for the spiral PIFA are attributed to its higher far–field gain value in the $(\theta, \varphi) = (90°, 270°)$ direction (–42.92 dB), as compared to that of the meandered PIFA in the same direction (–44.68 dB).

The dependence of P_r on the transmitter–receiver distance (d) is shown in Fig. 6(c). Net input powers of 2.402 mW ($P_t = 2.412$ mW or 3.823 dBm) and 2.367 mW ($P_t = 2.397$ mW or 3.798 dBm) are considered for the meandered and spiral PIFAs, respectively, in order to satisfy the strictest IEEE limitations for the SAR (Table 2). Assuming the most distal scenario of this study (d = 1 m), the exterior receiver should have enough sensitivity to detect the signal from the implanted device which is as weak as –63.04 dBm and –58.62 dBm for the meandered and spiral PIFAs, respectively.

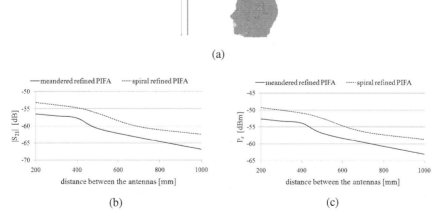

Fig. 6. (a) Transmission scenario simulation set–up, (b) transmission coefficient ($|S_{21}|$) of the refined PIFAs and (c) power delivered to the receiver antenna (P_r) versus distance between the antennas (d)

4   Conclusion

Meandering and spiraling techniques were applied to design two identically–sized MICS miniature PIFAs for integration into head–implantable biomedical devices. Phantom–related resonance detuning issues were addressed, and design modifications were proposed to refine tuning of the PIFAs when implanted inside a 15–tissue anatomical head model. Based on FDTD simulations, the performance of the antennas (exhibited far–field radiation pattern, SAR conformance with international guidelines, SAR distribution, and quality of up–link communication with exterior monitoring equipment) was assessed and found to be highly comparable.

Future work will include application of optimization algorithms to optimally tune the PIFAs inside the 15–tissue anatomical head model. Subsequent investigations will also include fabrication of the proposed antennas and experimental validation of the simulation results.

References

1. Kim, J., Rahmat–Samii, Y.: Implanted Antennas Inside a Human Body: Simulations, Designs and Characterizations. IEEE Transactions on Microwave Theory and Techniques 52, 1934–1943 (2005)
2. Medical implant communications service (MICS) federal register. Rules Reg. 64, 69926–69934 (1999)
3. Lee, C.M., Yo, T.C., Huand, F.J., Luo, C.H.: Bandwidth enhancement of planar inverted–F antenna for implantable biotelemetry. Microwave and Optical Technology Letters 51, 749–752 (2009)
4. Karacolak, T., Cooper, R., Topsakal, E.: Electrical properties of rat skin and design of implantable antennas for medical wireless telemetry. IEEE Transactions on Antennas and Propagation 57, 2806–2812 (2009)
5. Lee, V.M., Yo, T.C., Luo, C.H.: Compact broadband stacked implantable antenna for biotelemetry with medical devices. In: IEEE Annual Wireless and Microwave Technology Conference (2006)
6. Liu, W.C., Chen, C.H., Wu, C.M.: Implantable Broadband Circular Stacked PIFA for Biotelemetry Communication. Journal of Electromagnetic Waves and Applications 22, 1791–1800 (2008)
7. Liu, W.C., Yeh, F.M., Ghavami, M.: Miniaturized Implantable Broadband Antenna for Biotelemetry Communication. Microwave and Optical Technology Letters 50, 2407–2409 (2008)
8. XFDTD, Electromagnetic Solver Based on the Finite Difference Time Domain Method, Remcom Inc.
9. Kiourti, A., Christopoulou, M., Nikita, K.S.: Performance of a Novel Miniature Antenna Implanted in the Human Head for Wireless Biotelemetry. In: 2011 IEEE International Symposium on Antennas and Propagation (2011)
10. Kiourti, A., Christopoulou, M., Koulouridis, S., Nikita, K.S.: Design of a Novel Miniaturized Implantable PIFA for Biomedical Telemetry. In: Lin, J., Nikita, K.S. (eds.) MobiHealth 2010. LNICST, vol. 55, pp. 127–134. Springer, Heidelberg (2011)
11. Gabriel, C., et al.: The Dielectric Properties of Biological Tissues. Physics in Medicine and Biology 41, 2231–2293 (1996)

12. IEEE, Standard for Safety Levels with Respect to Human Exposure to Radio Frequency Electromagnetic Fields, 3kHz to 300GHz. IEEE Standard C95.1–1999 (1999)
13. IEEE, Standard for Safety Levels with Respect to Human Exposure to Radio Frequency Electromagnetic Fields, 3kHz to 300GHz. IEEE Standard C95.1–2005 (2005)
14. IEEE, Recommended Practice for Measurements and Computations of Radio Frequency Electromagnetic Fields with Respect to Human Exposure to such Field, 100 kHz to 300 GHz. IEEE Standard C95.3–2002 (2002)
15. Warty, R., Tofighi, M.R., Kawoos, U., Rosen, A.: Characterization of implantable antennas for intracranial pressure monitoring: reflection by and transmission through a scalp phantom. IEEE Transactions on Microwave Theory and Techniques 56, 2366–2376 (2008)

A Radio Channel Model for In-body Wireless Communications

Kamya Yekeh Yazdandoost

Dependable Wireless Laboratory, Wireless Network Research Center,
National Institute of Information and Communications Technology
3-4 Hikarino-oka, 239-0847 Yokosuka, Japan
yazdandoost@nict.go.jp

Abstract. Propagation model plays a very important role in designing wireless communication systems. Transmitting and receiving data from/to inside the body from tissue implanted medical devices are of great interest for wireless medical applications due to the promising of different clinical usage to promote a patient healthcare and comfort from one side and the most effective treatment for medical conditions from other side. The number of available electronic implantable devices is increasing every year. The complexity and functionality of these devices are also increasing at a significant rate. Hence, a reliable and efficient communication link is necessary to guarantee the best connection from/to an implanted device. In this paper we present a radio channel model for body implanted device over Medical Implant Communications Service (MICS) band in the frequency range of 402-405 MHz.

Keywords: Implant communication, radio channel model, electromagnetic wave, thermal effect.

1 Introduction

The use of implantable wireless communication device is growing at a remarkable rate, because the Medical Implant Communications Service (MICS) is replacing inductive communication for radio frequency implanted device. Therefore everybody can benefits of the best healthcare service irrespective of their geographic location.

Using MICS, a healthcare provider can set up a wireless link between an implanted device and a base station, allowing physicians to establish high-speed, easy-to-use, reliable, short-range access to the patient health data in real-time. Innovation in wireless communications aligned with the MICS band of frequencies is fueling the growth. The frequency band for MICS operations is 402-405 MHz [1], [2].

The 402-405 MHz band is well suited for in-body communication networks due to its international availability and compatibility with the incumbent users of the band (weather balloons). The maximum permitted output power for MICS devices is 25 μW EIRP (Effective Isotropic Radiated Power). The EIRP for an implanted device is defined as the signal power measured on external surface of human body and not at closed contact to the implanted device [1], [2].

K.S. Nikita et al. (Eds.): MobiHealth 2011, LNICST 83, pp. 88–95, 2012.

The use of implanted medical device is not without many significant challenges, particularly, the increasing of propagation losses in biological tissue. Therefore, to ensure the efficient performance of body implanted wireless communication the channel model need to be characterized and modeled for reliable communication system with respect to environment and antenna.

The paper discusses a radio propagation modeling, their characteristics, and human body as a medium for radio frequency propagation for medical implant communication service. The rest of this paper is as follows. The radio frequency and human body are discussed in section 2. Section 3 will describe the thermal effects of implant device. Then, description of the tissue interface and intrinsic impedance are provided in Section 4. Propagation model is discussed in section 5. Finally discussion and conclusion are expressed in Section 6.

2   Human Body and RF Wave

The human is partially conductive and consists of materials of different dielectric constants, thickness, and characteristic impedance. Therefore depending on the frequency of operation, the human body can lead to high losses caused by power absorption, central frequency shift, and radiation pattern destruction. The absorption effects vary in magnitude with both frequency of applied field and the characteristics of the tissue, which is largely based on water and ionic content. It is very difficult to determine the absorption of electromagnetic power radiated from an implanted source by the human body. Although quite a few investigations have been done to determine the effect of human body on radiated field [3], [4], and almost all of these studies have been based on external sources.

Prior to taking into consideration any in-body data communication, the effect of the human body on the RF signal must be understood. In order to construct a reliable wireless communication link from/to the human body, the electrical properties of the body tissues should be known for the frequency of interest. Table 1 shows the electrical properties of muscle, fat, and skin at frequency of 403.5 MHz [5]-[7]. Where ε is the dielectric constant, σ is the conductivity, δ is the penetration depth.

Table 1. The electrical properties of the body tissues at 403.5 MHz

Tissue	ε	$\sigma[S/m]$	$\delta[m]$
Muscle	57.100	0.797	0.052
Fat	5.578	0.041	0.308
Skin	46.706	0.689	0.055

3   Thermal Effects

The Specific Absorption Rate (SAR) is a standard measure of how much power is absorbed in the tissue. It will determine the amount of power lost due to heat dissipation, which depends upon E and H-fields strength.

The electromagnetic coupling into and/or out of the human body usually requires an antenna to transmit a signal into a body or pick up a signal from a body. The antenna operating environment for the implanted antenna is different from the traditional free-space communications, which is lossy environment. The implanted antennas may be classified in to two main groups: Electrical antennas, such as dipole antennas, and Magnetic antenna, for instance loop antennas.

The electrical antenna typically generates large components of E-field normal to the tissues interface, which overheat the fat tissue. This is because boundary conditions require the normal E-field at the interface to be discontinuous by the ratio of the permittivities, and since fat has a lower permittivity than muscle, the E-field in the fat tissue is higher [8].

SAR in the near field of the transmitting antenna depends on the H-field, whereas the SAR in the far field of the transmitting antenna depends mainly on the E-field. An important factor on the absorption characteristics of human body tissue layers is due to standing wave or impedance matching of the tissue types with high and low water content. The reflections of the propagation waves at different tissue layer interfaces, as shown in Fig. 1, can give rise to standing wave effect, which can increase the amount of local SAR.

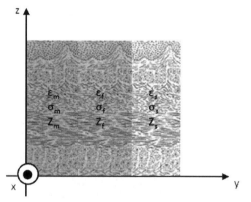

Fig. 1. Model of body tissues; muscle (m), fat (f), and skin (s)

The transmission from an antenna embedded in biological tissue is subject to a number of different electromagnetic phenomena including wavelength shortening due to dielectric loading, reflections from material transitions and absorption losses. The relationship between radiation and SAR is given by

$$SAR = \frac{\sigma |E|^2}{\rho} \ (W/Kg) \tag{1}$$

where E is the induced electric field and ρ is the density of tissue. For safety reason, in-body radiation is restricted to certain level. The Federal Communication Commission (FCC) regulations limit the Maximum Permissible Exposure (MPE) to non-ionizing radiation based on the amount of temperature rise that will occur. The regulation limits the temperature rise to 1 degree of Celsius. This limit being is determined by the specific heat of the tissue. The IEEE C95.1 [9] is defined that, the body exposure to radiation from implanted medical device is considered as partial body exposure in an uncontrolled environment. In such a case, the general provisions of the standard should not be violated, which is whole-body averaged SAR during localized exposure. The SAR averaged over the whole body is to be lower than 0.08 W/Kg and the spatial peak value of the SAR averaged over any 1 g of tissue (define as a tissue volume in the shape of cube) is to be less than 1.6 W/Kg. The spatial peak SAR shall not exceed 4 W/Kg over any 10 g of tissue in wrists, ankle, hands and feet. Experiments show exposure to an SAR of 8 W/Kg in any gram of tissue in the head or torso for 15 min may have a significant risk of tissue damage [10].

4   Tissues Interface

The reflections of propagation waves at different tissue layer interfaces can give rise to standing wave effects and impedance matching, which can, lead to local SAR increase. The intrinsic impedance (η) of a dielectric medium can be calculated from the material parameters by [11]:

$$\eta = \frac{\sqrt{\dfrac{\mu}{\varepsilon'}}}{\left[1 + \left(\dfrac{\sigma_{eff}}{\varpi \varepsilon'} \right)^2 \right]^{1/4}} e^{j(1/2)\tan^{-1}(\sigma_{eff}/\varpi \varepsilon')} \qquad (2)$$

where μ is the permeability, σ_{eff} is the effective conductivity, ω is the radian frequency, and $\acute{\varepsilon}$ is the real part of the complex relative permittivity. The intrinsic impedance of human tissues at 403.5 MHz is shown in Table 2 [11].

Table 2. Intrinsic impedance of tissues at 403.5 MHz

Tissue	$\eta(\Omega)$
Muscle	$43.5 \angle 13.0°$
Fat	$105.4 \angle 14.1°$
Skin	$47.7 \angle 14.4°$

Each tissue has own electrical properties which is different from other tissue. Therefore, there will be reflections of the propagation waves at different tissues layer interfaces. When a plane wave traveling in medium 1 strike a medium 2, the fraction that is reflected is given by the reflection coefficient (Γ), and the fraction is transmitted into medium 2 is given by the transmission coefficient (τ). A simplified model to compute the reflection and transmission coefficients are shown in Fig. 2.

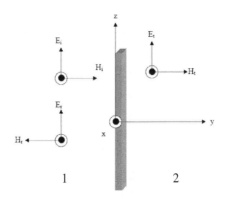

Fig. 2. Simple model to represent the reflection and transmission power by a planer interface

The reflection coefficient (Γ) and transmission coefficient (τ) at the interface are determined by [12]:

$$\Gamma = \frac{E_r}{E_i} = \frac{\eta_2 - \eta_1}{\eta_2 + \eta_1} \tag{3}$$

$$\tau = \frac{E_t}{E_i} = \frac{2\eta_2}{\eta_2 + \eta_1} \tag{4}$$

where E_i, E_r, and E_t are incident, reflected, and transmitted waves correspondingly. The reflection coefficient of field and power transmission factor at tissue boundaries at 403 MHz is shown in Table 3 [11].

Table 3. Field reflection coefficient and power transmission factor at tissue boundaries

Interface	Γ	τ (%)
Muscle to Fat	0.41	83.2
Fat to Skin	0.37	86.3
Skin to Air	0.78	39.2

5   Propagation Model

Optimization of the link efficiency and path loss must be quantified for expected radiation performance and link budget calculation due to effect of body tissues on output power and radiation pattern. The output power and radiation pattern are frequency dependent and strongly influenced by the electrical properties of the surrounding tissues. The foundation of any link budget is the Friis transmission equation

$$P_r = P_t G_t G_r \left(\frac{\lambda}{4\pi d} \right)^n \tag{5}$$

where P_r is the received power to the receive antenna, P_t is the transmitted power, G_t and G_r are gains of the transmit and receive antennas respectively, λ is the signal's wavelength, and r is the distance between two antennas.

To model the path loss of an implanted device, the field exited by an antenna can be expressed in terms of reactive wave and propagating wave with E-polarization or H-polarization with respect to the antenna type and body coordinates. This will considers both the near-field of the antenna, where reactive waves will be dominant, and the far-field of the antenna, which are determined by the propagating wave. Therefore the propagation loss between the transmitting and the receiving antenna, where one of them at least is placed inside a human body, as a function of frequency and distance, is dependent to: thermal attenuation due to conductivity, reflection losses at tissue boundaries, near-field losses and, far-field losses.

By determining the average SAR over the entire mass of the tissue between the transmitter and the receiver for near-field and far-field regions, we are able to compute the total power lost for human body part.

5.1   Near-Field

Kuster *et.al* shown in [13] that, SAR in the near-field is proportional to the square of H-field. Also they have shown the peak SAR is related to the antenna current, not to the input power. The SAR in the near field is

$$SAR = \frac{\sigma}{\rho} \frac{\mu\omega}{\sqrt{\sigma^2 + \varepsilon^2 \omega^2}} \left(\frac{Idl\sin\theta}{4\pi} e^{-\alpha R} \left(\frac{1}{R^2} + \frac{|\gamma|}{R} \right) \right)^2 \tag{6}$$

where I is current and R is distance. The power absorbed in the infinitely small volume is $\Delta P = SAR \times \Delta mass = SAR \times \rho \times dV$ where dV is $dV = R^2 \sin\theta dR d\theta d\phi$. The powered absorbed in the near-field (P_{nf}) is

$$P_{nf} = \int_{R=r}^{do} \int_{\theta=0}^{\pi} \int_{\phi=0}^{2\pi} \Delta P = \frac{\sigma}{\rho} \frac{\mu\omega}{\sqrt{\sigma^2 + \varepsilon^2 \omega^2}} \left(\frac{Idl\sin\theta}{4\pi} \right)^2 \times \int_{R=r}^{do} \int_{\theta=0}^{\pi} \int_{\phi=0}^{2\pi} R^2 \sin^3\theta e^{-2\alpha R} \left(\frac{1}{R^4} + \frac{|\gamma|^2}{R^2} + \frac{2\gamma}{R^3} \right) dR d\theta d\phi \tag{7}$$

5.2   Far-Field

The SAR in the far field of the transmitting antenna depends mainly on the *E*-field.

$$SAR = \frac{\sigma}{\rho} E_{rms}^2 = \frac{\sigma}{\rho} \left(|\eta||\gamma| \frac{Idl\sin\theta}{4\pi R} e^{-\alpha R} \right)^2 \tag{8}$$

The power absorbed in the infinitely small volume is $\Delta P = \sigma \left(|\eta||\gamma| \frac{Idl}{4\pi} \right)^2 \sin^3\theta e^{-2\alpha R} dR d\theta d\phi$
The powered absorbed in the far-field (P_{ff}) is

$$P_{ff} = \int_{R=do}^{do} \int_{\theta=0}^{\pi} \int_{\phi=0}^{2\pi} \Delta P = \sigma |\eta|^2 |\gamma|^2 \frac{I^2 dl^2}{12\pi\alpha} \left(e^{-2\alpha do} - e^{-2\alpha d} \right) \tag{9}$$

5.3   Received Power

From (7) and (9), the total power loss in tissue (P_{tl}) is $P_{tl} = P_{nf} + P_{ff}$. Hence the received power is

$$P_r = \frac{(P_t - P_{tl})\lambda_m^{\;2}}{(4\pi d)^2} G_t G_r \tag{11}$$

where λ_m is the wavelength in the biological tissue.

If one of the device is placed in free space and communicating with implanted one, path loss of free space should be counted for received power. In this case the total power loss is $P_{Tl} = P_{tl} + P_{fl}$, where the P_{Tl} is the total power loss and P_{fl} is the loss in free space.

6   Discussions and Conclusion

The path loss analysis at 403 MHz between two devices where one of them is placed outside the body at distance of 2 m from the body surface, while other one is implanted to the muscle tissue at 3cm deep inside the body is shown in Fig. 3.

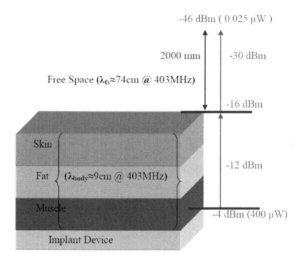

Fig. 3. Path Loss at 403 MHz

The paper presented a possible channel model for in-body communication. The study confirmed the importance of near-field and far-filed attenuation, which can affect not only radiation inside the body but can also determine the optimum distance of which good performance can be achieved in the surrounding environment.

The MICS technology delivers mobility, comfort, and higher levels of patient care. As designers develop new implanted medical devices taking advantage of RF technology to improve the quality of care for patients, propagation model is a key to this new system. Although the propagation models and RF system design is well understood for today's telecommunication systems, their application in medical systems offer unique challenges. A channel model performance assessment for implant device is more complex than for models in the free space, due to the conductivity and permittivity of an environment surrounded the implanted device.

The challenge in understanding of body implanted device is to make a propagation model in the environment which is extremely different from free space.

References

1. FCC, Medical implant communications (January 2003),
   `http://wireless.fcc.gov/services/index.htm?job=service_home`
   `&id=medical_implant`
2. ERC Recommendation 70-03 relating to the use of Short Range Device (SRD), European Conference of Postal and Telecommunications Administrations, CEPT/ERC 70-03, Tromsø, Norway (1997)
3. Scanlon, W.G., Evans, N.E.: RF performance of a 418 MHz radio telemeter packaged for human vaginal placement. IEEE Trans. Biomedical Eng. 44, 427–430 (1997)
4. Scanlon, W.G., Burns, J.B., Evans, N.E.: Radio wave propagation from a tissue-implanted source at 418 MHz and 916.5 MHz. IEEE Transaction on Biomedical Engineering 47, 527–534 (2000)
5. Gabriel, C., Gabriel, S.: Compilation of the dielectric properties of body tissues at RF and microwave frequencies. AL/OE-TR-1996-0037 (1996),
   `http://www.brooks.af.mil/AFRL/HED/hedr/reports/`
   `dielectricReport/Report.html`
6. Duney, C.H., Massoudi, H., Iskander, M.F.: Radiofrequency radiation dosimetry handbook. USAF School of Aerospace Medicine (1986)
7. Italian National Research Council, Institute for Applied Physics: Dielectric properties of body tissues, http://niremf.ifac.cnr.it
8. Gandhi, O.P.: Biological Effects and Medical Applications of Electromagnetic Energy. Prentice Hall, Englewood Cliffs (1990)
9. IEEE standard for safety levels with respect to human exposure to radio frequency electromagnetic field, 3 KHz to 300 GHz, IEEE Std C95.1 (1999)
10. Tang, Q., Tummala, N., Kumar, S., Gupta, S., Schwiebert, L.: Communication scheduling to minimize thermal effects of implanted biosensor networks in homogeneous tissue. IEEE Transaction on Biomedical Engineering 52, 1285–1294 (2005)
11. Scanlon, W.G.: Analysis of Tissue-coupled antennas for UHF intra-body communications. In: 12th International Conference on Antenna and Propagation, vol. 2, pp. 747–750 (2003)
12. Balanis, C.: Advance engineering electromagnetics. John Wiley & Sons, USA (1989)
13. Kuster, N., Balzano, Q.: Energy absorption mechanism by biological bodies in the near field of dipole antennas above 300 MHz. IEEE Transactions on Vehicular Technology 41, 17–23 (1992)

Parametric Study and Design of Implantable PIFAs for Wireless Biotelemetry

Asimina Kiourti, Michalis Tsakalakis, and Konstantina S. Nikita

National Technical University of Athens, School of Electrical and Computer Engineering
akiourti@biosim.ntua.gr, knikita@ece.ntua.gr
Abstract. The design parameters of a skin–implantable planar inverted–F antenna (PIFA) operating in the 402–405 MHz Medical Implant Communications Service (MICS) band are studied and their impact on the exhibited resonance characteristics is assessed. Based on the parametric results, two novel MICS PIFAs are proposed for skin–implantation. The study, thus, evaluates the tuning stability of a novel implantable PIFA on potential manufacturing inaccuracies. Significant guidance on implantable PIFA design is also provided. Numerical results based on Finite Element Method (FEM) numerical simulations are presented.

Keywords: Finite element method (FEM), implantable antenna, medical implant communications service (MICS) band, parametric study.

1 Introduction

Significant research is, nowadays, carried out on the design of medical implant–integrated antennas for biomedical telemetry in the Medical Implant Communications Service (MICS) frequency band (402–405 MHz) which is regulated by the Federal Communications Commission (FCC) [1] and the European Radiocommunications Committee (ERC) [2] for ultra–low–power active medical implants [3]–[9]. Stacked planar inverted–F antennas (PIFAs) are most commonly preferred because of their flexibility in size miniaturization, bandwidth enhancement, biocompatibility and conformability [5]–[9].

In this study, we evaluate the detuning caused by manufacturing inaccuracies in a stacked alumina 96%–PIFA with meandered patches, operating in the MICS band while implanted in skin tissue. Position of the shorting pin, meanders' lengths and widths, and dielectric thickness and material are parametrically studied to assess their impact on the antenna resonance frequency and its ability to operate in the MICS band. Significant guidance on implantable PIFA design is also provided. Based on the initial PIFA configuration and the derived parametric results, two novel skin–implantable Rogers 3210–PIFAs are proposed for biotelemetry in the MICS band. Finite Element Method (FEM) numerical simulations are performed [10].

The rest of the paper is organized as follows. In Section 2, a skin–implantable miniature alumina 96%–PIFA is proposed and parametrically studied. In Section 3,

K.S. Nikita et al. (Eds.): MobiHealth 2011, LNICST 83, pp. 96–102, 2012.
© Institute for Computer Sciences, Social Informatics and Telecommunications Engineering 2012

two novel skin–implantable Rogers 3210–PIFAs are designed based on the derived parametric results. The paper concludes in Section 4.

2   Parametric Study

The parametric PIFA model of Fig. 1(a)–(d) is considered [8], [9]. The PIFA consists of an R–radius ground plane and two R_p–radius vertically–stacked meandered patches, printed on h_1– and h_2–thick dielectric substrates, respectively. Meanders are equi–distant (d), have the same width (w), and their lengths are denoted by l_i, where i is the meander number (Fig. 1(b), (c)). Patches are fed by a 50 Ohm–coaxial cable centered at F (0 mm, 3 mm) and radiate, while a 0.2 mm–radius shorting pin connects the ground plane with the lower patch at S (s_x, s_y). An h_3–thick dielectric superstrate covers the structure to preserve its biocompatibility and robustness.

Since the PIFA is intended for skin tissue implantation, the simulation setup of Fig. 2 is considered, in which the antenna is assumed to be implanted in the center of a 100 mm–edge skin tissue simulating cube ($\varepsilon_r = 46.7$, $\sigma = 0.69$ S/m at 402 MHz [11]) [8], [9]. Iterative simulation tests are performed, and the variable values of Table 1 and 2 (PIFA I) are finally found to tune PIFA I at the desired frequency of 402 MHz (Fig. 3(a)), as demonstrated by the authors in [8].

Slight deviations around the original variable values of PIFA I are considered hereafter, and their effects on the exhibited resonance characteristics are discussed. Finite element (FE) simulations are conducted in which automatic iterative tetrahedron–meshing refinement is performed. Iterative refinement stops when the change in the magnitude of the reflection coefficient (in absolute value) between two consecutive passes is less than 0.02. Free–space surrounds the simulation setup of Fig. 2 by 200 mm ($\lambda_0/4 \approx 186.5$ mm, where λ_0 is the free–space wavelength at 402 MHz), while radiation boundaries extend radiation infinitely far into space.

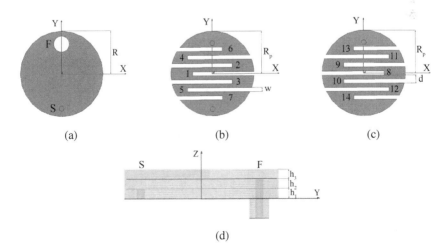

(a)   (b)   (c)

(d)

Fig. 1. Configuration of the proposed parametric PIFA model: (a) ground plane, (b) lower patch, (c) upper patch, and (d) side view

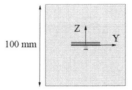

Fig. 2. Simulation set–up: PIFA implanted in the center of a 100 mm–edge skin tissue simulating cube [8], [9]

Table 1. Meanders' lengths (Fig. 1) for PIFA I, PIFA II and PIFA III (in [mm])

	l_1	$l_2 = l_3$	$l_4 = l_5$	$l_6 = l_7$	l_8	$l_9 = l_{10}$	$l_{11} = l_{12}$	$l_{13} = l_{14}$
PIFA I (alumina 96%)	7.4	5.8	4.5	0	7.4	5.8	4.5	0
PIFA II (Rogers 3210)	7.2	6.5	5.2	0	8.3	7.9	7.3	0
PIFA III (Rogers 3210)	4.9	5.1	6.4	5.9	7.9	7.8	6.6	5.8

Table 2. Design variable values (Fig. 1) for PIFA I, PIFA II and PIFA III (in [mm])

	R	R_p	$h_1 = h_2$	h_3	d	w	s_x	s_y
PIFA I (alumina 96%)	4	3.9	0.25	0.15	1	0.5	3	−1
PIFA II (Rogers 3210)	4.5	4.4	0.635	0.635	1	0.5	0.3	−3.2
PIFA III (Rogers 3210)	4.3	4.2	0.635	0.635	0.7	0.35	0	−3

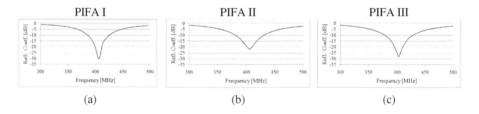

(a)   (b)   (c)

Fig. 3. Reflection coefficient frequency responses of (a) PIFA I, (b) PIFA II, and (c) PIFA III

2.1   Effect of Shorting Pin Position

Shorting the PIFA ground plane with the lower patch increases its effective size in a way which strongly depends on the exact position of the shorting pin. The design can be thought of as a modified monopole, since the shorting pin acts in much the same way as a ground plane on a monopole antenna, enhancing its electrical size. The further the shorting pin is placed on the "serpentine" current flow path which starts from the PIFA feed point (F), the more the effective size of the PIFA is enhanced, and, thus, the lower its resonance frequency becomes (Fig. 4(a), where positioning of the shorting pin at the end of meanders 1 to 5 is examined).

The values of the reflection coefficient achieved at the desired operation frequency of 402 MHz ($|S_{11}|_{@402\ MHz}$) are shown in Fig. 4(b) for each of the simulation scenarios. Assuming that satisfactory operation of the PIFA in the MICS band is guaranteed when $|S_{11}|_{@402\ MHz} < -10$ dB, then PIFA I is found to appropriately operate only in the case where the shorting pin is re–positioned at the end of meander 5.

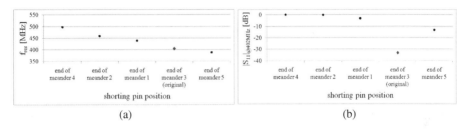

(a)   (b)

Fig. 4. Effect of shorting pin position on the (a) resonance frequency (f_{res}), and (b) reflection coefficient at 402 MHz ($|S_{11}|_{@402\ MHz}$) of PIFA I

2.2   Effect of Meanders' Length and Width

Longer (wider) meanders increase the length of the "serpentine" current path on the radiating patches, or, equivalently, the effective size of the PIFA, thus decreasing its resonance frequency, as shown in Fig. 5(a) (Fig. 6(a)). As expected, more intense deviations are observed in PIFA resonance frequency when all meanders are simultaneously lengthened (widened), rather than when the meanders of only the lower or upper patch are modified.

Simultaneous change in all meanders' lengths and widths by around ±0.2 mm (Fig. 5(b)) and ±0.1 mm (Fig. 6(b)), respectively, is found to preserve adequate operation of the antenna in the MICS band. The aforementioned values increase to approximately ±0.3 mm (Fig. 5(b)) and ±0.2 mm (Fig. 6(b)), respectively, when the lengths and widths of only a single patch (either the lower or upper) are modified.

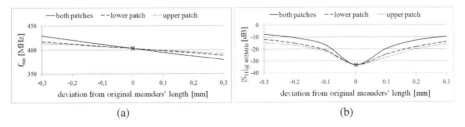

(a)   (b)

Fig. 5. Effect of meanders' length on the (a) resonance frequency (f_{res}), and (b) reflection coefficient at 402 MHz ($|S_{11}|_{@402\ MHz}$) of PIFA I

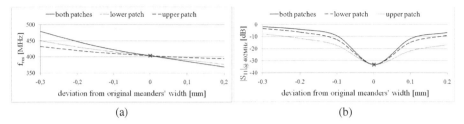

(a)   (b)

Fig. 6. Effect of meanders' width on the (a) resonance frequency (f_{res}), and (b) reflection coefficient at 402 MHz ($|S_{11}|_{@402\ MHz}$) of PIFA I

2.3   Effect of Dielectric Thickness

Thicker dielectric layers isolate the PIFA from the high–dielectric constant skin tissue, thus decreasing its effective dielectric constant and electrical length, while increasing its resonance frequency. This phenomenon is apparent in Fig. 7(a) in the cases where changes in the upper substrate (h_2)– and superstrate (h_3)–thicknesses are considered. Modifying the lower substrate thickness (h_1) also affects the shorting pin–related resonance effect, overall resulting in lower resonance frequencies for higher values of h_1.

As shown in Fig. 7(b), only slight deviations (of the order of less than ±0.05 mm) from the original dielectric material thicknesses are allowed in order to preserve satisfactory operation of the PIFA in the MICS band.

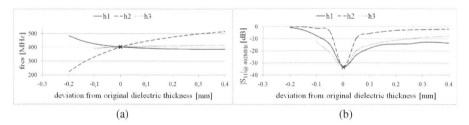

(a)   (b)

Fig. 7. Effect of dielectric thickness on the (a) resonance frequency (f_{res}), and (b) reflection coefficient at 402 MHz ($|S_{11}|_{@402\ MHz}$) of PIFA I

2.4   Effect of Dielectric Material

Resonance frequencies achieved when substituting the dielectric material of PIFA I (alumina 96%, $\varepsilon_r = 9.4$) with other dielectric materials commonly used in implantable antenna design (teflon ($\varepsilon_r = 2.1$), macor ($\varepsilon_r = 6.1$), alumina 92% ($\varepsilon_r = 9.2$) and Rogers 3210 ($\varepsilon_r = 10.2$) [5]–[9], [12]) are shown in Fig. 8(a). Higher dielectric constant–materials shorten the PIFA's wavelength, increase its electrical length and are, thus, found to decrease its resonance frequency.

Adequate performance of PIFA I in the MICS band is achieved only in the case where the dielectric is to be substituted with a material of highly similar dielectric constant, ε_r, such as alumina 92% (Fig. 8(b)). The choice of substrate and superstrate materials is thus proved to be highly critical in the design and performance of miniature implantable PIFAs.

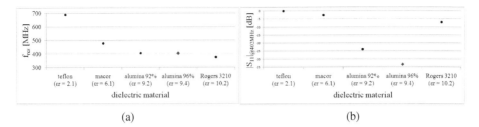

(a)   (b)

Fig. 8. Effect of dielectric material on the (a) resonance frequency (f_{res}), and (b) reflection coefficient at 402 MHz ($|S_{11}|_{@402\ MHz}$) of PIFA I

3   Design of Novel Implantable PIFAs

The parametric results of Section 2 provide significant guidance on the design of implantable PIFAs. For example, if the dielectric material of PIFA I is to be replaced with Rogers 3210 (ε_r = 10.2, versus ε_r = 9.4 for alumina 96%) [5]–[7], a decrease in resonance frequency is expected (Fig. 8(a)). However, the minimum available thickness of Rogers 3210–sheets (0.635 mm, versus 0.15 mm for alumina 96%) weakens tissue–loading on the PIFA, overall increasing its resonance frequency (Fig. 7(a)). PIFA design variables can, thus, be suitably selected (Table 1 and 2, PIFA II) to enhance its physical and/or effective size, and achieve an adequately low reflection coefficient at the desired operating frequency of 402 MHz (Fig. 3(b)).

Another example would be attempting to decrease the physical size (R) of PIFA II, while enhancing its effective size instead. Re–design consists of reducing the meanders' width (w) and separation distance (d) to allow space for extra meanders, and adding two extra meanders to each of the patches. In this way, lengthening of the "serpentine" current path is achieved. Iterative tests can be performed to identify those variable values (Table 1 and 2, PIFA III) which achieve the desired resonance characteristics (Fig. 3(c)).

4   Conclusion

Design variables of a skin–implantable MICS PIFA were parametrically studied through FEM simulations to assess their impact on antenna resonance and the ability of the antenna to adequately operate in the MICS band (defined as $|S_{11}|_{@402\ MHz}$ < −10 dB). Even minor deviations from the original parameter values were found to considerably alter the PIFA resonance frequency, and degrade its performance at the desired operating frequency of 402 MHz.

Significant guidance on implantable PIFA design was also provided. Based on the parametric results, two novel PIFAs were proposed for skin–implantation and wireless biotelemetry in the MICS band. Several other implantable PIFAs can be designed to suit specific size constraints, material–availability limitations, bandwidth requirements and resonance detuning sensitivity values. Finally, the parametric results can be generalized to the design of PIFAs operating in other frequency bands and implanted in any type of human tissue.

References

1. Medical implant communications service (MICS) federal register. Rules Reg. 64, 69926–69934 (1999)
2. ERC recommendation 70–03 relating to the use of short range devices (SRD). Conf. Eur. Postal Telecomm. Admin (EPT), CEPT/ERC 70–03, Annex 12 (1997)
3. Sani, A., Alomainy, A., Hao, Y.: Numerical Characterization and Link Budget Evaluation of Wireless Implants Considering Different Digital Human Phantoms. IEEE Transactions on Microwave Theory and Techniques 57, 2605–2613 (2009)
4. Abadia, J., Merli, F., Zurcher, J.-F., Mosig, J.R., Skrivervik, A.K.: 3D–Spiral Antenna Design and Realization for Biomedical Telemetry in the MICS Band. Radioengineering 18, 359–367 (2009)
5. Liu, W.-C., Chen, S.-H., Wu, C.-M.: Bandwidth Enhancement and Size Reduction of an Implantable PIFA Antenna for Biotelemetry Devices. Microwave and Optical Technology Letters 51, 755–757 (2009)
6. Karacolak, T., Cooper, R., Topsakal, E.: Design of a Dual-Band Implantable Antenna and Development of Skin Mimicking Gels for Continuous Glucose Monitoring. IEEE Transactions on Microwave Theory and Techniques 56, 1001–1008 (2008)
7. Liu, W.-C., Chen, C.-H., Wu, C.-M.: Implantable Broadband Circular Stacked PIFA for Biotelemetry Communication. Journal of Electromagnetic Waves and Applications 22, 1791–1800 (2008)
8. Kiourti, A., Christopoulou, M., Nikita, K.S.: Performance of a Novel Miniature Antenna Implanted in the Human Head for Wireless Biotelemetry. In: 2011 IEEE International Symposium on Antennas and Propagation (2011)
9. Kiourti, A., Christopoulou, M., Koulouridis, S., Nikita, K.S.: Design of a novel miniaturized implantable PIFA for biomedical telemetry. In: Lin, J., Nikita, K.S. (eds.) MobiHealth 2010. LNICST, vol. 55, pp. 127–134. Springer, Heidelberg (2011)
10. Sadiku, M.N.O.: Numerical Techniques in Electromagnetics, 2nd edn. CRC Press (2001)
11. Gabriel, C., et al.: The Dielectric Properties of Biological Tissues. Physics in Medicine and Biology 41, 2231–2293 (1996)
12. Sootornpipit, P., Furse, C.M., Chung, Y.C.: Design of Implantable Microstrip Antenna for Communication with Medical Implants. IEEE Transactions on Microwave Theory and Techniques 52, 1944–1951 (2004)

In Vitro and In Vivo Operation of a Wireless Body Sensor Node

Francesco Merli[1], Léandre Bolomey[2], François Gorostidi[3], Yann Barrandon[3],
Eric Meurville[2], and Anja K. Skrivervik[1]

[1] Laboratory of Electromagnetics and Acoustics
[2] Laboratory of Microengineering for Manufacturing 2,
Ecole Polytechnique Fédérale de Lausanne
[3] Laboratory of Stem Cell Dynamics,
Ecole Polytechnique Fédérale de Lausanne and Department of Experimental Surgery,
Centre Hospitalier Universitaire Vaudois and University of Lausanne
Switzerland
{francesco.merli,eric.meurville,anja.skrivervik}@epfl.ch,
{francois.gorostidi,yann.barrandon}@epfl.ch,
leandre.bolomey@a3.epfl.ch

Abstract. A wireless Body Sensor Node (BSN) and its operations are presented. The BSN comprises all the necessary components (i.e., antenna, electronics, batteries and bio-sensor) to allow continuous monitoring of physiological data. In vitro characterization validates the simulated performances, while in vivo experiment shows the capability of the system for real life telemedicine applications.

Keywords: Body Sensor Node (BSN), implantable antennas, implantable telemetry system, Medical Device Radiocommunications Service (MedRadio), Telemedicine.

1 Introduction

Wireless implantable systems promise large improvements in patients' care and quality of life. For this purpose, small biocompatible devices have been recently presented for different applications such as pH monitoring [1], gastro-intestinal tract exploration [2, 3] or cardiovascular pressure control [4]. In this work we present a complete Body Sensor Node which has been tested in vitro and applied in vivo in a porcine animal for local temperature monitoring. The system performs data telemetry with an external Base Station in the Medical Device Radiocommunications Service band (MedRadio, 401-406 MHz) [5] and the Industrial, Scientific and Medical band (ISM, 2.4-2.5 GHz).

2 Body Sensor Node

The proposed BSN aims at a high system integration of all its components, namely: the Multilayered Spiral Antenna (MSA), the electronics (the RF transceiver and

K.S. Nikita et al. (Eds.): MobiHealth 2011, LNICST 83, pp. 103–110, 2012.

the Digital Signal Processor), the batteries and the bio-sensor. Fig. 1 depicts the complete packaging of the device. All the elements fit in a biocompatible cylindrical housing which measures 10 x 32 [mm].

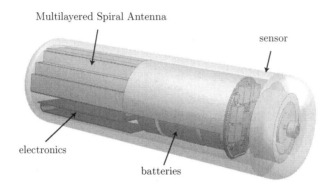

Fig. 1. Complete packaging of the proposed BSN. The biocompatible casing is in light gray. As a possible sensor, the driver and rotor for glucose microviscometer described in [6] are illustrated.

The MSA consists in a conformal radiator whose main resonating element is a three dimensional spiral metallization developed on a *pyramidal* assembly as described in [7, 8]. Polyetheretherketones (PEEK) was used for the biocompatible housing, as it is very easy to machine and its biocompatible characteristics are well established [9, 10]. The housing thickness (0.8 mm) was selected to improve the electromagnetic radiation of the implanted radiator in agreement with the results reported in [11]. The MSA has dual band capabilities working in both the MedRadio and the ISM bands and its radiation performances (maximum gain equal to -29.4 and -17.7 dBi in the lower and higher frequency ranges, respectively) provide a robust communication link for applications targeting a minimum working range of 2 m.

The electronics components were assembled on a flexible Printed Circuit Board (PCB) [12] to fit in the small available volume. The RF communication is provided by an ultra-low power Integrated Circuit (IC), the ZL70101 [13] manufactured by Zarlink, operating in both the MedRadio and ISM bands. The BSN operations are executed by an ultra-low power digital signal processor: the Ezairo 5900 manufactured by ON Semiconductor. Four coin type 377 batteries (1.5 V/ 27 mAh), manufactured by Energizer, were selected to provide the required power supply.

The conception of the proposed BSN gives a broad freedom regarding the monitoring device or bio-actuator to be included. The front-end electronics to drive the glucose microviscometer presented in [6] and the potentionstat described in [14] are just examples of possible sensors.

3   Power Consumption

In order to reduce the power requirements and to extend the life time of the BSN, the system is kept in a *sleeping* state. In this condition only the 2.45 GHz *wake-up* part of the transceiver is active and the power consumption is 2 μW. A signal received in the ISM band wakes up the IC; subsequently, the measurements are performed and the bidirectional communication occurs in the MedRadio frequency spectrum. Considering the TMP112 temperature sensor (from Texas Instruments (used in the in vivo experiment), the total active phase lasts 430 ms and consumes 15.7 mW. If one measurement is taken every 5 minutes, the embedded batteries provide a life time of 284 days. The power ratio of each part of the BSN is depicted in Fig. 6 (RF 91.9%, DSP 8.0%, sensor 0.1%). The RF portion can be separated in the actual power used to transmit data (RF_{TX}=2.7%), to establish the MedRadio communication channel (RF_{est}=14.6%) and to reserve the channel until the data are ready to be sent (RF_{ch}=74.6%), as illustrated in Fig. 2.

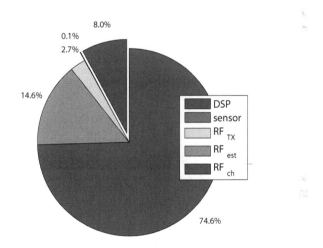

Fig. 2. Power consumption repartition among the different components of the BSN during the active phase

Comparing the performances of implantable devices with telemetry capabilities is not an easy task as many different conditions (working frequency, data rate, duty cycle dimension, implant location, purpose, power supply) can be considered. In Fig. 3 once can notice the good compromise of the proposed BSN among volume occupation, power consumption and life time when compared to other implantable systems[1].

---

[1] The selected implantable sensors are only chosen to show examples of different applications as the list is by no means exhaustive.

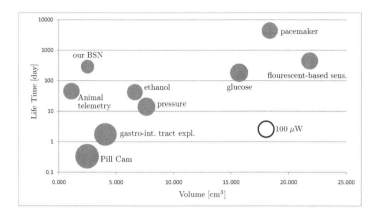

Fig. 3. Graphical comparison of active implantable devices with far field (blue) and near field (red) communication capability: flourescent-based sens. [15], pacemaker [16], glucose [17], pressure [18], ethanol [19], gastro-int. tract expl. [3], Pill Cam [2], Animal telemetry [20] and our BSN. The size of the bubble reflects indicated the power consumption computed as the ration between life time and the battery supplies.

4   In Vitro Characterization

In vitro tests were performed in order to check the functioning of the realized device and validate the MSA radiation characteristic. The BSN was inserted in a liquid body phantom. The latter has dielectric properties equivalent to the human muscle tissue and has a cylindrical shape (80 x 110 [mm]). Outdoor test were carried out to assess the capability of the MedRadio communication link (as illustrated in Fig. 4-(a)), while the *wake-up* performances in the 2.45 GHz ISM band were verified in an anechoic chamber.

Maximum working ranges, reported in Fig. 4-(b), confirmed the simulated performances of the MSA. Considering the power link budget characteristics of the IC ZL70101 and the Base Station provided by Zarlink, the maximum registered ranges correspond to antenna gain values equal to -30.5 and -18.6 dBi in the MedRadio and ISM bands, respectively. These values, which take into account the mismatch and the losses within the electronics assembly, closely agree with the predicted characteristics, i.e., -29.4 and -17.7 dBi.

5   In Vivo Experiment

Two BSNs were implanted in a large animal model (Göttingen minipigs) chosen for the similarity between their and the human tissues. A temperature sensor was included into the BSNs to study the correlation between local temperature and the healing process of a deep wound in the settings of a cultured epidermal autograft. In fact, it has been observed that the temperature has an important impact on epidermal stem cell behavior in vitro [21].

Frequency	Range [m]
403 MHz	14.0
2.47 GHz	4.8

(a)   (b)

Fig. 4. In vitro characterization: (a) outdoor communication tests and (b) maximum registered ranges for both working frequencies

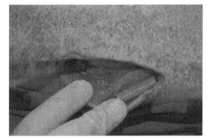

Fig. 5. Implantation, in accordance to all ethical considerations and the regulatory issues related to animal experiments, of the two BSNs at different depths. A subcutaneous location (5 mm deep) and an intra-muscular one (30 mm deep), were chosen.

In vivo experiment lasted for 15 consecutive days. During this period the animal dwelled most of the time in a farm and periodically, it was taken away for a follow-up at the hospital. Fig. 6 shows the registered temperatures during one of these follow-ups. Values measured by a rectal probe are also reported to appreciate the effect of the placement of the two BSNs.

While being at the farm, the animal was maintained indoor in a cage of dimensions 1.3 x 2.7 [m]. In order not to interfere with the farm daily life (feeding of the animals, cleaning, etc.), the Base Station was placed in the mansard above the cage room at a distance of 2.5 m.

The overall performances during the entire test period are reported in Table 1. During the 2696 interrogations between each BSN and the external Base Station the relative number of failed communications are 5.82% and 16.10% for the subcutaneous and intra-muscular BSN, respectively. From the point of view of communication protocol, three sources of error are identified in Table 1: MedRadio, ISM and firmware. One can appreciate that the data transmission in the MedRadio was found to be the most critical on for the deepest

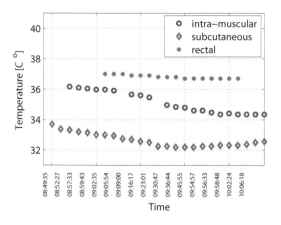

Fig. 6. Measured temperature values at the hospital

sensor ($err_{MedRadio}$=12.46%), while the *wake-up* communication showed a relative number of errors err_{ISM} lower than 5% for both BSNs. Almost negligible problems were caused by the driving firmware ($err_{firmware}$ <0.2%). Explanations for the registered errors have been found observing the pig activity in the cage and the consequent relative positions between the BSNs and the Base Station.

Table 1. Communication Performances during the In Vivo Experiment

BSN	Number of Measurements	$err_{MedRadio}$ [%]	$err_{firmware}$ [%]	err_{ISM} [%]	err_{tot} [%]
intra-muscular	2696	12.46	0.00	3.63	16.09
subcutaneous	2696	1.67	0.11	4.04	5.82

6   Conclusion

This work presented the operation of a complete wireless Body Sensor Node. The node integrates all the necessary components in a cylindrical volume (10 x 32 [mm]). Communication with an external Base Station placed in a few meters working range (up to 14 m) has been proven. The results obtained by the in vitro and in vivo tests confirm the promising capabilities of the proposed BSN and pave the way for future research oriented to the making of complete telemedicine systems.

Acknowledgments. The authors would like to thank J.-F. Zürcher and G. Corradini for the help provided during the realization of the proposed BSN.

References

[1] Gonzalez-Guillaumin, J.L., Sadowski, D.C., Kaler, K.V.I.S., Mintchev, M.P.: Ingestible capsule for impedance and pH monitoring in the esophagus. IEEE Trans. Biomed. Eng. 54(12), 2231–2236 (2007)

[2] Pill cam, Given Imaging,
http://www.givenimaging.com/en-us/Pages/GivenWelcomePage.aspx

[3] Johannessen, E.A., Wang, L., Wyse, C., Cumming, D.R.S., Cooper, J.M.: Biocompatibility of a lab-on-a-pill sensor in artificial gastrointestinal environments. IEEE Trans. Biomed. Eng. 53(11), 2333–2340 (2006)

[4] Chow, E.Y., Chlebowski, A.L., Chakraborty, S., Chappell, W.J., Irazoqui, P.P.: Fully wireless implantable cardiovascular pressure monitor integrated with a medical stent. IEEE Trans. Biomed. Eng. 57(6), 1487–1496 (2010)

[5] Medical Device Radiocommunications Service (MedRadio), Federal Communication Commission (FCC) Std. CFR, Part 95.601-673 Subpart E, Part 95.1201-1221 Subpart I, formerly. Medical Implanted Communication System (MICS) (2009),
http://wireless.fcc.gov/services/
index.htm?job=service_home&id=medical_implant

[6] Kuenzi, S., Meurville, E., Ryser, P.: Automated characterization of dextran/ concanavalin a mixtures–a study of sensitivity and temperature dependence at low viscosity as basis for an implantable glucose sensor. Sensors and Actuators B: Chemical 146(1), 1–7 (2010)

[7] Merli, F., Bolomey, L., Meurville, E., Skrivervik, A.K.: Dual band antenna for subcutaneous telemetry applications. In: Proc. IEEE Antennas and Propagation Society Int. Symp. (APSURSI), pp. 1–4 (2010)

[8] Merli, F., Bolomey, L., Zürcher, J.-F., Corradini, G., Meurville, E., Skrivervik, A.K.: Design, realization and measurements of a miniature antenna for implantable wireless communication systems. IEEE Trans. Antennas Propag. 59(10), 3544–3555 (2011)

[9] Johannessen, E.A., Wang, L., Cui, L., Tang, T.B., Ahmadian, M., Astaras, A., Reid, S.W.J., Yam, P.S., Murray, A.F., Flynn, B.W., Beaumont, S.P., Cumming, D.R.S., Cooper, J.M.: Implementation of multichannel sensors for remote biomedical measurements in a microsystems format. IEEE Trans. Biomed. Eng. 51(3), 525–535 (2004)

[10] Kurtz, S.M., Devine, J.N.: PEEK biomaterials in trauma, orthopedic, and spinal implants. Biomaterials 28(32), 4845–4869 (2007)

[11] Merli, F., Fuchs, B., Mosig, J.R., Skrivervik, A.K.: The effect of insulating layers on the performance of implanted antennas. IEEE Trans. Antennas Propag. 59(1), 21–31 (2011)

[12] Bolomey, L., Meurville, E., Ryser, P.: Implantable ultra-low power DSP-based system for a miniature chemico-rheological biosensor. In: Proceedings of the Eurosensors XXIII Conference, vol. (1), pp. 1235–1238 (2009)

[13] Bradley, P.D.: Implantable ultralow-power radio chip facilitates in-body communications. RF Design (online magazine) (2007),
http://rfdesign.com/next_generation_wireless/
short_range_wireless/706RFDF1.pdf

[14] De Micheli, G., Ghoreishizadeh, S., Boero, C., Valgimigli, F., Carrara, S.: An integrated platform for advanced diagnostics. In: Proc. Design, Automation and Test in Europe (DATE 2011), pp. 2995–2999 (2011)

[15] Valdastri, P., Susilo, E., Forster, T., Strohhofer, C., Menciassi, A., Dario, P.: Wireless implantable electronic platform for chronic fluorescent-based biosensors. IEEE Trans. Biomed. Eng. 58(6), 1846–1854 (2011)

[16] Audet, S., Herrmann, E.J., Receveur, R., et al.: Medical applications. Comprehensive Microsystems, 421–474 (2008)

[17] Valdastri, P., Susilo, E., Förster, T., Strohhöfer, C., Menciassi, A., Dario, P.: Wireless implantable electronic platform for blood glucose level monitoring. In: Proceedings of the Eurosensors XXIII Conference, vol. (1), pp. 1255–1258 (2009)

[18] Valdastri, P., Menciassi, A., Arena, A., Caccamo, C., Dario, P.: An implantable telemetry platform system for in vivo monitoring of physiological parameters. IEEE Trans. Inf. Technol. Biomed. 8(3), 271–278 (2004)

[19] Cheney, C.P., Srijanto, B., Hedden, D.L., Gehl, A., Ferrell, T.L., Schultz, J., Engleman, E.A., McBride, W.J., O'Connor, S.: In vivo wireless ethanol vapor detection in the wistar rat. Sens. Actuators B Chem. 138(1), 264–269 (2009)

[20] DSI PhysioTel PA-C10 for Mice, data sciences international (2005)

[21] Brouard, Barrandon: Private Communication

Recent Results in Computer Security
for Medical Devices

Shane S. Clark and Kevin Fu

Department of Computer Science
University of Massachusetts Amherst
{ssclark,kevinfu}@cs.umass.edu

Abstract. The computer security community has recently begun research on the security and privacy issues associated with implantable medical devices and identified both existing flaws and new techniques to improve future devices. This paper surveys some of the recent work from the security community and highlights three of the major factors affecting security and privacy solutions for implantable medical devices: fundamental tensions, software risks, and human factors. We also present two challenges from the security community with which the biomedical community may be able to help: access to medical devices and methods for *in vitro* experimentation.

Keywords: security, privacy, imd, wireless, risk, human factors.

1 Introduction

The computer security community has shown significantly increased interest in implantable medical devices (IMDs) in the last few years. Security researchers have identified a number of security and privacy flaws in devices that are widely implanted in patients and have begun to suggest technologies to mitigate the associated risks [19,17,11,12]. Many of the issues identified are attributable to the recent widespread adoption of networked, and especially wireless, interfaces in IMDs. Paul Jones of the U.S. Food and Drug Administration has said:

> "The issue of medical device security is in its infancy. This is because, to date, most devices have been isolated from networks and do not interoperate. This paradigm is changing now, creating new challenges in medical device design." (personal communication, Aug. 2007)

IMDs behave like any other networked computing devices in many ways, and, as such, many existing security and privacy risks apply to them. However, computer scientists find IMDs to have unconventional peripherals (e.g., electrical connections to control cardiac tissue). These unique characteristics demand that device designers take care in the adoption of security and privacy mechanisms to this domain.

This paper summarizes some of the work done by security researchers in the area of IMDs, with the intention of fostering effective collaborations between the

K.S. Nikita et al. (Eds.): MobiHealth 2011, LNICST 83, pp. 111–118, 2012.

Table 1. References to major subtopics in medical device security

Medical Device Security Subtopic	References
Access Control	[5,8,9,10,11,19,20]
Emergent Threats	[12]
Encryption	[5,10,11,17]
Failures	[11,14,15,17,19]
Foundations & Design Principles	[8,10,26]
Hardware	[9,11,14,20]
Human Factors	[4,23,26]
Policy	[6,7,8]
Privacy	[4,6,9,10,11,17,19,20]
Software	[7,8,12,15,18,26]
Specifications	[6,7,8,18]
Wireless	[5,8,9,10,11,17,19,20]

security and biomedical communities. To this end, we discuss three major security and privacy issues that are vital to future research in the area of implantable medical devices: fundamental tensions, software risks, and human factors. We also describe two major challenges for the security community: access to medical devices and methods for *in vitro* experimentation. While not an exhaustive survey, this paper provides the biomedical engineer with several footholds to search for literature about security and privacy for medical devices.

2 Fundamental Tensions: Security, Privacy, Utility, Safety

One of the first key issues with IMDs recognized by security researchers is the existence of fundamental tensions between security and privacy goals and traditional goals such as utility and safety [10]. The strong application of access control and cryptography could endanger patients in the case of an emergency if healthcare professionals are unable to gain access to a device. The status quo, on the other hand, leaves patients vulnerable to malicious parties who could potentially disable an IMD or even use it to induce a life-threatening condition [11,17,19]. A balance must be struck between these two extremes.

Security researchers have already proposed some approaches that seek a balance. One proposal is the use of proximity-based access control [20]. By using a technique known as distance bounding, new IMDs could determine how close a device programmer is and only allow access to those nearby. This approach mitigates the effects of malicious behavior because it would require the adversary to be close to the potential victim, potentially alerting the patient to the presence of a threat and allowing him to seek safety. The likelihood that a patient would recognize an unsafe situation and react appropriately, however, is unknown. The main advantage of an approach like distance bounding is that it would not complicate interactions for medical staff because physical proximity is not a barrier in a clinical setting.

There is no "one size fits all" solution to the tensions between security and privacy and utility and safety. Other proposals include the addition of a second, removable device intended to enforce security and privacy goals [5,9], a battery-less proxy to handle access control [11], and ultraviolet-ink tattoos to store device keys for emergency access [21]. See Figure 1. Different IMDs bear different risks, so the appropriate balance to strike depends upon the particular class of device under consideration.

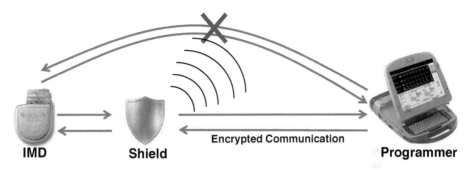

Fig. 1. An illustration of the shield device proposed by Gollakota et al. The shield device jams any radio communication with the IMD unless the programmer has first authenticated with the shield. Removing the shield provides unauthenticated access in case of emergency. Image from [9] used with permission.

3 Hardware vs. Software Security Risks

The increasing size and complexity of the software used in IMDs is another key concern in security and privacy research [13]. As software complexity increases, so do the interactions amongst software artifacts and thus the likelihood of vulnerabilities. For instance, software complexity combined with promiscuous communication can lead to an emergent risk of malware. The recent addition of network interfaces to IMDs massively expands the complexity of the software system that must be protected. Not only must every device designed to communicate with the IMD be trustworthy, but, "any component capable of communication with the device [must] be trustworthy [8]." While it is tempting to write off malware as a concern only for PCs, these devices are increasingly likely to communicate with IMDs and this communication can be used as an infection vector.

There are at least two real-world examples that emphasize the potential impact of malware on networked IMDs. The Stuxnet worm is a computer worm that propagates through Windows computers and seeks out a specific model of Programmable Logic Controller (PLC) used to control some industrial centrifuges for nuclear enrichment. Once the PLCs are infected, the worm causes the attached centrifuge to spin out of control [3,24]. The Stuxnet worm is the most widely known example of a computer worm capable of destroying real-world systems, but it is not the only one. More recently, Hanna et al. successfully loaded custom firmware on the Cardiac Science G3 Plus Automated External Defibrillator (AED) because the AED did not verify the authenticity and freshness of a software update. This type of vulnerability is an indication that a malicious party could potentially create a self-replicating worm that spreads to many AEDs. Such a worm could prevent infected AEDs from delivering life-saving shocks or cause them to deliver shocks of arbitrary strength while appearing to function normally [12].

As the complexity and size of a software system increases, the task of ensuring its security and privacy becomes both more important and more difficult to accomplish. There is evidence that the increasing complexity of IMD software is already taking a toll on the biomedical industry. From 1983 to 1997, 6% of the recalls issued for medical devices containing software were attributable to software failures [25]. From 1999 to 2005, 11.3% of the recalls were attributable to software failures, a near doubling of the software-related recall rate [2].

To combat the rising rate of software faults, traditional software-engineering tools such as requirements specification and static analysis should be applied to IMDs [8]. These techniques are used to control complexity and gain confidence in software. Meaningful requirements specification necessitates the consideration of security and privacy at early stages and static analysis tests the final software artifact. An end-to-end approach is essential to developing trustworthy software because it provides confidence in both the design decisions made, and in their correct implementation [16].

4   Security and Human Factors

Security mechanisms must account for human factors. Users must be willing and able to both understand and enforce their own security and privacy goals. If a security mechanism frustrates a user or lacks an intuitive interface, it will serve only to increase the complexity of the system.

The tendency of users to ignore or incorrectly apply security features should not be discounted. In a seminal work in the area of human factors [26], Whitten and Tygar found that only half of the users they tested were able to success-fully encrypt an email message—even with access to software manuals and hints from the experimenters. Other work in human factors has demonstrated that users are also likely to ignore security warnings, especially those that become commonplace. Sunshine et al. found that in nine out of their ten user tests, the majority of the test subjects chose to ignore web browser warnings about website certificate validity and proceed to potentially harmful websites [23]. In half of these tests, at least 90% of the subjects chose to ignore the security warnings.

As Whitten and Tygar emphasize, one major obstacle to the adoption and effective use of security and privacy mechanisms is the fact that security is generally a secondary goal. Just as the users in Whitten and Tygar's study simply wanted to send email, IMD patients may simply want to receive treatment for their existing medical conditions. Convincing users to think about and consciously manage a secondary aspect of their IMDs is no small task.

Denning et al. carried out a user study with 13 IMD patients in which the patients were asked about their attitudes toward IMD security and privacy, as well as whether they liked or disliked a variety of security and privacy solutions [4]. The participants' responses revealed that most were concerned about security and privacy in general, but they showed comparatively little concern about specific scenarios. 10 out of 12 participants agreed that they were concerned about the safety and privacy of their electronic information and 9 out of 11 agreed that they were concerned about their physical safety. When asked more detailed questions about security and safety, 10 out of 12 participants disagreed that they were concerned about someone changing their IMD settings without their permission and 7 out of 10 disagreed that they were concerned about medical staff being unable to change settings on their IMDs in the case of an emergency. Despite these somewhat contradictory results, 7 out of 9 participants agreed that something should be done to protect the security of future IMDs [4].

The adoption and usefulness of any new security or privacy technology hinges on user understanding and willingness to participate. Human factors in computer security are still a lively area of research because of the complexities involved with providing understandable options and motivating users to adopt sound security practices. How human factors will affect security and privacy for IMDs is still unclear. More work that specifically addresses this issue is necessary, but any solution that is proposed in the future must take human factors under careful consideration.

5   Challenges for Computer Science

Two major challenges are that (1) computer security researchers seldom have access to real medical devices for experimentation, and (2) the computer security community is largely disjoint from the biomedical engineering community. While security researchers have made some recent progress in understanding the security and privacy issues associated with networked IMDs, the area is still largely unexplored and there are significant barriers to entry.

The ICDs and pacemakers used in all of the published research from the security community were explanted devices obtained from patients or healthcare providers. The explanted devices that our lab uses for experimentation are generally older models, often have aging batteries entering the elective replacement indicator, and none have intact leads. Tests of phenomena such as RF interference are difficult or impossible to carry out in a repeatable way without access to the complete system. It is also difficult to acquire a large number of identical, or even similar, devices.

To make devices more available for reproducible experiments, the Open Medical Device Research Library (OMDRL) now provides explanted medical devices for research in trustworthy computing [1]. Researchers also need open access to hardware-software platforms (e.g., open source pacemakers) to innovate. Otherwise researchers will likely focus on identifying anecdotal vulnerabilities in devices found on eBay rather than innovating new technologies that improve security and privacy.

The security community also faces major obstacles in designing and completing reproducible *in vitro* experiments because of its isolation from the biomedical

(a)   (b)

Fig. 2. (a) The human analogue used by Halperin et al. to prototype defenses [11]. The plastic bag was filled with ground beef and bacon in which a hardware prototype was embedded. (b) Our own FDA-inspired prototype, which is a carefully calibrated saline bath with electrode plates.

community. Security and privacy are issues that must be addressed at a variety of layers. While most of the published vulnerabilities are the result of digital communication interfaces that do not require *in vitro* experiments to validate, many of the proposed defenses rely on physical-layer properties [11,20,9]. Halperin et al. were the first security researchers to attempt a realistic *in vitro* experiment. They chose to use a plastic bag full of hamburger and bacon to approximate a human torso (see Figure 2) [11]. Despite a lack of scientific rigor, bags of meat remain the state of the art testing methodology for computer scientists working with IMDs. Lacking familiarity with the literature from the biomedical community, it is difficult for security researchers to determine proper procedures, or even to find the appropriate standards to follow. Our own group has recently begun prototyping new testing setups based on published literature from the biomedical community (see Figure 2) [22]. The state of *in vitro* testing in the security community is improving, but the biomedical community still has an opportunity to improve the testing methodology used by actively engaging in collaborations with security researchers.

6   Conclusion

This paper presents three major considerations that must be addressed by researchers working on security and privacy for IMDs and outlines two challenges from the computer science community. Our hope is that future research can leverage the strengths of both the computer science and biomedical communities to produce new and effective approaches to IMD security and privacy.

Acknowledgments. We thank Wayne Burleson, Gesine Hinterwalder, and Ben Ransford for their feedback on early drafts. This research is supported by NSF CNS-0831244, a Sloan Research Fellowship, an NSF graduate research fellowship, and Cooperative Agreement No. 90TR0003/01 from the Department of Health and Human Services. Its contents are solely the responsibility of the authors and do not necessarily represent the official views of the DHHS or NSF.

References

1. Open Medical Device Research Library, http://www.omdrl.org/
2. Bliznakov, Z., Mitalas, G., Pallikarakis, N.: Analysis and Classification of Medical Device Recalls. World Congress on Medical Physics and Biomedical Engineering (2006)
3. Broad, W.J., Markoff, J., Sanger, D.E.: Israeli Test on Worm Called Crucial in Iran Nuclear Delay (2011),
   http://www.nytimes.com/2011/01/16/world/middleeast/16stuxnet.html
4. Denning, T., Borning, A., Friedman, B., Gill, B.T., Kohno, T., Maisel, W.H.: Patients, Pacemakers, and Implantable Defibrillators: Human Values and Security for Wireless Implantable Medical Devices. In: International Conference on Human Factors in Computing Systems (2010)
5. Denning, T., Fu, K., Kohno, T.: Absence Makes the Heart Grow Fonder: New Directions for Implantable Medical Device Security. In: USENIX Workshop on Hot Topics in Security (2008)
6. Fu, K.: Inside Risks, Reducing the Risks of Implantable Medical Devices: A Prescription to Improve Security and Privacy of Pervasive Health Care. Communications of the ACM 52(6), 25–27 (2009)
7. Fu, K.: Software Issues for the Medical Device Approval Process. Statement to the Special Committee on Aging, United States Senate, Hearing on a Delicate Balance: FDA and the Reform of the Medical Device Approval Process (2011)
8. Fu, K.: Trustworthy Medical Device Software. Public Health Effectiveness of the FDA 510(k) Clearance Process: Measuring Postmarket Performance and Other Select Topics, Workshop Report (2011)
9. Gollakota, S., Hassanieh, H., Ransford, B., Katabi, D., Fu, K.: They Can Hear Your Heartbeats: Non-Invasive Security for Implanted Medical Devices. ACM SIGCOMM (2011)
10. Halperin, D., Heydt-Benjamin, T.S., Fu, K., Kohno, T., Maisel, W.H.: Security and Privacy for Implantable Medical Devices. IEEE Pervasive Computing 7, 30–39 (2008)
11. Halperin, D., Heydt-Benjamin, T.S., Ransford, B., Clark, S.S., Defend, B., Morgan, W., Fu, K., Kohno, T., Maisel, W.H.: Pacemakers and Implantable Cardiac Defibrillators: Software Radio Attacks and Zero-Power Defenses. In: IEEE Symposium on Security and Privacy (2008)
12. Hanna, S., Rolles, R., Molina-Markham, A., Fu, K., Song, D.: Take Two Software Updates and See Me in the Morning: The Case for Software Security Evaluations of Medical Devices. In: USENIX Workshop on Health Security and Privacy (2011)
13. Israel, C.W., Barold, S.S.: Pacemaker Systems as Implantable Cardiac Rhythm Monitors. American Journal of Cardiology (2001)
14. Lee, S., Fu, K., Kohno, T., Ransford, B., Maisel, W.H.: Clinically Significant Magnetic Interference of Implanted Cardiac Devices by Portable Headphones. Heart Rhythm Journal 6(10), 1432–1436 (2009)
15. Leveson, N.G., Turner, C.S.: An Investigation of the Therac-25 Accidents. Computer 26(7), 18–41 (1993)

16. Leveson, N.G.: Safeware: System Safety and Computers. Addison-Wesley (1995)
17. Li, C., Raghunathan, A., Jha, N.K.: Hijacking an Insulin Pump: Security Attacks and Defenses for a Diabetes Therapy System. In: IEEE International Conference on e-Health Networking, Applications and Services (2011)
18. Networking and Information Technology Research and Development Program: High-Confidence Medical Devices: Cyber-Physical Systems for 21st Century Health Care (2009)
19. Paul, N., Klonoff, D.C.: Insulin Pump System Security and Privacy. In: USENIX Workshop on Health Security and Privacy (2010)
20. Rasmussen, K.B., Castelluccia, C., Heydt-Benjamin, T.S., Capkun, S.: Proximity-Based Access Control for Implantable Medical Devices. In: ACM Conference on Computer and Communications Security (2009)
21. Schechter, S.: Security that is Meant to be Skin Deep: Using Ultraviolet Micropigmentation to Store Emergency-Access Keys for Implantable Medical Devices. In: USENIX Workshop on Health Security and Privacy (2010)
22. Seidman, S.J., Ruggera, P.S., Brockman, R.G., Lewis, B., Shein, M.J.: Electromagnetic Compatibility of Pacemakers and Implantable Cardiac Defibrillators Exposed to RFID Readers. International Journal on Radio Frequency Identification Technology and Applications (2007)
23. Sunshine, J., Egelman, S., Almuhimedi, H., Atri, N., Cranor, L.F.: Crying Wolf: An Empirical Study of SSL Warning Effectiveness. In: USENIX Security Symposium (2009)
24. The Stuxnet Worm, http://www.symantec.com/business/outbreak/index.jsp?id=stuxnet/
25. Wallace, D., Kuhn, D.: Failure Modes in Medical Device Software: An Analysis of 15 Years of Recall Data. International Journal of Reliability Quality and Safety Engineering (2001)
26. Whitten, A., Tygar, J.: Why Johnny Can't Encrypt: A Usability Evaluation of PGP 5.0. In: USENIX Security Symposium (1999)

Wearable System for EKG Monitoring - Evaluation of Night-Time Performance

Antti Vehkaoja[1], Jarmo Verho[1], Alper Cömert[2],
Markku Honkala[3], and Jukka Lekkala[1]

[1] Tampere University of Technology, Department of Automation Science and Engineering
[2] Department of Biomedical Engineering
[3] Institute of Fibre Materials Science,
Korkeakoulunkatu 3, 33720 Tampere, Finland
{antti.vehkaoja,jarmo.verho,alper.comert,
markku.honkala,jukka.lekkala}@tut.fi

Abstract. We evaluated the night-time heart rate recognition performance of our wireless EKG measurement system with four persons during fourteen nights. The system uses fabric electrodes, which rely on moisture of the skin as the electrolyte for optimal operation. Even with the small amount of sweating during night, we achieved an average of 99.8 % R-peak recognition rate when data transmission failures of the wireless network were not considered. Based on our test the performance of the textile electrodes is on par with commercial disposable gel EKG electrodes that were used as a comparison.

Keywords: Wearable EKG, wireless network, textile electrode, night-time heart rate.

1 Introduction

Interest towards the night-time heart rate (HR) and heart rate variability (HRV) monitoring has been growing when understanding of the heart rate control mechanisms of the autonomous nervous system has increased. Besides medical applications, also wellness and security applications, such as elderly monitoring, may utilize the results of the night-time HR and HRV research.

Variety of methods exists and is currently being developed for night-time HR measurement. When developing and validating new methods for monitoring night-time physiological signals, it is important to have a reliable source of reference data. Traditionally contact EKG recorded with regular gel electrodes have been considered as the most reliable source of the reference data.

As a part of our earlier research, we developed a wearable and wireless system for measurement and on-line analysis of EKG during group rehabilitation training. The initial version of the system was presented in [1]. Our system that relies on the use of fabric electrodes made from silver coated polyamide yarn has proven to work very well in its aerobic exercise monitoring application where sweating provides the necessary electrolyte between the electrodes and skin. We wanted to test how reliable the system is in night-time measurement, where usually the amount of sweat is limited, but on the

K.S. Nikita et al. (Eds.): MobiHealth 2011, LNICST 83, pp. 119–126, 2012.
© Institute for Computer Sciences, Social Informatics and Telecommunications Engineering 2012

other hand, the amount of movement artifacts is also small. We are planning to use our system as a reference method when developing and validating other night-time HR measurement techniques and therefore its reliability needs to be known well.

2   Background and Related Work

Night-time EKG and HR information have been measured for diagnosing cardiac malfunctions and for other medical purposes for a long time. These measurements are normally done with Holter cardiac monitors that usually record from two, up to twelve EKG measurement leads. Currently, these devices are becoming smaller and more comfortable to wear because memory cards have replaced C-cassettes as storage medium. In spite of this trend, Holter devices are cardiac data loggers and the data must always be offloaded from the memory card for viewing or post-processing. This is acceptable in a case of an occasional medical test, but for more regular use, this would be too cumbersome. In our measurement system, all data is relayed in real-time to a computer from which it can be automatically sent further to a hospital server if desired or analyzed on-site and reported directly to the user.

Cerutti *et al* investigated wearable monitoring of biomedical signals in an EU project called "My-Heart". In a sub-project of My-Heart, a wearable monitoring and analysing system, "Take-Care", was developed and used to collect physiological information during sleep [2]. The same group has also further studied possibilities to use HRV and respiratory signals recorded with wearable devices in sleep stage classification [3]. In spite of the work reported in [2] and [3] that utilizes the signals recorded with a wearable systems and textile electrodes, we have not been able to find a study where night-time HR recording performance of textile electrodes has been compared with gel electrodes.

All the aforementioned systems, including ours, require measurement devices to be worn by the user, which may disturb the sleep of a sensitive person. Another possibility for monitoring night-time EKG and HR is to attach the electrodes to the bed. This method has been investigated by Devot *et al* in [4], where they achieved an 81.8 % average recognition rate for heart beats. Peltokangas *et al* [5] received an average of 95.1 % recognition coverage during 22 measurement nights. Their system uses seven bipolar channels that measure EKG with textile electrodes. The electrodes are sewn on a bed sheet and the measurement is done from the area of upper torso.

An important benefit of the wearable HR measurement systems, when compared to those installed in the bed, is that the origin of the data can be known with high confidence also when more than one person is sleeping in the bed. Another benefit of wearable measurement is that the data can be collected also when the test person gets up from the bed e.g. for going to the bathroom during the night. This enables a longer time HRV analysis to be done from the uninterrupted signals.

Night-time HR information can also be recorded by using a regular heart rate belt. Modern belts also measure beat-to-beat HR, which can be used for HRV analysis. However, the data does not contain the original EKG signal and therefore the reliability of the HRV information or possible EKG irregularities cannot be verified afterwards.

3   Materials and Methods

3.1   Monitoring System

Our monitoring system consists of a wearable unit that includes the electrodes and the measurement hardware, a wireless network, and signal processing software running on a PC. The monitoring system also includes HR display modules that could, in addition to showing the HR, be used for relaying other information related with the quality of sleep to the person after awakening. Our system was originally designed for EKG monitoring and analysis during cardiac group rehabilitation training. The system has been designed to conform to the European medical safety standards and it has received a CE approval as a class IIa medical device.

Wearable Unit. The electrodes as well as the detachable EKG measurement unit are worn with a special measurement shirt shown on the right side of Fig. 1 The electrode setup includes three textile electrodes, two at the chest area (attached onto a chest belt) and one around the right arm attached to the shirt sleeve. All electrodes are made from 275 dtex silver plated multifilament nylon (polyamide 6.6) yarn embroidered onto a textile base. The chest electrodes are oval shaped, whereas the arm electrode has a square wave shape embroidered on a stretchable ribbon. This special shape was selected based on work presented in [4]. The left side chest electrode is placed at the location of V4 electrode of the 12-lead EKG system, the right one being placed similarly but mirrored to the right side. We have been using both regular silicone covered copper cables and braided textile cables as electrode wires. The conductive core of the textile cables is made from the same conductive silver yarns as the electrodes. The type of the cable has not had any significant effect to the result in this application.

The shirt provides high wearing comfort when compared to the other current commercial Holter devices. Other benefits of using an EKG shirt are the easy and exact attachment of the electrodes onto right positions when using personalized shirts. Movement artifacts are also decreased and the fabric electrodes irritate the skin less than the regular EKG electrodes that contain adhesives.

Wireless Network and Measurement Hardware. Our wireless network is based on the IEEE 802.15.4 standard and is realized as a star network using TDMA channel access. Eight measurement units and displays are supported in one network. The measurement units have two EKG channels with 250 Hz/channel sampling rate and 16-bit dynamics.

Measurement channel 1 records EKG between the two chest electrodes and measurement channel 2 records EKG between the left chest and the arm electrode. The EKG amplifiers have 0.2 – 40 Hz pass-band that is realized with 1st order analog high-pass and 2nd order analog low-pass filters. The amplifier input has 250 mV DC voltage tolerance in order to withstand the potentially high difference of the half cell potentials of the textile electrodes.

The measurement units include a buffer memory of eight data packets, which equals 0.8 seconds, to cover transient data losses. Fig. 1 shows the components of the measurement system.

Fig. 1. Night time HR measurement hardware. Measurement unit marked as 1, display unit as 2, and network coordinator as 3.

Signal Processing. Even though the measurement units have their own analog high and low pass filters, the measured signals are further filtered at the PC. The baseline wandering is removed by a nonlinear filtering method presented by Keselbrener *et al* in [7]. This very simple method, which removes a 100 ms median value from the data, has shown to be especially efficient. This technique leaves the R-peaks untouched and therefore the R-R intervals unaltered. The data is also low-pass filtered with 10th order Butterworth filter with a cut-off frequency of 40 Hz to remove the 50 Hz powerline noise and other high frequency interference sometimes seen in the signal. The filter is implemented in forward-backward fashion to avoid any phase shifting.

QRS complexes are recognized simply by their falling edge. Voltage level drop during 40 ms is compared to a threshold value. When falling edge has been found the preceding local maximum which is the R peak is the searched. Recognition threshold for the voltage drop is adaptive. It is set to a half of the peak-to-peak amplitude of the signal during the previous second. The R-peak candidates found are then filtered based on their temporal separation and amplitude. The minimum separation of two consecutive peaks is 250 ms. The minimum amplitude requirement prevents finding false R-peaks from noise signal in case of loss of the electrode contact.

3.2   Test Setup

We tested the performance of our measurement system with four test subjects. All subjects were male and none of them had known medical conditions that could have affected the results. Demographical data of the subjects is shown in Table 1.

The subjects were given instructions about wearing the measurement shirt and fastening the electrodes. They were also instructed to place the network coordinator inside two meter radius from the bed. The recordings with textile electrodes were made during a total of 14 nights. During eleven nights, the subjects were also wearing another measurement unit that recorded EKG with standard gel electrodes. In gel electrode measurements the electrodes were placed below the chest belt textile electrodes and the one corresponding the arm electrode under the right clavicle. The textile electrodes were not artificially moistened during the recordings.

Table 1. Demographical data of the test subjects

Parameter	Person 1	Person 2	Person 3	Person 4
Age	31	23	29	25
Height (cm)	173	184	175	178
Body mass index	26.7	20.7	20.6	24.9
Recorded nights	6	3	3	2
Nights with reference	4	3	3	1

4   Results and Discussion

Fig. 2 shows an example EKG signals recorded with textile and gel electrodes. The signals have been picked from the middle of a night where the person has been asleep. As seen from the signals the quality of EKG recorded with the textile electrode equals the quality of the gel electrode EKG. Fig. 2 also shows the rapid HR variation observed with one test subject. The natural variations of this magnitude make it more difficult to automatically point out all the incorrect QRS detections from the R-R interval data without losing the sensitivity of the detection.

Table 2 shows the results of the test measurements. The average recognition sensitivities, calculated as TruePositives / (TruePositives + FalseNegatives), for textile and gel electrodes are very close to each other. The measurement channel 1 seems to give slightly higher sensitivity in textile recordings which suggests that the channel is less prone to movement artifacts, which are the prevalent cause of both false negative and false positive R-peak detections in textile electrode measurements. In gel recordings the higher QRS amplitude received from EKG channel 2 may be the cause of its slightly better performance. Overall the grand average calculated by weighting the amount of measurement nights of each person, shows the sensitivity of 99.8 % for both electrode types and measurement channels. This is considerably higher than what has been received with electrodes installed in the bed in [4] and [5].

When calculating the false negative and false positive detection rates we do not have had a reliable reference data available because the detection errors occur with both electrode types. Therefore the classification of the beats has been done in part manually and the resulting numbers may contain small errors.

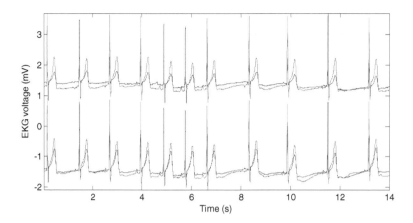

Fig. 2. An example of raw EKG signals recorded with our system. The upper curves are received from the textile electrodes and the lower from the gel electrodes. Blue curves are recorded between the chest electrodes and green between the right arm and the left chest electrode.

Table 2. Average heart rate recognition and data receiving rates for the test subjects

Parameter		Person 1	Person 2	Person 3	Person 4	Grand average
HR recognition sensitivity with textile electrodes (%) *	ch1	99.90	99.70	99.66	99.92	99.81
	ch2	99.89	99.71	99.63	99.86	99.79
HR recognition sensitivity with gel electrodes (%) *	ch1	99.89	99.28	99.96	99.99	99.75
	ch2	99.80	99.62	99.96	100.0	99.81
False positive recognitions, ch 1 / ch 2, textile electrodes**		0/2	3/19	48/53	8/17	12.1 / 18.7
False positive recognitions, ch 1 / ch 2, gel electrodes**		1/3	38/14	17/12	1/5	15.6 / 9.2
Data loss due to radio interruptions, textile recorder (%)		0.59	0.60	0.43	0.49	0.54
Data loss due to radio interruptions, gel recorder (%)		0.12	0.84	0.21	0.01	0.33
Total length of the record with textile electrodes (h)		34.3	20.7	25.5	16.1	
Total length of the record with gel electrodes (h)		23.9	20.7	25.2	8.0	

*   The sections with data loss due to radio interruptions are not taken into account.

** The average number of false positives per night.

Fig. 3 shows an R-R interval series from one measurement night of the test subject 1. The quality and accuracy of the data is sufficient for HRV analysis when sections with no missing QRS-complexes are selected. The waveform in the end of the series, called microarousal, relates to sleep fragmentation and can be used as one measure in sleep quality analysis.

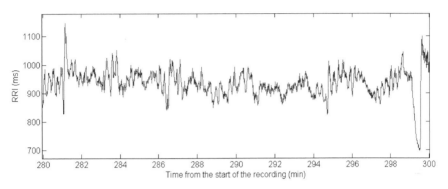

Fig. 3. A 20 minute R-R interval signal produced by our system

As seen from the data loss rates presented in the Table 2, an average of 0.54 % and 0.33 % of the EKG data from textile and gel electrode records were lost during the radio transmission. This is a relatively high value especially when considering that the measurement units include 0.8 second data buffers and that the network allows an average data transmission loss of 25 % while still being able to cover the packet dropouts with retransmissions. The system has performed without complications during the group rehabilitation sessions at various gyms where the transmission distances are much longer as in our tests. The interruptions in night-time measurements mainly occur in bursts at certain time periods during the night. We believe that they are related to decreased radio signal strength due to certain sleeping postures, e.g. sleeping in prone position and having the measurement unit under the body. Unfortunately we do not have any data source for evaluating the correlation of missing data and sleeping posture.

5   Future Work

In the future we are planning to develop the signal processing algorithms so that the QRS detection results of both measurement channels are combined. The target is to further enhance the detection coverage and especially increase the confidence of the correct QRS recognitions, which is important when using the system as a reference method in night-time HR measurements.

The performance of our system still needs to be evaluated with female subjects and people with higher body mass index. The operation of the wireless network also has to be improved.

Acknowledgments. This work was funded by the Finnish Cultural Foundation, Heli and Paavo V. Suominen special fund of Pirkanmaa regional Fund.

References

1. Vehkaoja, A., Verho, J., Cömert, A., Aydogan, B., Perhonen, M., Lekkala, J., Halttunen, J.: System for ECG and heart rate monitoring during group training. In: Proceedings of the 30th Annual International Conference of the IEEE Engineering in Medicine and Biology Society, EMBC 2008, pp. 4832–4835 (2008)
2. Cerutti, S., Bianchi, A.M., Mendez, M.O.: My-Heart Project: Analysis of Sleep and Stress Profiles from Biomedical Signal. In: International Conference on Health Informatics, HEALTHINF 2008, pp. 273–278 (2008)
3. Bianchi, A.M., Mendez, M.O., Cerutti, S.: Processing of signals recorded through smart devices: sleep-quality assessment. IEEE Transactions on Information Technology in Biomedicine 14(3), 741–747 (2010)
4. Devot, S., Bianchi, A.M., Naujokat, E., Mendez, M.O., Brauers, A., Cerutti, S.: Sleep Monitoring Through a Textile Recording System. In: Proceedings of the 29th Annual International Conference of the IEEE Engineering in Medicine and Biology Society, EMBC 2007, pp. 2560–2563 (2007)
5. Peltokangas, M., Verho, J., Vehkaoja, A.: Unobtrusive Night-Time EKG and HRV Monitoring System. In: 10th International Conference on Biomedical Engineering, BiomedEng (2011) (accepted)
6. Cömert, A., Honkala, M., Aydogan, B., Vehkaoja, A., Verho, J., Hyttinen, J.: Comparison of Different Structures of Silver Yarn Electrodes for Mobile Monitoring. In: Proceedings of the 4th European Conference of the International Federation for Medical and Biological Engineering, IFMBE 2009, vol. 22, Part 10, pp. 1204–1207. Springer, Heidelberg (2009)
7. Keselbrener, L., Keselbrener, M., Akselrod, S.: Nonlinear high pass filter for R-wave detection in ECG signal. Medical Engineering & Physics 19(5), 481–484 (1997)
8. Bianchi, A.M., Villantieri, O.P., Mendez, M.O., Cerutti, S.: Signal Processing and Feature Extraction for Sleep Evaluation in Wearable Devices. In: Proceedings of the 28th Annual International Conference of the IEEE Engineering in Medicine and Biology Society, EMBC 2006, pp. 3517–3520 (2006)

A Smartphone-Based Healthcare Monitoring System—PHY Challenges and Behavioral Aspects

Nabil Alrajeh[1], Ezio Biglieri[2,4], Bouchaïb Bounabat[3], and Angel Lozano[4]

[1] Biomedical Technology Department, College of Applied Medical Sciences, King Saud University, Riyadh, KSA
nabilksu@yahoo.com

[2] King Saud University, Riyadh, KSA
e.biglieri@ieee.org

[3] École Nationale Supérieure d'Informatique et d'Analyse des Systèmes, Rabat, Morocco
bouchaib.bounabat@gmail.com

[4] Universitat Pompeu Fabra, Barcelona, Spain
angel.lozano@upf.edu

Abstract. Within the broad theme of wireless healthcare systems, this paper focuses on emergency interventions. Specifically, we address one of the aspects of this problem, namely sensing and transmitting to a health center data regarding an emergency status. No dedicated infrastructure is assumed—rather, the premise is that existing wireless networks be utilized. Our goal is to enable quick and reliable access whenever an emergency arises. For system design, we describe an integrated solution based on the use of wireless/mobile technologies and of smartphones integrated by ad hoc sensors. We examine the possible use of licensed or unlicensed spectrum through two separate techniques: Interweaved Cognitive Radio and a novel version of Interference Alignment, that we call Interference Priority and is tailored to the specific problem of this paper. Validation based on the behavioral aspects of the system is also discussed.

Keywords: Wireless healthcare, Smartphone, Emergency communication, Cognitive radio, Interference alignment, Multi-agent reactive decisional systems.

1 Introduction

Following [1], we can categorize healthcare applications according to the following rubrics: (i) Prevention, (ii) Healthcare maintenance and checkup, (iii) Short-term (or home healthcare) monitoring, (iv) Long-term (or nursing home) monitoring, (v) Personalized healthcare monitoring, (vi) Incidence detection and management, and (vii) Emergency intervention. This paper focuses on the problems posed by last item, and aims at suggesting an integrated solution based on the use of wireless/mobile technologies and of smartphones integrated by *ad hoc* sensors. Our goal is to provide a design rationale and analysis tools for a

K.S. Nikita et al. (Eds.): MobiHealth 2011, LNICST 83, pp. 127–134, 2012.

healthcare system whereby a patient's vital parameters can be constantly sensed and updated round-the-clock via a wearable wireless device, compared with patient data stored in a sensor memory, and transmitted to a center (along with supplementary data as needed) when an emergency status is detected. Aspects of this problem are:

① Choice of vital parameters to be monitored (heart beat rate, position of the patient to detect a fall, ...), and sensors to be used.
② Need for quick detection of the emergency status. This could be done using results from "quickest detection" theory [2].
③ Transmission to the health center of data regarding the emergency status. Transmission should be fast and reliable, and adapted to the level of emergency.
④ Secrecy/security of information should be guaranteed.

The system we advocate operates as follows: The patient's vital parameters are periodically or constantly sensed and compared with stored data. If an anomaly is detected, a preliminary decision is made about its degree of urgency. Then, the data to be transmitted to the health center are selected, along with their transmission protocol, according to the urgency degree detected—the minimum amount of data to be transmitted is an alarm message, supplemented by the last reading of the vital parameters, and might require two-way communication.

This paper focuses on two design and analysis aspects: PHY-level challenges dictated by the system (Section 2), and its behavioral aspects (Section 3).

1.1 Sensing Vital Parameters

Health-care monitoring systems that are compact, lightweight, low-cost, have low power consumption, and are able to identify physiological signals reliably and stably, have long been advocated (see, e.g., [3] and references therein, and [4] for a recent advanced application). Continuous remote monitoring can be achieved by external or implanted sensors, which communicate with the external world in two steps. The first one uses a very-short-range communication technology (e.g., ZigBee- or Bluetooth-based—see also [5]) to connect with a smartphone, and the second transmits sensed data to a remote health data center using standard wireless infrastructure. Within such scheme, pure relay-based functionality of the smartphone may be inappropriate, as it may cause a large energy consumption. For this reason, context-aware processing and data filtering have been advocated to use sensors episodically rather than continuously, and hence reduce the amount of transmitted data [6].

1.2 Prioritization Aspects

Taking into account that reliability is the key factor for any vital-parameter monitoring system suitable for emergency interventions, transmission should be prioritized according to the effects of its delivery failure [7]. Several reliability-based prioritization systems exist, classifying failure consequences from "very low" to "very high" (the effect of a delivery failure might cost the loss of human life). Here, alarm signals will be given the highest priority. Real-time monitoring traffic for patient conditions are the second highest priority, while other medical applications are given a lower priority. One may further categorize priorities as *relative* and *absolute* [7]. Under relative priority, transmission resources is allocated proportionally to priority, while, under absolute priority, high-priority transmissions are granted exclusive access to the channel regardless of the possible presence of low-priority traffic. We can observe that, under relative priority, quality-of-service (QoS) cannot be guaranteed to highest-priority messages if there is a heavy load of low-priority traffic, and hence absolute priority should be chosen whenever an emergency message must be transmitted.

1.3   QoS

The QoS level must be calibrated to the specific system within medical applications (see, e.g., [7]). For example, life-critical emergency information transmission requires a latency shorter than other types of traffic (transmission of abnormal— but not life-threatening—vital parameters, or normal vital parameters) can tolerate. Along with conventional QoS metrics such as delay, throughput, and error rate, unconventional metrics have also been advocated, like medical diagnosability of the data (see, for example, the *weighted diagnostic distortion* discussed in [7]).

2   PHY Challenges

Once no dedicated infrastructure is assumed, two options are available: unlicensed or licensed use of the spectrum. We examine separately the two options, namely:

① Unlicensed use of the spectrum may be realized by Cognitive Radio, operating according to the "interweaving" paradigm [8]. Through spectrum sensing, secondary users can detect an empty spectrum space in which transmission is achieved with a modicum of interference from adjacent spectrum users.

② For primary utilization of licensed spectrum, we advocate a technique we call *Interference Priority (IP)*. This borrows notions from the recent concept of Interference Alignment (IA) [9, 10]. Its guiding principle is to prioritize transmissions by enhancing the spectral efficiency of emergency transmissions while keeping to a minimum the penalty on nonemergency transmissions. IP results in: (i) Higher bit rates, (ii) Broader coverage for critical emergency transmissions, and/or (iii) Lower power consumption, which in turn maps to longer battery lives.

2.1   Cognitive Radio

The version of interweaved cognitive radio we are advocating is based on a *spectrum pooling* concept, whereby the secondary user isolates a portion of the radio spectrum that is not being utilized by the primary user [11, 12]. The spectrum owned by the primary user is divided into subchannels, each assumed to be active at any time with probability less than one. The secondary user, which has the ability of sensing all subchannels, forms a link by combining a number of subchannels depending on their quality and bandwidths, and on the QoS requirements. In principle, results from channel sounding may dictate the number of subchannels to be used by the secondary user, and the adaptive choice of transmission parameters (transmitted power, use of multiple antennas, modulation format, bit rate, coding scheme, etc.). However, for the system examined here we focus on a simpler solution, based on channel sensing, which classifies each subchannel that can be used for emergency communication as a *white space* when it is completely empty—except for noise and adjacent-channel interference—and hence can be used by the secondary user without restriction, or a *black space* when it is fully occupied by primary-communication signals, interfering signals, and noise, and hence cannot be used. When N_s primary-user subchannels are white, they are ordered according to a suitable simple metric (for example, the energy measured in each of them), and the N best are picked to form the secondary link.

2.2   Coding Aspects

We may categorize two types of effects reducing the transmission throughput:

① Presence of thermal noise, fading, and channel impairments.
② Nonzero probability of mistaking a black space for a white one.

Now, while standard error-control coding of the information data can reduce the first effect, nonstandard solutions should be envisaged to compensate for the other one. A simple model for the system examined here, which holds when the transmitted messages are protected by powerful error-control codes but not by an ARQ mechanism (which would require a feedback subchannel and hence introduce further delay), is based on the concept of "block-erasure channel" [13]. As observed in [14, 15], if the packet transmitted over a secondary-user subchannel is received erroneously (i.e., the internal checksum of the packet does not match), then it may not be passed to the higher layers in the protocol stack. Consequently, the application layer sees this as an *erasure*. A packet is received erroneously in a subchannel if the joint effects of noise, fading, and interference from the primary user cause a detection error. This error may or may not be detected, so that an erasure may or may not be declared. A common simplifying assumption is that an erasure occurs if and only if there is a collision, while the packet is so protected by its error-control code that it is received correctly whenever there is no collision (this assumption is valid for high-enough SNR and powerful codes). Thus, under this channel model, a packet is either received correctly, or is lost. The validity of this model leads to advocacy of *erasure-correcting codes*. A family of erasure-correcting codes allowing simpler decoding, and one often advocated for cognitive-radio applications, is that of

Digital Fountain (DF) codes. The use of DF codes involves a computational complexity much lower than for block codes. Concatenation of a powerful error-control code (used within each packet) with an erasure-correcting code (across packets) yields a possible solution, as advocated and analyzed in [16]. Another solution, which does not require N to be large, is based on the use of low-density parity-check codes designed for the block-fading channel, and hence suitable for channels affected by erasures and independent errors. [17]

2.3   Interference Priority

For primary utilization of licensed spectrum, what we advocate a technique we shall refer to as *Interference Priority (IP)*, which borrows notions from the recently proposed concept of Interference Alignment (IA) [9, 10] as well as from more established spatial processing schemes [18]. The guiding principle in IP is to prioritize emergency transmissions, if necessary at the expense of nonemergency transmissions.

In contrast with IA, which is a symmetric strategy in the sense that the interference is aligned at each of the participating receivers, *in IP the interference is aligned only at the receiver for which the emergency transmission is intended.* The constraints on the emergency transmitter are relaxed, allowing it to focus on maximizing the performance of its own link, while the other participating transmitters not only respect the constraint of aligning at the emergency receiver but further minimize the strength of their interference thereon. Altogether, the emergency transmission enjoys a hefty boost, while the other transmissions do not. This translates into broader coverage for those critical events and/or lower power consumption (which, in turn, maps to longer battery lives). Moreover, the asymmetric nature of IP reduces the need for channel-state information; only the channel realizations from each transmitter to the health-care center receiver need be tracked by the transmitters. In IA, alternatively, all channel realizations between each participating transmitter and receiver must be tracked.

3   Behavioral Aspects—An Analysis

Due to its complex nature, also due to its reactive aspects, our system may be hard to specify and validate. To this end, we propose a formal model for its specification and validation. Our approach is based on the concept of a Multi-Agent Reactive Decisional System [19, 20, 21, 22, 23]. Indeed, agents and sensors are elements with a similar nature [24]: they are independent, stand-alone, do not contain information about their whole environment, and finally the creation of a number of agents in necessary quantity is possible. Thus, a sensor can be formally specified and checked as a Decisional Reactive Agent (DRA), and the whole system as a Multi-Agent Reactive Decisional System (MARDS).

The objective here is to use the agent based techniques to formally model and verify WSN behavior and especially its temporal properties. Such modeling and checking should reduce the uncertainty in the system, and ensure the early correction in the development process (prior to the implementation), as well as a higher quality error-free product.

3.1   WSN Behavior Formal Specification and Healthcare System

Formal wireless sensor networks (WSN) behavior specification and verification are essential to critical systems as pervasive healthcare systems, and especially to the network subsystem composed of environmental sensors deployed around mobile or nomadic devices belonging to the patient. Such modeling and checking is particularly needed by the self-organization between these nodes, as it helps to provide rich contextual information about the people to be monitored, as well as about necessary alerting mechanisms [25]. It is also essential for end-user healthcare monitoring applications [25] that the system should prove to be doing the job it was designed to accomplish. Faulty system components and exceptions must not result in system misbehavior.

3.2   DRA Based Approach

Numerous results have shown that the agent paradigm proves useful in the development of complex WSN. The approach we advocate is based on the Decisional Reactive Agent (DRA) paradigm, which is among the most useful concepts used for reactive system modeling, especially in modeling and checking of mobile systems. In reaction to a detected external action, the DRA is able to adapt its behavior in an autonomous way and generate a set of adequate decisions that have to be undertaken within adequate deadlines. DRA-related concepts aim at modeling WSN behavior as a set of interacting reactive agents composing an event-driven real-time system and identifying temporal properties to check its logical correctness, i.e., the adequacy of it responses to dynamics. These concepts can especially improve the robustness of the whole system, in case of failure or malfunctioning of one or several sensors.

A major issue in future research concerns the extension of a DRA-based approach to the Autonomic System view of WSN behavior, integrating other self-management properties as the automatic configuration of components, the automatic monitoring and control of resources to ensure the optimal functioning with respect to the defined requirements.

Acknowledgments. This work was supported by the Project CONSOLIDER-INGENIO 2010 CSD2008-00010 "COMONSENS" and by a Grant from King Saud University, Riyadh, KSA.

References

1. Varshney, U.: Pervasive healthcare and wireless health monitoring. Mobile Netw. Appl. 12, 113–127 (2007)
2. Poor, H.V., Hadjiliadis, O.: Quickest Detection. Cambridge University Press (2009)

3. Chen, C.-M.: Web-based remote human monitoring system with intelligent data analysis for home health care. Expert Systems with Applications 38, 2011–2019 (2011)
4. Kromhout, W.W.: Got flow cytometry? All you need is five bucks and a cell phone. UCLA News (July 26, 2011), `newsroom.ucla.edu/portal/ucla/ucla-engineers-create-cell-phone-210982.aspx`
5. Abouei, J., Brown, J.D., Plataniotis, K.N., Pasupathy, S.: Energy efficiency and reliability in wireless biomedical implant systems. IEEE Trans. Inf. Technol. Biomed. 15(3), 456–466 (2011)
6. Mohomed, I., Misra, A., Ebling, M., Jerome, W.: Context-aware and personalized event filtering for low-overhead continuous remote health monitoring. In: 9th IEEE Int. Symp. on a World of Wireless, Mobile and Multimedia Networks (WoWMoM 2008), Newport Beach, CA, June 23-26, pp. 1–8 (2008)
7. Lee, H., Park, K.-J., Ko, Y.-B., Choi, C.-H.: Wirless LAN with medical-grade QoS for e-healthcare. Journal of Communications and Networks 13(2), 149–159 (2011)
8. Biglieri, E., Goldsmith, A., Greenstein, L.J., Mandayam, N., Poor, H.V.: Principles of Cognitive Radio. Cambridge University Press, Cambridge (2011) (to be published)
9. Cadambe, V.R., Jafar, S.A.: Interference alignment and degrees of freedom of the K-user interference channel. IEEE Trans. Inform. Theory 54(8), 3425–3441 (2008)
10. Maddah-Ali, M., Mohatari, A., Khandani, A.: Communication over MIMO X channels: Interference alignment, decomposition, and performance analysis. IEEE Trans. Inform. Theory 54(8), 3457–3470 (2008)
11. Čabrić, D., Mishra, S.M., Willkomm, D., Broderson, R.W., Wolisz, A.: A cognitive radio approach for usage of virtual unlicensed spectrum. In: 14th IST Mobile Wireless Summit 2005, Dresden, Germany, June 19-22 (2005)
12. Weiss, T., Jondral, F.: Spectrum pooling: An innovative strategy for the enhancement of spectrum efficiency. IEEE Commun. Mag. 42, S8–S14 (2004)
13. Guillén i Fàbregas, A.: Coding in the block-erasure channel. IEEE Trans. Inform. Theory 52(11), 5116–5121 (2006)
14. Kushwaha, H., Xing, Y., Chandramouli, R., Heffes, H.: Reliable multimedia transmission over cognitive radio networks using fountain codes. IEEE Proc. 96(1), 155–165 (2008)
15. Kushwaha, H., Xing, Y., Chandramouli, R., Subbalakshmi, K.P.: Erasure tolerant coding for cognitive radios. In: Mahmoud, Q.H. (ed.) Cognitive Networks: Towards Self-Aware Networks, pp. 315–331. J. Wiley & Sons, Chichester (2007)
16. Berger, C.R., Zhou, S., Wen, Y., Willett, P., Pattipati, K.: Optimizing joint erasure- and error-correction coding for wireless packet transmission. IEEE Trans. Wireless Commun. 7(11), 4586–4595 (2008)
17. Boutros, J.J., Guillén i Fàbregas, A., Biglieri, E., Zémor, G.: Low-density parity-check codes for nonergodic block-fading channels. IEEE Trans. Inform. Theory 56(9), 4286–4300 (2010)
18. Rashidi-Farrokhi, F., Foschini, G.J., Lozano, A., Valenzuela, R.A.: Link-optimal space–time processing with multiple transmit and receive antennas. IEEE Commun. Letters 5(3), 85–87 (2001)
19. Aaroud, A., Labhalla, S.E., Bounabat, B.: A new formal approach for the specification and the verification of multi–agent reactive system operating modes. International Journal for Information Processing and Technology (March 2001)
20. Aaroud, A., Labhalla, S.E., Bounabat, B.: Modelling the handover function of Global System for Mobile Communication. International Journal of Modelling and Simulation 25(2) (2005)

21. Bounabat, B.: Méthode d'Analyse et de Conception des Systèmes Orientée Objet Décisionnel. Application aux langages synchrones et aux systèmes répartis. Doctoral Dissertation, Cadi Ayyad University, Faculty of Sciences, Marrakech, Morocco (2000)
22. Bounabat, B., Romadi, R., Labhalla, S.E.: User's behavioural requirements specification for a reactive agent. In: CESA 1998, Nabeul-Hammamet, Tunisia, April 1-4 (1998)
23. Furbach, U.: Formal specification methods for reactive systems. Journal of Systems and Software (21), 129–139 (1993)
24. Podkorytov, D., Rodionov, A., Sokolova, O., Yurgenson, A.: Using Agent-Oriented Simulation System AGNES for Evaluation of Sensor Networks. In: Vinel, A., Bellalta, B., Sacchi, C., Lyakhov, A., Telek, M., Oliver, M. (eds.) MACOM 2010. LNCS, vol. 6235, pp. 247–250. Springer, Heidelberg (2010)
25. Alemdar, H., Ersoy, C.: Wireless sensor networks for healthcare: A survey. Computer Networks 54(15), 2688–2710 (2010)

Phone Based Fall Risk Prediction

Vânia Guimarães[1], Pedro M. Teixeira[1], Miguel P. Monteiro[2], and Dirk Elias[1]

[1] Fraunhofer AICOS Portugal
Rua Alfredo Allen, 455/461, 4200-135 Porto, Portugal
{vania.guimaraes,pedro.teixeira,dirk.elias}@fraunhofer.pt
[2] Faculdade de Engenharia da Universidade do Porto
Rua Dr. Roberto Frias, s/n, 4200-465 Porto, Portugal
apm@fe.up.pt

Abstract. Falls are a major health risk that diminishes the quality of life among older people and increases the health services cost. Reliable and earlier prediction of an increased fall risk is essential to improve its prevention, aiming to avoid the occurrence of falls. In this paper, we propose the use of mobile phones as a platform for developing a fall prediction system by running an inertial sensor based fall prediction algorithm. Experimental results of the system, which we still consider as work in progress, are encouraging making us optimistic regarding the feasibility of a reliable phone-based fall predictor, which can be of great value for older persons and society.

Keywords: Fall Prevention, Fall Risk Prediction, Inertial Sensors, Older people, Gait analysis, Smartphone.

1 Introduction

The progressive ageing of population is creating new social and economic challenges, concerning people's health and well-being. Particularly, falling is a serious and common problem facing older people that frequently leads to injury, suffering, fear, depression, loss of independence, reduced quality of life and death [1]. This is further problematic for older people living in the community, where help or medical assistance can be provided late [2].

To give a faster assistance when a fall occurs, several strategies to alert its occurrence have been developed. These are essentially reactive and don't prevent fall occurrences and some of their related consequences [2].

"Fall prevention" is therefore becoming increasingly important. Besides external risk factors (e.g. slippery floor), medicaments intake, chronic diseases, gait or balance disorders and hazardous activities also contribute to the occurrence of a fall [1]. Consequently, older people presenting some of these risk factors can be considered a high risk target group. Multi-factorial interventions are then applied to modify/eliminate those risks [3].

Nowadays, the risk is based on questionnaires and the assessment of gait and balance disorders, which are among the most consistent predictors of future fall [3]. These tests are typically administered by experts in a clinical environment

K.S. Nikita et al. (Eds.): MobiHealth 2011, LNICST 83, pp. 135–142, 2012.

and are only accessible when a visit to the clinic is necessary, which frequently happens after an injurious fall, so that the application of preventive strategies can be already too late. Also, the equipments used are usually costly, not portable and time-consuming, limiting their use as routine. These clinical-centric models are therefore becoming increasingly unsatisfactory.

1.1 Why Using a Mobile Phone as Fall Predictor

A proactive community-based strategy is necessary in order to earlier recognize increased risks and improve prevention strategies [4]. Some systems based on the use of wearable inertial sensors like accelerometers and gyroscopes have been proposed in recent studies for unsupervised long-term fall risk screening, through the evaluation of functional ability and mobility. These systems have the advantage of being portable, low-cost and easy-to-use. In contrast to clinical tests, these are self-administrable and can be used outside clinical environments, being able to anticipate the detection of problems and therefore to administer/modify prevention strategies at an earlier stage [4].

The popularity of mobile phones is likely to continuously increase in the near future due to decreasing prices, thus projecting an overall acceptance regarding it as a fall prediction platform. Based on these principles, a smartphone is adapted to be used as a fall risk screening tool, using its inertial sensors.

To make maximum explore of phone strengths and to improve fall prevention strategies, other risk factors besides gait and mobility problems were considered. The same questionnaires currently used by doctors were therefore adapted to the phone, so that several risk factors for falling could be identified and monitored over time. Since the information is stored in the smartphone, the historical can be used by the user and/or automatically transmitted to the doctor by gateway capabilities in order to evaluate the risk over time and earlier apply/modify preventive schemes [4].

2 Related Work

In order to detect subtle problems with gait or balance and provide objective measurements of individuals fall risk, instrumented assessments, such as force platforms and cameras, have been developed [12].

Nowadays, the studies focus on the measurement of parameters from wearable inertial sensors signals, which demonstrate advantages in relation with previous methods. Sensors are used to quantify validated medical tests for fall risk screening [2], directed routines with a series of movements/assessment tasks [4] and walking patterns/gait [5,11].

Then, these parameters are proposed to be used to measure the risk of falling, discriminating those at no risk and at risk. Results from [2] suggested that an accelerometer could be used not only to measure the performance of Timed Up and Go Test, but also to extract further data from gait, providing a more depth analysis of the individual's risk. Machine learning techniques can also be used to combine parameters, so that risk stratification can be done [4].

Literature suggests a global quantification of patient's fall risk, not only through measurements of physical performance, but also with other risk factors (e.g. fall history, medication use and balance confidence), emphasizing the multi-factorial nature of falls [6]. As a result, the likelihood of falling is assessed by screening multi-factorial risk factors, and the intervention should be made when a risk factor enters a warning zone [3].

3   Risk Prediction Method

Given the multi-factorial nature of falls and the current problems of solution scalability, we propose a smartphone-based solution which comprises three main modules: the gait analysis test, the clinic questionnaires and the feedback module, as illustrated in Fig. 1.

(a) Gait test   (b) Clinical Questionnaires   (c) Results

Fig. 1. Purposed Smartphone-based Fall Risk Analysis Solution

Several authors have identified that gait speed alone could be used as a simple and quick option to measure fall risk, possibly as an alternative to more complex mobility tests performed in clinical environments [5]. According to [11], a gait speed lower than $70cm/s$ is associated with an increased risk. In our first approach, the mobile phone was placed at the lower back of trunk, attached at the belt (Fig. 1a). This position is stable and near the centre of mass of the body, moving parallel to it, and has been frequently used in the literature [7].

Acceleration data is read from an Android based mobile phone, and the axis are adjusted according to ISB recommendations [13]. After reading the signals, foot contacts detection is done. Foot contacts are obtained through readings of forward acceleration peaks, preceding a change of signal polarity (from positive to negative values) [9].

After foot contacts detection, discrimination between right and left foot contacts was done by analysis of the medio-lateral acceleration profiles of the trunk. As recognized by [9], the major part of the medio-lateral acceleration is to the left during the right support phase, and vice versa.

Toe offs were detected on the vertical acceleration signal, already excluding the static gravity component.

The minimums after heel strikes were considered the toe offs [10].

Since the time stamps of all events were recorded, all gait phases (i.e. stance, swing, single and double support phases) could be properly delimited, both from a left and right perspective.

The calculation of step length was based on two pendulum models: the first relative to swing phase and the second, with an unknown radius, to the double support phase, as described by [10] and [9] (Equation 1). Step length was therefore calculated as the sum of displacement during swing phase (S_1) and the displacement during double stance (S_2). S_1 was derived from leg length (l) and vertical displacement between the time of toe off and heel strike events (h_1). S_2 was set as a constant equal to foot length [10].

$$Steplength = S_1 + S_2 = 2\sqrt{2h_1l - h_1^2} + S_2 \tag{1}$$

Vertical displacement was calculated by double integration of vertical accelera-tion signal, using the trapezoidal rule. To eliminate the problems related with the lack of initial conditions and the presence of acceleration drift, intermedi-ate steps of high pass filtering were required before and after integration [14]. Fast Fourier Transform filtering was used to eliminate the frequencies below the frequency of foot contacts.

Step duration was calculated as the time between two consecutive foot con-tacts [9]. Mean step length and duration was calculated from all available strides. Mean step length divided by mean step duration was used to estimate walking speed.

Table 1. Risk profile

Risk Factor	Risk profile
Fall History (*Have you fallen during the past 12 months?*)	Y / N
ADL difficulties (*Katz ADL score $\leq$ 2?* [15])	Y / N
IADL difficulties (*IADL score $<$ 8?* [16])	Y / N
Gait/Mobility difficulties (*Velocity $<$ 70cm/s?* [11])	Y / N
Balance confidence (*ABCS score $<$ 67%?* [17])	Y / N
Medication Use	
Polypharmacy (*Do you take 4 or more medications?*)	Y / N
Cardiovascular system medications (*diuretics, anithypertensives*)	Y / N
Psychoactive medications (*sedatives, antidepressants*)	Y / N
Musculoskeletal system medications (*narcotics, corticosteroids*)	Y / N
Other (*hypoglycaemics, allergy, cold medications*)	Y / N
Medical Conditions	
Musculoskeletal (*arthritis, ...*)	Y / N
Neurological (*stroke, Parkinson's, ...*)	Y / N
Heart diseases (*postural hypotension, arrhythmias, unstable, ...*)	Y / N
Diabetes	Y / N
Dizziness	Y / N
Psychological function (*FSQ score $\leq$ 70?* [18])	Y / N
Social activities (*FSQ score $\leq$ 78?* [18])	Y / N
Quality of interactions (*FSQ score $\leq$ 69?* [18])	Y / N

Our risk prediction approach also tried to include other risk factors for falling, in order to take the first steps on a multidimensional risk screening method, which would include not only an evaluation of gait, but also other risk factors for falling. Several risk factors which could be self perceived by the person through the use of validated questionnaires were selected. Table 1 summarizes risks and tools that could be included on risk profile.

This profile would give rise to an overall health status score (Fig. 1c) and later the time-dependent risk factors could be used to estimate a likelihood of falling changing over time.

4 Evaluation Method

A group of 14 participants (mean age 26 ± 3.6, height $1.74 \pm 0.1cm$ and weight $73.5 \pm 11.3Kg$) without any visible gait problem and able to walk unassisted without using walking aids participated on the test. The aim of this test phase was to evaluate the algorithm's performance on detecting step length, duration and velocity from acceleration signals during normal gait.

The experimental setup comprised a walkway $5m$ long, with distance markers placed on the ground. The phone was placed inside a case and adjusted around the pelvis using a belt. It was positioned with a known orientation relative to the ground and to the walking direction at the lower back of trunk of each participant. Subjects were asked to walk along the walkway at three different self-selected speeds: comfortable normal pace, slower pace and a faster pace. Each test was repeated one time. During each test, a simultaneous recording of a digital video camera parallel to the ground and of phone sensors was done.

Foot length and leg length were measured experimentally. Leg length was measured as described by [8], from the difference between standing and sitting height. Foot length was measured with the shoes on.

Each video was analysed in order to obtain an estimative of the mean step length, duration and velocity. A frame by frame analysis was necessary to determine the time of each heel contact. Each step length was determined using the information of the distance markers on the ground. Video information was used as a reference to evaluate the results obtained from sensors data.

5 Data Analysis and Discussion

Acceleration signals similar to those reported on the literature could be obtained using phone's sensors. On Fig. 2, an example of an acceleration signal with detected foot contacts is provided.

The maximum deviation from expected (i.e. from video results) and estimated (i.e. from sensors signal analysis) mean step length was 17% and the mean deviation was $7 \pm 5\%$. Deviations of the measured mean step length compared with the expected values are either positive either negative and a tendency to over or underestimation is not observed.

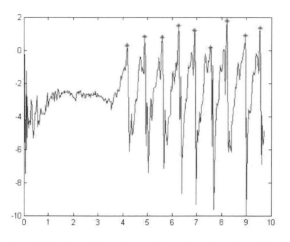

Fig. 2. Foot contacts detection. *blue:* forward acceleration signal (m/s^2) *vs.* time (s); *red:* foot contacts.

The deviations encountered on estimated values seem to be acceptable even if we consider all the sources of error that can be present on step length estimation. First, the model used to estimate this value is more rigid than the real displacement patterns during gait, which presents some variability. For example, leg is not always maintained as a rigid pendulum, like considered, and the displacement during double support not always correspond to foot length, which can be tilt on the direction of progression. Second, errors on acceleration and position signals can be present, so that others are introduced on step length estimation.

The maximum deviation from expected and estimated mean step duration was 3%. No difference was observed in 41% of the considered tests. The mean deviation between measured and real values was very low and equal to $1 \pm 1\%$.

The velocity, calculated as the mean step length divided by mean step duration, is affected by errors at these two estimated values. The maximum velocity deviation from real values is 15%, and the mean deviation is $7 \pm 5\%$. A plot of the comparison between expected and estimated velocities is shown on Fig. 3.

Although errors on velocity estimation are present, they are not very high, so that using phone's sensors to estimate gait velocity seems to be a reliable option when the main purpose is to discriminate high and low risk persons based on velocity. As indication, all the young persons walked at a normal speed higher that $70cm/s$, so that they may not be at risk of falling, as expected.

As a general comment of the adopted method, it can be stated that the walkway had a short size, so that few steps were available, and possibly no enough time was available so that gait was stabilized still within the walkway during each test. In spite of uniform clinical protocols to detect gait abnormalities are lacking, representative measures of gait variability (which are considered as significant fall risk predictors [11]) could be extracted if a larger walkway was available.

For now, nothing can be concluded regarding the kind of signals and quality of detections that are expected to occur when old people (presenting or not

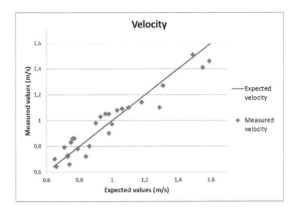

Fig. 3. Expected vs. Estimated velocity. Red line represents the expected velocity.

gait abnormalities) walk. It would be important to validate data from phone's sensors signals on these persons in comparison with the traditional methods of gait analysis, including cameras and force-plates systems.

From the described results, evidences exist that phone's inertial sensors can be used to analyse gait, which would also include the extraction of variability parameters. However, further exploration is necessary.

6   Conclusion

In this paper, a study was done regarding the use of mobile phones as fall risk screening tools, aiming to improve the current fall prevention strategies.

From the results, evidences exist that phone's sensors signals can be used to quantify gait or other movements, by extracting parameters with a relation with the risk of falling. At present, this relation is not well known, but strong evidences exist that several parameters can be combined to screen for fall risk.

Other risk factors can also be assessed using the phone, by using the same questionnaires currently used by doctors at clinics. Evidences exist that in the future all these risk factors can be combined, by attributing time-varying weights to each one, enabling the calculation of a global likelihood of falling. All the assessments can be made in an unsupervised manner and centralized on the phone, so that a complete history of risk factors can be built and transmitted to the doctor.

A greater frequency of assessments would therefore be encouraged, earlier alerting the persons for higher risks and providing new insights about fall prevention strategies.

References

1. National Institute for Health Excellence and Clinical: Clinical practice guideline for the assessment and prevention of falls in older people (CG21), vol. 2. Royal College of Nursing, London (2004)

 2. King, R.C., Atallah, L., Wong, C., Miskelly, F., Yang, G.Z.: Elderly Risk Assessment of Falls with BSN. In: 2010 International Conference on Body Sensor Networks, pp. 30–35 (2010)
 3. Ganz, D.A., Bao, Y., Shekelle, P.G., Rubenstein, L.Z.: Will my patient fall? The Journal of the American Medical Association 297(1), 77–86 (2007)
 4. Narayanan, M.R., Redmond, S.J., Scalzi, M.E., Lord, S.R., Celler, B.G., Lovell, N.H.: Longitudinal falls-risk estimation using triaxial accelerometry. IEEE Transactions on Bio-Medical Engineering 57(3), 534–541 (2010)
 5. Montero-Odasso, M., Schapira, M., Soriano, E.R., Varela, M., Kaplan, R., Camera, L.A., Mayorga, L.M.: Gait velocity as a single predictor of adverse events in healthy seniors aged 75 years and older. The Journals of Gerontology. Series A, Biological Sciences and Medical Sciences 60(10), 1304–1309 (2005)
 6. Laessoe, U., Hoeck, H.C., Simonsen, O., Sinkjaer, T., Voigt, M.: Fall risk in an active elderly population–can it be assessed? Journal of Negative Results in Biomedicine 6(2) (2007)
 7. Senden, R., Grimm, B., Heyligers, I.C., Savelberg, H., Meijer, K.: Acceleration-based gait test for healthy subjects: reliability and reference data. Gait & Posture 30(2), 192–196 (2009)
 8. Bogin, B., Varela-Silva, M.I.: Leg length, body proportion, and health: a review with a note on beauty. International Journal of Environmental Research and Public Health 7(3), 1047–1075 (2010)
 9. Zijlstra, W.: Assessment of spatio-temporal parameters during unconstrained walking. European Journal of Applied Physiology 92, 39–44 (2004)
10. Alvarez, D., Gonzalez, R.C., Lopez, A., Alvarez, J.C.: Comparison of Step Length Estimators from Wearable Accelerometer Devices. In: Proceedings of the 28th IEEE EMBS Annual International Conference, New York City, USA, pp. 5964–5967 (2006)
11. Verghese, J., Holtzer, R., Lipton, R.B., Wang, C.: Quantitative gait markers and incident fall risk in older adults. The Journals of Gerontology. Series A, Biological Sciences and Medical Sciences 64(8), 896–901 (2009)
12. Culhane, K.M., O'Connor, M., Lyons, D., Lyons, G.M.: Accelerometers in rehabilitation medicine for older adults. Age and Ageing 34(6), 556–560 (2005)
13. Wu, G., Cavanagh, P.R.: ISB Recommendations in the Reporting for Standardization of Kinematic Data. Journal of Biomechanics 28(10), 1257–1261 (1995)
14. Slifka, L.D.: An Accelerometer based approach to measuring displacement of a vehicle body. PhD thesis, Horace Rackham School of Graduate Studies of the University of Michigan (2004)
15. Réseau Francophone de Prévention des Traumatismes et de Promotion de la Sécurité. Good Practice Guide - Prevention of falls in the elderly living at home. Éditions inpes (2005)
16. Graf, C.: The Lawton Instrumental Activities of Daily Living Scale. American Journal of Nursing 108(4), 52–62 (2008)
17. Hill, K.: Activities-specific and Balance Confidence (ABC) Scale. Australian Journal of Physiotherapy 51(3), 197 (2005)
18. Society of Hospital Medicine. Website (2011),
    http://www.hospitalmedicine.org/

Infusing Image Processing Capabilities
into an RFID-Based Personal Mobile Medical Assistant

Konstantinos Sidiropoulos[1], Pantelis Georgiadis[2],
Nikolaos Pagonis[2], Nikolaos Dimitropoulos[3], Pantelis Asvestas[2],
George Matsopoulos[4], and Dionisis Cavouras[2]

[1] School of Engineering and Design, Brunel University West London, UK
Konstantinos.Sidiropoulos@brunel.ac.uk
[2] Department of Medical Instruments Technology, Technological Educational Institute of
Athens, Greece
{pasv,cavouras}@teiath.gr, pgeorgiadis@med.upatras.gr,
npagonis@sch.gr
[3] Medical Imaging Department, EUROMEDICA Medical Center,
2 Mesogeion Avenue, Athens, Greece
n.dimitropoulos@euromedica.gr
[4] School of Electrical and Computer Engineering,
National Technical University of Athens, Greece
gmatso@esd.ece.ntua.gr

Abstract. The technological fusion of modern handheld devices, like Personal Digital Assistants (PDAs), wireless networking, and Radio Frequency Identification (RFID) technology, is considered capable of providing the solution to the healthcare community's increasing need to enhance patient safety and reduce medication-dispensing errors by rapid and precise delivery of medical information within an all-wireless digital hospital environment. Therefore, the aim of the current study is to propose a wireless solution that utilizes RFID technology to enable physicians and healthcare professionals to automatically identify patients and easily access their medical information through PDAs, remotely. The developed PDA-based application is capable to identify each patient's RFID tag, retrieve his/her medical data (images, bio-signals, reports, pharmaceutical treatment etc.) and display them by means of a user friendly graphical interface that suites the needs of the healthcare professional. Additionally, the application is further enhanced by incorporating advanced image processing capabilities to improve the diagnostic potential of medical images.

Keywords: PDA, RFID, Medical Assistant.

1 Introduction

According to the 1999 Institute of Medicine report, entitled "To Err Is Human: Building a Safer Health System", medical errors constitute one of the major threats in

K.S. Nikita et al. (Eds.): MobiHealth 2011, LNICST 83, pp. 143–149, 2012.

modern healthcare application resulting in 44.000 to 98.000 deaths in America alone [1]. In order to conceive the magnitude of the imposed threat, the aforementioned report states that deaths due to medical errors in 1998 exceeded the number of deaths attributed to motor vehicle accidents (43.458), breast cancer (42.297), and even AIDS (16.516). What's more, total annual costs associated with medical errors resulting in injury are estimated to be between 17 and 29 billion dollars.

Although, a couple decades ago, the major cause of medical errors was misbelieved to be inadequate training, or even negligence on the part of the healthcare professionals, recent studies have proven that the predominant cause of medical errors is the lack of detailed and timely information concerning the patient. Such information includes patient's medical history and existing conditions, results of clinical tests and examinations, medical images, allergies, and current prescriptions [1-3].

Millions of people suffer from medical conditions that should be made known to healthcare practitioners prior to treatment. Doctors and nurses can not provide optimal care without sufficient knowledge of a patient's medical history. Without access to vital patient information, healthcare professionals are prone either to delay treatment or rely on erroneous or deficient data [3].

Unfortunately, timely access to this information is in most of the cases not practical or even unfeasible, especially in emergency situations [3]. The main reason of this deficiency it that patient medical records are maintained, traditionally, in paper-based format at the offices of individual healthcare providers.

The initial step towards the solution of this problem was made with the growth of electronic medical record (EMR) systems over the past decades. Furthermore, the increased use of mobile devices, Personal Digital Assistants (PDAs), and wireless devices in general, during the past few years, has metamorphosed the way we perceive things around us and has had a radical impact on our working environment. The healthcare sector, where the need for precise and rapid delivery of information is of vital importance, could not remain unaffected by these developments. Therefore, the use of portable PDAs along with wireless LANs can satisfy the need of healthcare professionals for remote access to information.

Currently, the small size and weight of PDA devices provide tremendous convenience and portability. Additionally, with the rapid evolution of electronic technology, PDAs are now capable of accomplishing more challenging tasks, such as reproduction of video sequences and processing of static high quality medical images [4]. Moreover, the technological fusion of modern PDA handheld devices, wireless networking, and RFID technology, is considered capable of providing the solution to the healthcare community's increasing need to enhance patient safety and reduce medication-dispensing errors by rapid and precise delivery of medical information within an all-wireless digital hospital environment.

Hence, in the current study, a previously proposed wireless solution [5] that utilizes RFID technology to enable physicians and healthcare professionals to automatically identify patients and easily access their medical information through PDAs, remotely, is further enhanced by incorporating advanced image processing capabilities in order to improve the diagnostic potential of the system.

2   Methods

2.1   System Design and Implementation

The hardware platform selected for the final prototype of this project was the Qtec 9000. The device features a 520 MHz Intel Bulverde Processor (ARM compatible), 64 MB SDRAM, 128 MB Flash ROM, a Secure Digital Card expansion slot for optionally adding extended memory capabilities and a 3.6" / 89-millimetre diagonal 640x480 64-bit color transflective TFT-LCD with backlight LEDs that supports both landscape and portrait mode. Furthermore, the specific model incorporates a full QWERTY keyboard. It also offers WiFi connectivity through its integrated IEEE 802.11b compliant wireless network card. Bluetooth and IrDA connectivity are also supported. Finally, regarding the operating system, Qtek 9000 arrives with Microsoft Windows Mobile 5.0 preloaded.

In order to add RFID capabilities to the PDA, Wireless Dynamics' SDiD 1020 RFID reader card was utilized. The particular card reads and writes to all ISO 15693, ISO 14443A/B and many proprietary 13.56Mhz RFID tags being used or deployed for asset tracking, access control, process control and healthcare/medical/pharmaceutical applications [6]. Moreover, due to its battery friendly design, the RFID card draws only a few milliamps (11 mA) of current in standby mode, extending the battery life of the PDA. The RFID transponders-tags employed in this study were a member of class 0 (passive, read only), operated at the frequency of 13.56 MHz and were mounted on the patient's wrist bracelet. This choice was based on the fact that RFID systems using this frequency have a large user base and are supported by many RFID manufacturers such as Sony and Phillips [7].

A PDA-based application was developed in C# (C sharp) programming language taking full advantage of the .NET Compact Framework. The software packages used during the implementation phase include: (a) Microsoft Visual C# 2005 Express Edition (Integrated Development Environment), (b) Microsoft .NET Compact Framework SDK (Software Development Kit) 2.0 (c) Microsoft Pocket PC SDK 2003.

It should be noted that the application was developed on a typical desktop PC (Intel Pentium 4 / 1.8GHz with 1GB RAM) running Microsoft Windows XP, while for database management Microsoft SQL Server 2005 Express Edition was employed. An overview of the proposed system is illustrated in Fig. 1.

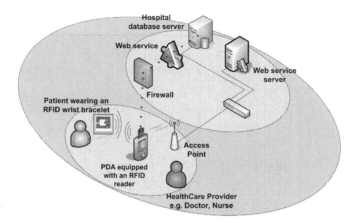

Fig. 1. An overview of the proposed system

2.2   System Functionality

The aforementioned application is capable to control the RFID reader connected to the PDA, and therefore to read each patient's RFID tag (interrogation phase). During this interrogation phase, which lasts only less than a second, the RFID reader should be within a couple of centimetres away from the patient's tag (Fig. 2). Successful completion of this phase results in the acquisition of the tag's unique ID, which is an 8 byte stream.

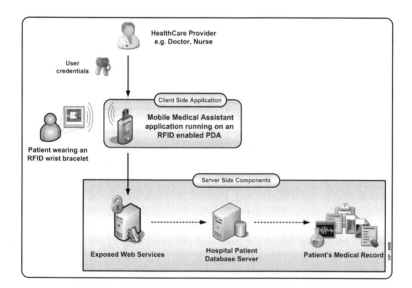

Fig. 2. A typical use case scenario involving all key components of the proposed system

Following the correct identification of the patient through his/her RFID tag, the developed application can query the hospital's database and retrieve his/her medical data comprising medical history, past conditions, medical images, vital signs (temperature, pressure, pulse, etc) presented in graphical view, clinical reports, current pharmaceutical treatment etc.

All the retrieved information is then displayed by means of a user-friendly graphical interface that suits the needs of the healthcare professional. In order to accomplish this, special effort was made during the design of the graphical user interface (GUI), so that the application provides for easy information navigation. Additionally, information is categorized into logically related units for more effective consumption. Moreover, once the patient's medical record has been retrieved from the database, the healthcare professional can easily update it with the latest information.

As far as image viewing is considered, the developed application can be used to download and display a wide variety of images, whether they are color (RGB) or grayscale of the following formats: DICOM, BMP, JPG, GIF, PNG and TIFF. Additionally, it can open practically any size of image, but the maximum visible resolution is limited to 640 x 480 pixels. Larger images can also be viewed using a special 'Shrink to Fit' function that employs the appropriate zoom factor, so that the original image fits in a 640 x 480 frame. The user has also the ability to scroll the image (using the stylus) in its original size. This function is called 'Tap & Scroll'. Zooming and rotation, in angles multiples of 90°, are also featured. Additionally, standard mirroring and flipping image transformations are supported. Finally, the information contained in the DICOM header can be accessed to obtain valuable information regarding the displayed image.

The application can process the loaded images using special-purpose algorithms to enhance image quality. One of the most commonly used image enhancement method is the "windowing correction" technique [8], which is used in the application in two ways: (a) by window-width and window-level adjustment using two slider bars and (b) by stylus movement, for adjusting image brightness and contrast. In addition, the application uses image enhancement techniques for (a) contrast enhancement by means of histogram modification (Cumulative Density Function based Histogram Equalization), (b) typical 2-dimensional convolution filtering, including smoothing, laplacian, high emphasis and unsharp, and (c) adaptive median filtering for de-speckling of ultrasound images.

The whole application was designed in a robust and compact way, in order to utilize the PDA's CPU and memory resources as optimally as possible.

Web services technology was employed in order to accommodate every transaction between the remote medical database and the client application running on the handheld device. Hence, the main role of the implemented web service is to act as an abstraction layer operating above the database, and return data to the client application. Furthermore, the task of user authentication is also assigned to the web service. All web functions were described (name, return types and input parameters) using the WSDL (Web Services Description Language) [9].

Finally, regarding security, Microsoft's implementation of UsernameToken was used to perform direct authentication at the message layer. This process can be summarised as follows. The application passes the credentials to the web service as part of a secure message exchange. A password is sent in the message as plaintext, which is data in its unencrypted or decrypted form. The Web service, then, decrypts the message, validates the credentials, verifies the message signature, and then sends an encrypted response back to the application [10].

3   Results and Discussion

The solution was evaluated by an experienced healthcare professional in a real healthcare environment (EUROMEDICA Medical Center) in terms of mobility, usability, stability, and performance, and was found to be a useful modality for healthcare providers that could further enhance their diagnostic capabilities by providing timely access to patient's information. Moreover the use of RFID tags guaranteed correct patient identification.

The application's response times fell within acceptable limits depending on the type and volume of data exchanged. The use of XML web services for access to the database promoted the system's interoperability, at the cost of slightly higher response times, caused by an extra overhead in data exchange. The use of new generation wireless network technologies, such as IEEE 802.11g (55Mbps) instead of 802.11b (11Mbps), can further improve the response times.

Furthermore, both static image viewing and graphical illustration of patient's vital signs were evaluated as adequate by the expert physician. Full DICOM transferring and decoding support for image files rendered the application plausible for a modern hospital environment. Most of the integrated image filtering algorithms exhibited acceptable performance regarding their processing times.

The application's GUI was found to be simple and intuitive, while it facilitated a pleasant and rich user experience, even to users unfamiliar with PDA-based applications.

Regarding future work, the use of active RFID tags – that support both read and write access- should be considered. Active RFID tags could be employed in order to store the patient's own medical record.

Because it is considered crucial for enhancing the user experience and enabling the effective consumption of information content, design of an improved GUI is therefore among the major issues that need further study and effort.

Another idea for future work could be the implementation of more advanced image enhancement algorithms that could be integrated in the application.

Finally, security of the proposed system should be further enhanced. Hence, more reliable authentication methods and up-to-date, powerful encryption algorithms could be adopted in order to fortify patients' sensitive medical data against tampering or interception.

4   Conclusion

By exploiting state-of-art RFID technology, a PDA-based application was designed for identification and accessing of each patient's medical record and through evaluation was proved to be plausible for application in a modern hospital environment. In addition, integration of image processing algorithms was a valuable feature that could potentially upgrade the role of the proposed system from a tool providing access to essential patient information into an important preliminary diagnostic asset.

References

1. Leape, L.L.: Error in medicine. JAMA 272, 1851–1857 (1994)
2. Lesar, T.S., Briceland, L., Stein, D.S.: Factors related to errors in medication prescribing. JAMA 277, 312–317 (1997)
3. Vawdrey, D.K., Hall, E.S., Knutson, C.D., Archibald, J.K.: A self-adapting healthcare information infrastructure using mobile computing devices. In: 5th International Workshop on Enterprise Networking and Computing in Healthcare Industry, Santa Monica, USA, pp. 91–97 (2003)
4. Banitsas, K.A., Georgiadis, P., Tachakra, S., Cavouras, D.: Using handheld devices for real-time wireless teleconsultation. In: 26th Annual International Conference of the IEEE Engineering in Medicine and Biology Society, San Francisco, USA, pp. 3105–3108 (2004)
5. Sidiropoulos, K., Georgiadis, P., Dimitropoulos, N., Cavouras, D.: Personal Mobile Medical Assistant employing RFID technology. In: 5th European Symposium on Biomedical Engineering, Patras, Greece (2006)
6. Bhatt, H., Glover, B.: RFID Essentials. O'Reilly (2006)
7. Taimur, H., Samir, C.: A Taxonomy for RFID. In: 39th International Conference on System Sciences, Hawaii (2006)
8. Gonzalez, R.C., Woods, R.E.: Digital Image Processing. Addison-Wesley, New York (1992)
9. Ballinger, K.: .NET Web Services: Architecture and Implementation. Addison Wesley, Boston (2003)
10. Stamos, A., Stender, S., Araujo, R.: Web Service Security, Scenarios, Patterns, and Implementation Guidance for Web Services Enhancements (WSE) 3.0. Microsoft Press, Redmond (2006)

Securing Medical Sensor Network with HIP

Dmitriy Kuptsov[1,2], Boris Nechaev[1,2], and Andrei Gurtov[1,3]

[1] Helsinki Institute for Information Technology HIIT
[2] Department of Computer Science and Engineering, Aalto University
[3] Centre for Wireless Communications, University of Oulu
firstname.lastname@hiit.fi

Abstract. Recent developments of embedded wireless technologies, such as low-cost low-power wireless sensor platforms, uncovered big potential for novel applications. Health care and well-being are examples of two applications that can have large impact on society. Medical sensor networks via continuous monitoring of vital health parameters over a long period of time, can enable physicians to make more accurate diagnosis and provide better treatment. Such network allow emergency services to react fast to dangerous patient's conditions and perhaps save more lives. For such applications to become viable, their design has to consider fail-safe mode of operation, protection of sensitive user data, and especially provide solution for efficient access control. Given the specifics of these applications, in this work we identify communication pattern that will guarantee the most secure way to exchange medical data, propose a standard based security protocol enabling authentication and data protection, and introduce a mechanism for access control—a crucial building block in privacy sensitive applications. To validate our design we implement a prototype on a wireless sensor platform.

Keywords: medical sensor network, security, embedded wireless Internet, network architectures, energy consumption.

1 Introduction

With proliferation of advanced wireless sensor platforms they found numerous applications in medicine. The vast list of applications includes invasive and non-invasive monitoring of blood pressure, ECG, temperature, oxygen saturation, etc. Wireless medical sensors worn by patients can form a *medical sensor network* (MSN) accessible to medical personnel. The patients need not be at the hospital, monitoring can be done remotely with patients staying at home or traveling in a vehicle. Specifics of health care field impose several requirements on such networks: reliability, ease of deployment and maintenance, device mobility, security and privacy. Without underestimating importance of other requirements, one can argue that security and privacy are crucial for modern medical applications.

Medical ethics requires that all medical records are kept private. This lays strict security requirements on all hardware and software used for medical purposes. Not only data collection and transfer in an MSN must be kept private, but there should also be strict access control. Breaches in security may have grievous consequences: from leaks of

K.S. Nikita et al. (Eds.): MobiHealth 2011, LNICST 83, pp. 150–157, 2012.

personal medical records to fatal outcomes. Since life of a patient may depend on adequate doctor's response time to abnormal sensor readings, a malicious person may want to try to undermine timely transmission of sensor measurements or replace alarming readings with benignly looking ones, thus potentially putting patient's life in danger.

In this paper we propose a medical sensor network framework which focuses on security and privacy. We consider several usage scenarios described in detail in the next section, involving sensors having and lacking Internet connection, different locations of doctor and patient, etc. The framework is aimed at providing confidentiality of data transmission over insecure network, access control and user authorization including temporary role delegation, authenticity and integrity of communications between patients and medical personnel.

Our framework heavily relies on Host Identify Protocol (HIP) [1,2,3]—a protocol proposed to overcome the problem of using IP addresses both for host identification and routing. HIP defines a new cryptographic *Host Identity* name space, thereby splitting the double meaning of IP addresses. In HIP, Host Identities (HI) are used instead of IP addresses in the transport protocol headers for establishing connections. Prior to communication over HIP, two hosts must establish a HIP association. This process is known as HIP base exchange (BEX) [2] and it consists of four messages transferred between initiator (I) and responder (R). A successful BEX authenticates hosts to each other and generates a Diffie-Hellman shared secret key used in creation of two IPsec Encapsulated Security Payload (ESP) Security Associations (SAs), one for each direction. All subsequent traffic between communicating nodes is encrypted by IPsec.

The task of designing a medical sensor network was already explored by other researchers. In [4] the authors propose a system which allows post cardiac surgery patients to be monitored remotely. In their system ECG signals collected from patients staying at home are uploaded to a server and automatically analyzed for possible arrhythmia. [5] and [6] describe medical sensors developed by the authors and routing, discovery and query protocols together forming an MSN. Neither of the above projects are concerned with security. [7] proposes a security scheme relying on Elliptic Curve Cryptography (ECC) which only focuses on protecting communication between sensors and a base station. In [8] authors go further and define access control for secure pervasive health care systems. The work in [9], which is the closest to ours, proposes a new protocol for communication between all parties in MSNs. Our solution is different from all the above in considering various realistic communication modes and proposing more general solution which covers all essential aspects of secure medical applications. Another advantage of our approach is in using a standardized protocol (HIP) which is more reliable, versatile and flexible than ad-hoc custom solutions.

The rest of the paper is structured as follows. In Section 2 we outline communication model, assumptions and requirements for our system. Section 3 describes our protocol in detail. Feasibility evaluation of our system is discussed in Section 4. Finally, Section 5 concludes the paper.

2   Model and Requirements

Our system for medical monitoring comprises several key components: (a) **Personal Area Network (PAN)** which includes multiple low-power **medical sensors** placed on or implanted into a patient's body and performing long-term reading of vital health parameters (e.g., blood pressure, pulse, etc.) and a single **on-body gateway** serving as a full-functional node for medical sensors and which has two wireless interfaces (one short range wireless interface, e.g., 802.15.4 for maintaining connection with medical sensors, and one long-range wireless interface, e.g., UMTS or 802.11, for maintaining Internet connection); (b) a **trusted authority (TA)** which is a node trusted by all other nodes belonging to the system and responsible for managing identities, revocation statuses, and access rights; (c) a **backend server** responsible for storing collected patient's sensor readings (a PAN gateway when connected to the Internet always establishes a secure channel to a backend server and uploads sensor readings to it periodically); and (d) a **backend terminal** with a graphical interface used by accredited personnel (this does not necessary need to be only patient's doctor, but may include any emergency services such as paramedics or police, having various access rights for reading patient's data); a backend terminal can retrieve patients' sensor readings from backend service, or accept readings directly from the gateway node after establishing security association with the sensor. The last is needed in emergency situations when no Internet connection is available.

In our medical application we consider the following communication patterns. Initially, medical sensors and patient's gateway node should establish a long term security association, or perform **initial pairing**. In its basic form this is achieved with a HIP handshake between a medical sensor node and the gateway. However, mutual authentication between two nodes can differ and depends on how the protocol is configured (we discuss two alternatives in Section 3). No node, other than the patient's gateway, can have a security association with medical sensors. Depending on the logic implemented, gateway node may perform various sophisticated tasks with sensor readings, such as preprocessing, adaptations, anomaly detection analysis, etc. However, the basic and most important functionality is: (a) ability to establish **security associations** with either backend service or directly with a backend terminal; (b) enforce **access control** with or without TA being online. Furthermore, an entity that belongs to the system— patient's gateway, backend service, or backend terminal—can access TA when it is online (meaning that there exist a communication channel between accessing node and TA) and either check certificate status or ask for a new certificate.

Next we discuss the requirements we impose on our system. Medical sensor nodes performing long-term health parameters reading will have a **limited processing power** and very **limited battery capacity**. Therefore, such nodes should be stressed with cryptography as little as possible and should not perform any sophisticated tasks with data processing. We thus require that medical sensor nodes perform HIP handshake only once during initial pairing to produce a symmetric keys common to a particular sensor and gateway and store such keys permanently for the entire time of sensor operation.

All monitored patient's data that is transmitted between medical sensor nodes, gateway, backend service or backend terminal should be encrypted to ensure **privacy**. This data should also be labeled with authentication information to ensure **data authenticity** and prevent forgery attacks on patient's data. However, we don't discuss how the data should be stored on a backend server or handled by backend terminals, since it is out of scope of this work.

Finally, we require that strong **access control** mechanisms are implemented and deployed on medical sensor nodes and gateway to prevent unauthorized access to sensitive information.

3   Architecture

Next we present the architecture description which is based on the model specified in Section 2. Specifically, in this section we will describe in detail the following operational phases and components: (i) initial medical sensor to gateway pairing, (ii) gateway to backend service pairing, (iii) backend terminal to gateway pairing, and (iv) access control mechanism. We present general architecture view in Figure 1.

Sensor to Gateway Pairing. When PAN is being deployed for the first time, e.g. when a new patient receives wearable sensors at the hospital, it requires bootstrapping of security associations between all sensors and a gateway. In our application this operational phase is called *initial pairing*.

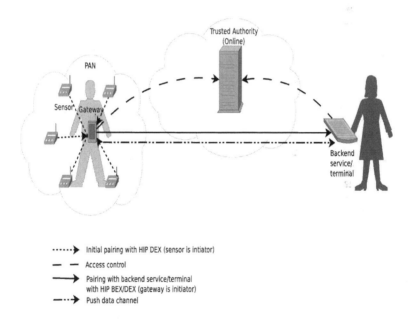

Fig. 1. System architecture

Initial pairing resembles HIP DEX [10]—a HIP handshake protocol which is a variant of HIP BEX [2]—where the public key is fixed and no signature algorithms are used. These two key differences allow HIP DEX to perform better than HIP BEX, and a lot better than SSL or its variants, thus making it a perfect solution for securing communication between low-powered medical sensors. In our architecture HIP DEX between medical sensors and a gateway is running directly on a MAC layer which allows to save space (indeed this is can be implemented as a variant of 6lowpan to allow header compression and save space). Because HIP DEX does not use signature algorithms, and certificates are not suitable for medical sensors due to limited processing power, and 128 bytes frame size in 802.15.4 radio which is too small to fit a certificate, a two factor authentication process is required to guarantee secure sensor to gateway communication. We suggest several possible ways to achieve this goal.

First, and the most simple approach is to use the link button scheme: prior to initiating HIP DEX a special button should be pressed on a gateway node to allow HIP instance to accept any unknown I1 packets for a short time interval (e.g., of 5-10 seconds). Once the button was pressed medical sensor can be turned on and send its I1 packet. The second approach is considered to be more secure, because it does not leave any opportunity window for an attacker who is in range of patient's 802.15.4 radio. This idea is based on HIP DEX ability to perform mutual authentication with passwords, i.e., HIP DEX allows to perform a challenge-response type authentication. The usage of this feature requires the passwords to be configured externally from HIP. We foresee that the password selection on medical sensor nodes will be implemented in several alternative ways. One way is to implement it as jumpers, or switches, and with a graphical user interface on the gateway. Another way to configure the passwords on sensor and gateway in a secure way is to exchange it via a visual channel [11]. The idea is simple, yet elegant and secure: After boot up the sensor generates a random password of arbitrary length, and conveys the bits to the gateway using series of blinks of a light-emitting diode (LED). Indeed the blinks represent a frequency modulated signal. The gateway in turn captures the blinks using camera and demodulates the signal and reconstructs the password. For the latter two cases above, after the password is stored in the memory (on both, sensor and gateway), the sensor node triggers HIP DEX with gateway (both use the password for mutual authentication).

Irrespectively of what type of authentication is used, the pairing procedure should be repeated for all sensors that will participate in a particular PAN. Moreover, the secret key produced after HIP DEX completion will be stored permanently in hardware on the sensor node and gateway. This will allow to reuse the keys even if the sensor nodes or the gateway reboots, or turn on after battery replacement. However, since the entropy of the secret key is used up every time a new packet is encrypted and sent out, it is preferable that the initial pairing is repeated every now and then, for instance once per month. The advantage of our initial pairing over simple presharing of keys is that it allows to pair nodes from different vendors, which is a necessary feature for commercial products.

Gateway to Backend Service Pairing. The default security rule installed on the gateway node does not allow any communication to be accepted from the Internet, including HIP packets. As such, after completing the initial pairing a gateway will attempt to connect to the backend service by establishing a security association with an instance of HIP BEX protocol. The key difference from the initial pairing is that certificates are mandatory in such pairing. Moreover, both gateway and backend service should exchange their certificates to perform mutual authentication. Detailed description on format and usage of certificates in HIP BEX can be found in [12]. Below we will further discuss how secure access control can be implemented with certificates. Another difference is that instead of sending I1 packet directly to a backend service the gateway sends it to rendezvous server first (who's location is known to the gateway), which redirects it to a proper backend server. More information on HIP rendezvous mechanism can be found in [13].

Backend Terminal to Gateway Pairing. Another communication pattern can be needed when Internet connection is not available and someone from emergency personnel tries to retrieve critical readings from the patient's sensor nodes. Such situation can occur, for instance, when the patient is transported from home in a rural area to a hospital. As we have mentioned earlier, the security rules on the gateway are configured so that the amount of traffic arriving on UMTS or 802.11 interfaces is limited to avoid denial of service attacks. To make this scenario work flawlessly, the gateway after establishing a connection on a physical layer with a 802.11 access point placed e.g. inside the ambulance, will trigger HIP BEX with certificates similarly to gateway to backend server pairing case. Though the key difference from the previous scenario here is that the gateway will broadcast I1 packet with null value used for destination HIT. This fallback mechanism will allow any backend terminal that receives such an I1 packet to reply with R1 packet. Despite that the destination HIT would be a null value, responder still should include a mandatory certificate in R1 packet which will allow the gateway to enforce access control rules that we discuss next.

Access Control. An important aspect of medical sensor network application is an ability to provide means for nodes to enforce access control by roles, cancel previously granted access rights by revoking issued certificates, and in certain cases to provide anonymous identity to the certificate holder. Due to space limitations we will merely discuss certificate revocation status verification procedure and role-based access control enforcement implemented on a gateway node.

If the certificate was revoked from the system, its further usage should be prohibited. If the nodes, especially gateways, do not have an ability to verify the status of the certificate there is a chance that an intended attacker can receive access to confidential information. Since the TA in our case is not guaranteed to be online all the time, the gateway may not have the ability to verify the status of the received certificate. To combat this, in our system every entity is granted two certificates: (i) *permanent membership certificate (PMC)* which a node receives when it joins the network for the first time (such certificate can for instance be installed on the node by TA) and (ii) *on-demand short term certificates (OSTC)* are granted by TA for a short period of time (e.g., an hour or a day) based on the status of PMC of the node. Consequently, a gateway node will receive HIP BEX packets if and only if such packets contain a valid OSTC certificate. On the other hand, it is responsibility of a particular node to request the next OSTC certificate in a timely manner. For instance, before an ambulance departures to the patient's premises, the backend terminal should be supplied with a valid OSTC certificate based on the PMC certificate permanently installed on the same terminal.

Once the revocation mechanism is in place the gateway can verify OSTC certificate and do a role-based access control to sensor node reading. Each gateway is configured with a set of roles and corresponding access rules for each role. For example, the gateway node may have rules that allow the role "public service" to read only basic identification information, and rules that allow to read all available information to roles "physician" and "paramedic".

4   Evaluation

In this section we present measurement results for cryptographic primitives that impact the performance of our medical WSN architecture. Table 1 contains key characteristics of one of the most basic operations in our network—HIP handshake. In our experiments we used Imote2 sensor node.

In Table 1 we consider two network types: PAN of the patient, which includes communication between sensors and the gateway, and Internet, when collected data is transmitted from the gateway to a doctor's machine or a central server. PAN communication should be much lighter than Internet communication due to computational and energy constraints of embedded devices. HIP allows to use DEX instead of BEX mode for this and tune handshake procedure by using a shorter key, disabling signatures, certificates and puzzle, and using a different MAC algorithm. This allows the total size of packets transmitted during DEX handshake to be much smaller than in BEX, also making it considerably faster and less energy demanding.

Table 1. Resource consumption per HIP handshake

Protocol components	Network types		
	PAN	Internet	
	ECC HIP DEX	HIP BEX ECC	HIP BEX
Key exchange	Fixed ECDH	ECDH	DH
	160 bit	160 bit	1536 bit
Signatures	None	ECDSA	RSA
Certification method	None	ECDSA	RSA
Puzzle difficulty	0	≥ 10	≥ 10
MAC	CMAC (AES-CBC)	SHA-1	SHA-1
Message sizes (bytes)			
I1	40	40	40
R1	92	916	1544
I2	148	944	1568
R2	102	108	188
	DEX Duration (ms)	BEX Duration (ms)	
	72.396	151.26	1115.96
Energy (mj)			
For Initiator (I)	17.0	53.8	471.5
For Responder (R)	17.0	34.1	222.5
Total w/ transmission, I	26.14	73.74	560.1
Total w/ transmission, R	26.14	61.19	443.1

5   Conclusion

In this paper we proposed a secure architecture for medical sensor networks. Our approach is based on the standardized HIP protocol. We considered several use case scenarios of accessing sensor data: when trusted authority used for user authentication is available and when authentication has to be done by the PAN gateway. We achieve this using a 2-tiered certificate infrastructure which involves issuing permanent and short-term certificates. After authentication is complete, the gateway is able to perform role-based access control to determine what information should the requesting entity be allowed to access. In order to assess feasibility of the proposed architecture we implemented and deployed it on Imote2 sensors, which also allowed us to measure main characteristics of HIP handshake in various scenarios.

References

1. Moskowitz, R., Nikander, P.: Host Identity Protocol architecture. IETF RFC 4423 (May 2006)
2. Moskowitz, R., Nikander, P., Jokela, P., Henderson, T.: Experimental Host Identity Protocol (HIP). IETF RFC 5201 (April 2008)
3. Gurtov, A.: Host Identity Protocol (HIP): Towards the Secure Mobile Internet. Wiley and Sons (2008)
4. Kyriacou, E., Chimonidou, P., Pattichis, C., Lambrinou, E., Barberis, V.I., Georghiou, G.P.: Post Cardiac Surgery Home-Monitoring System. In: Lin, J. (ed.) MobiHealth 2010. LNICST, vol. 55, pp. 61–68. Springer, Heidelberg (2011)
5. Shnayder, V., Chen, B., Lorincz, K., Fulford, J., Welsh, M.: Sensor networks for medical care. In: Proceedings of the 3rd International Conference on Embedded Networked Sensor Systems, SenSys 2005 (November 2005)
6. Shnayder, V., Chen, B., Lorincz, K., Fulford, J., Welsh, M.: Sensor networks for medical care. Harvard University Technical Report TR-08-05 (April 2005)
7. Malasri, K., Wang, L.: Design and implementation of a secure wireless mote-based medical sensor network. In: Proceedings of the 10th International Conference on Ubiquitous Computing, UbiComp 2008, pp. 172–181 (2008)
8. Venkatasubramanian, K., Gupta, S.K.S.: Security solutions for pervasive healthcare. In: Xiao, Y. (ed.) Security in Distributed, Grid, Mobile, and Pervasive Computing, pp. 443–464. Auerbach Publications, CRC Press (April 2007)
9. Garcia-Morchon, O., Flack, T., Heer, T., Wehrle, K.: Security for pervasive medical sensor networks. In: MobiQuitous 2009: Proc. of the 6th Annual International Conference on Mobile and Ubiquitous Systems. ICST/IEEE (July 2009)
10. Moskowitz, R.: HIP Diet EXchange (DEX): draft-moskowitz-hip-rg-dex-05 (March 2011), Expires in September 2011(work in progress)
11. Saxena, N., Ekberg, J.-E., Kostiainen, K., Asokan, N.: Secure device pairing based on a visual channel: Design and usability study. IEEE Transactions on Information Forensics and Security 6, 28–38 (2011)
12. Heer, T., Varjonen, S.: HIP Certificates: draft-ietf-hip-cert-12 (March 2011) (work in progress)
13. Laganier, J., Eggert, L.: Host Identity Protocol (HIP) rendezvous extension. IETF RFC 5204 (March 2008)

Development of an mHealth Open Source Platform for Diabetic Foot Ulcers Tele-consultations

George E. Dafoulas[1,2,*], Stylianos Koutsias[1], Joachim Behar[3], Juan Osorio[4], Brian Malley[5], Alexander Gruentzig[5], Leo Anthony Celi[5], Pantelis Angelidis[6], Kyriaki Theodorou[1], and Athanasios Giannoukas[1]

[1] Faculty of Medicine-University of Thessaly, Greece
{gdafoulas,ktheodor,giannouk}@med.uth.gr,
skoutsia@otenet.gr
[2] e-trikala SA, Municipality of Trikala, Greece
[3] Engineering Department-University of Oxford, UK
joachim.behar@gmail.com
[4] Research Group in Biomedical Engineering (GIBEC), Antioquia School of Engineering and CES University, Colombia
juan.s.osorio@gmail.com
[5] Sana, Computer Science and Artificial Intelligence Laboratory, Massachusetts Institute of Technology, USA
{Bemalley,a.gruentzig}@gmail.com, leoanthonyceli@yahoo.com
[6] Faculty of Engineering Informatics and Telecommunications, University of Western Macedonia, Greece
paggelidis@uowm.gr

Abstract. Diabetes is one of the foremost causes of death in many countries and a leading cause of blindness, renal failure, and non-traumatic amputation. Therefore, diabetic foot ulceration and amputation cause extensive burden on individuals and health care systems in developed and developing countries. Due to the multi-disciplinary requirements for the treatment of diabetic foot ulceration, telemedicine has been introduced to facilitate the access of the patients to specialized health professionals. In this paper the development of an open source mobile health platform is presented, able to support diagnostic algorithms, with the use of a smartphone.

Keywords: Telemedicine, m-health, diabetic foot ulcers.

1 Introduction

The prevalence of diabetes mellitus is growing at epidemic proportions worldwide and it is predicted to rise due to longer life-expectancy and changing dietary habits. Type 2 diabetes, formerly non-insulin dependent diabetes mellitus, accounts for the majority of the diagnosed cases. The greatest rise of type 2 diabetes is likely to be in developing countries such as Africa, Asia, and South America [1].

---

* Corresponding author.

K.S. Nikita et al. (Eds.): MobiHealth 2011, LNICST 83, pp. 158–164, 2012.
© Institute for Computer Sciences, Social Informatics and Telecommunications Engineering 2012

Diabetes is associated with many complications related to microvascular, macro-vascular, and metabolic etiologies. These include cardiovascular, cerebrovascular and peripheral arterial disease; neuropathy; retinopathy and nephropathy [2]. One of the most common complications of diabetes in the lower extremity is the diabetic foot ulcer. The lifetime risk of a person with diabetes developing a foot ulcer could be as high as 25% [3], though the prevalence of foot ulcers varies per continent, country and specific ethnics minorities within one country [4].

The burden of the diabetic foot ulcers in particular is high since it includes reduced functioning, increased sick leaves, increased financial costs for both the health care system and the societal perspectives. In addition to the direct costs of treatment of the diabetic foot complications, there are also indirect costs relating to loss of productivity, individual patient and family's cost and deterioration of the quality of life [5]. Diabetic foot continues to be the most common underlying cause of non-traumatic lower extremity amputations in the US and Europe [6].

It is possible to dramatically reduce the incidence of amputation through appropriate management and prevention programs. Considering the vast personal, social, medical, and economic costs, many countries have adopted policies to this direction [7].The treatment of the diabetic foot patients is however challenging and requires a systematic approach to the complete assessment of these patients. The multidisciplinary team-approach to diabetic foot disorders has been demonstrated as the optimal method to achieve better results [8]. The provision of such integrated care varies from country to country due to many reasons and it is difficult to be achieved [5].

Due to the multi-factorial pathology of diabetic foot ulceration, telemedicine has been suggested as a way of enabling the experts to monitor an increasing number of patients more regularly [9-13]. The aim of this research was to adapt an existing pilot mobile telemedicine platform, called Sana [14], that supports audio, images, location-based data, text, video and allows for clinical decision pathways through tele-consultation based on the diabetic foot tele-monitoring.

2   Methods

The open-source Sana platform (released under a permissive free software license, the Berkeley Software Distribution - BSD license) [15], is being used in the first diabetic foot tele-monitoring prototype. Sana is a mHealth project based at the Massachusetts Institute of Technology (MIT) that offers an end-to-end system that connects healthcare workers to medical professionals [16]. The tool allows healthcare workers to transmit medical files such as notes, audio and video through a cell phone to a central server for archiving, incorporation into an electronic medical record and reviewing by a remote specialist for real-time decision support [17].

The complete Sana system consists of at least one phone and a web-connected server. The server runs the medical records system of choice and the Sana Dispatch

Server program (MDS). The Sana Dispatch Server is responsible for communication to and from phones registered in the system. The data is received via lower-level synchronization, packetization and multimodal transfer that the Sana-enabled phones perform, three strategies to ensure reliable, low-cost data transfer. In addition to this, the Sana Dispatch Server has plug-ins that allows it to interface with different medical records systems. Sana is currently fully-compatible with OpenMRS [18], using an OpenMRS plug-in for the Sana Dispatch Server and a custom-patched version of OpenMRS, which extends it to have a queue of pending diagnoses in addition to allowing data such as images to be tagged to a patient record.

Moreover, the medical records system also runs on the server or a separate machine if desired. To sum up, the system infrastructure and design allows for modularity and interoperation.

Sana is highly customizable and, with the branching capability (that show different questions and results based on previous selections), allows physicians to make their own decision-tree diagnostic utilities for common procedures. Sana's front-end for data and media capture is accessible through a fully programmable workflow interface, that runs on the phone and doesn't need remote doctor's review. These workflows can be dynamically loaded onto phones running Sana. The back-end provides an intuitive user interface for management of medical data.

Procedures are step-by-step workflows, and are at the core of Sana. In most scenarios, a procedure is a set of pages that have questions or prompts. For the needs of the specific procedure, the available diagnostic algorithms project [12,19] can be used. The prototype procedure developed is being called "Diabetes Mellitus Follow-up" and appears in the Sana encounters stored on client devices. Procedures are defined in a simple XML format. All the information regarding installation and deployment of new procedures can be found at the Sana Wiki [17].

3   Results

The developed m-health platform, creates and records a personal ID for each patient with his medical history on the server. During the visit of the health workers at the patient house, the online medical record of the patient will be updated with the upload of high-resolution images and/or video of the ulcer as well as obtaining measurements such as peripheral blood pressure, blood sugar level and foot-skin temperature. The interactive telemedicine consultations with a home health-nurse can be supported by a decision pathway running on the Android phone, assisting the management of diabetic foot ulcers.

More specific, the developed application includes a three-step system that a health worker can use during a tele-consultation of a diabetic foot patient. In the first step of the procedure the health worker enters patient ID in order to connect with the personal records of the patient (or to create one if the patient is not registered), runs the diagnostic algorithms, take pictures, review the photographs, validates the procedure and sends the data to the server (Fig. 1).

Fig. 1. Screenshots of the smart phone running the application

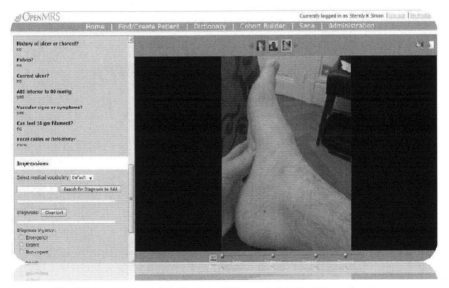

Fig. 2. Image viewer of the back-end Electronic Medical Record system

The server allows the specialized doctor to log-on and see the data, review the patient's information and images. Clicking on the procedure or on the patient's name, a new window opens, which displays all the information to be sent to the server. In particular, the doctor can display the pictures zoom in/out, vary the contrast and download the picture if further image analysis needs to be performed (Fig. 2).

After reviewing the information the doctor fills a form in order to give his/her feedback on the data that have been sent. As soon the review is completed, the doctor can send back his feedback to the phone so that the health worker can read through it. The platform can support both synchronous and non-synchronous operation.

4   Discussion

Telemedicine platforms can allow the treatment to be carried out by non-experts such as visiting nurses and general practitioners under the tele-guidance of specialized experts. These alternatives can reduce the high costs of out-patient treatment and the lack of specialized personnel in both developing and developed countries. Both synchronous and non-synchronous telemedicine platforms are available.

The past few years, innovative mobile information services that improve patient access to medical specialists for faster, high quality, and more cost effective diagnosis and intervention have been developed thanks to open-source and free to use platforms.

In order to develop Sana system, technological choices have been made. In particular choices such as using Smart phone and not simple cell phones, and the choice of Android for the Operating System (OS) and of the OpenMRS for patient's health record.

Smart phones run complete operating system (OS) that provide a platform for application developers. There is a number of different OS available. In particular the Android OS was released in 2008 and observed a growth of 615.1% between 2009 and 2010 [20]. This highlights the penetration of Android OS into the market and justifies its choice by Sana which aims to choose the future open source most widespread operating system in order to render its application available for the most people in the world currently. There are however alternative OS in the market able to support a similar m-health platforms. Interoperability of the telemedicine platforms is a major challenge for the market, and to this end the Continua Health Alliance has been established aiming towards interoperability of ehealth platforms. [21]

OpenMRS is an Open Medical Record System that was created in 2004 for developing countries and in order to provide them with an efficient information management system [18].

The tele-monitoring of diabetic foot, as many other telemedicine applications, is a challenging procedure but it has its limitations. In particular, it does not always guarantee accurate diagnosis or successful outcomes. For this reason the platform should be used and follow proposed clinical guidelines [22].

There is limited experience in the international literature on mobile systems used in diabetic foot tele-monitoring. A Danish qualitative study (13) indicated that that it is possible for specialized doctors in a hospital to conduct clinical examinations and decision making at a distance, in close cooperation with the visiting nurse and the patient, using a mobile phone.

The study reported that that visiting nurse experienced increased confidence with the treatment of the foot ulcer while the patients expressed satisfaction and felt confidence with this new way of working.

At the moment, a large RCT study -Tele Ulcer project within the Renewing Health large scale RCT EU funded project [23], takes place in the Region of Southern Denmark, aiming to test the clinical, economic, organisational and patient-related consequences of home treatment of diabetic foot ulcers via tele-consultations. However the platform used does not involve a m-health platform supporting the study nurse with an diagnostic algorithm.

5   Conclusion

The proposed diabetic foot tele-monitoring platform has the potential to support a tele-consultation between specialized doctor and the health worker visiting the patient, enabling the clinicians to provide an alternative tele-care service. However, intense validation of the developed platform is required via a clinical trial to evaluate its clinical validity and cost-effectiveness.

Acknowledgements. The authors would like to thank the principal advisor of Sana, Peter Szolovits, Professor of Computer Science and Engineering in the MIT Department of Electrical Engineering and Computer Science (EECS), Professor of Health Sciences and Technology in the Harvard/MIT Division of Health Sciences and Technology (HST), and head of the Clinical Decision-Making Group within the MIT Computer Science and Artificial Intelligence Laboratory (CSAIL). In addition the development advisor of Sana, Dr G. D. Clifford University Lecturer & Associate Director Centre for Doctoral Training in Healthcare Innovation Department of Engineering Science University of Oxford, UK.

The specific application for diabetic foot telemonitoring, was developed during an internship of Dr George E. Dafoulas in CSAIL-MIT, co-sponsored by the Municipal Enterprise e-trikala SA of the Municipality of Trikala.

The Android smartphone used for the testing of the propotype of the application developed, was donated by Google Inc, Cambridge MA, USA.

References

1. Wild, S., Roglic, G., Green, A., Sicree, R., King, H.: Global prevalence of diabetes: estimates for the year 2000 and projections for 2030. Diabetes Care 27, 1047–1053 (2004)
2. American Diabetes Association. Diabetes Facts and Figures, American Diabetes Association, Alexandria, VA (2000)
3. Singh, N., Armstrong, D.G., Lipsky, B.A.: Preventing foot ulcers in patients with diabetes. JAMA 293, 217–228 (2005)
4. Centers for Disease Control and Prevention. Diabetes: Disabling, Deadly, and on the Rise: At-a-Glance, Centers for Disease Control and Prevention, Atlanta (2005)

5. Boulton, A.J., Vileikyte, L., Ragnarson-Tennvall, G., Apelqvist, J.: The global burden of diabetic foot disease. Lancet 366, 1719–1724 (2005)
6. Jeffcoate, W.J.: The incidence of amputation in diabetes. Acta Chir. Belg. 105, 140–144 (2005)
7. International Diabetes Federation and International Working Group on the Diabetic Foot. Diabetes and Foot Care: Time to Act, International Diabetes Federation, Brussels (2005)
8. Diabetic Foot Disorders: A clinical practice guideline. J. Foot Ankle Surg. 45(suppl.) (September/October 2006)
9. Marg, M., et al.: Integrating Telemedicine into a National Diabetes Footcare Network. Pract. Diabetes Int. 17, 235–238 (2000)
10. Visco, D.C., Shalley, T., Wren, S.J., et al.: Use of telehealth for chronic wound care: a case study. J. Wound Ostomy Continence Nurs. 28, 89–95 (2001)
11. Horswell, R.: The use of telemedicine in the management of diabetes-related foot ulceration: a pilot study. Adv. Skin Wound Care 17, 232–238 (2004)
12. Wilbright, W.A., Birke, J.A., Patout, C.A., Varnado, M., Horswell, R.: The Use of Telemedicine in the Management of Diabetes-Related Foot Ulceration: A Pilot Study. Adv. Skin Wound Care 17, 232–238 (2004)
13. Clemensen, J., Larsen, S.B., Kirkevold, M., Ejskjaer, N.: Treatment of Diabetic Foot Ulcers in the Home: Video Consultations as an Alternative to Outpatient Hospital Care. Int. J. Telemed. Appl. (2008)
14. Celi, L.A., Sarmenta, L., Rotberg, J., Marcelo, A., Clifford, G.: For the Moca Team: Mobile care (Moca) for remote diagnosis and screening. J. Health Inform Dev. Ctries. 3, 17–21 (2009)
15. http://en.wikipedia.org/wiki/BSD_licenses (accessed May 1, 2011)
16. http://www.sanamobile.org/about.html (accessed May 1, 2011)
17. http://www.sanamobile.org/wiki/index.php/Overview (accessed May 1, 2011)
18. http://openmrs.org/about/ (accessed May 1, 2011)
19. Patout, C.A., Birke, J.A., Wilbright, W.A., Coleman, W.C., Mathews, R.E.: A Decision Pathway for the Staged Management of Foot Problems in Diabetes Mellitus. Arch. Phys. Med. Rehabil. 82, 1724–1728 (2001)
20. http://www.canalys.com/pr/2011/r2011013.html (accessed May 1, 2011)
21. http://www.continuaalliance.org/index.html (accessed August 20, 2011)
22. American Telemedicine Association's Practice Guideline for Teledermatology (December 2007)
23. http://www.renewinghealth.eu (accessed August 20, 2011)

Towards Continuous Wheeze Detection Body Sensor Node as a Core of Asthma Monitoring System

Dinko Oletic, Bruno Arsenali, and Vedran Bilas

University of Zagreb, Faculty of Electrical Engineering and Computing,
Unska 3, HR-10000 Zagreb, Croatia
{dinko.oletic,bruno.arsenali,vedran.bilas}@fer.hr
http://www.fer.unizg.hr/aig

Abstract. This article presents a wheeze detection method for wearable body sensor nodes used in management of asthma. Firstly, a short review of current state of telemonitoring in management of chronic asthma is given. A concept of the asthma monitoring system built around a body sensor node analysing respiratory sounds is proposed, with a smart phone as a self-management center and additional sensor nodes for environment monitoring. In search for a wheeze detection algorithm suitable for low power continuous operation on wireless sensor node, a simple algorithm based on the 4-th order linear prediction coefficients (LPC) method is presented. Predictor error energy ratio of Durbin's algorithm is used as the only feature. Algorithm is implemented on low power digital signal processor (DSP) to evaluate its performance. Sensitivity (SE) of 70.9%, specificity (SP) 98.6% and accuracy (ACC) of 90.29% are achieved using pre-recorded test signals. Program complexity is analysed in order to identify possibilities of lowering power consumption.

Keywords: asthma telemonitoring, body sensor networks, wheeze detection, LPC, Durbin's recursion.

1 Introduction

Rising prevalence of asthma increases the workload of medical staff and costs to healthcare systems. The oldest and simplest form of telemonitoring in respiratory disease management is upload of home peak-flow-meter (PFM) data. Clinical deployments and reviews of automated PFM [1], [2] conclude that although patients generally express positive attitude towards this type of management, PFM is unable to capture the moments of the worst airflow obstruction, and requires patient cooperation (nocturnal monitoring). Also, it is common to experience the fall of patient's interest during a long-term monitoring. This stresses the need to automate the management of the disease and intervene in patient's daily routine as little as possible.

On the other hand, maturity of personal wireless area networks such as Bluetooth and IEEE 802.15.4 variants, blooming of the smart phone market, rise of

K.S. Nikita et al. (Eds.): MobiHealth 2011, LNICST 83, pp. 165–172, 2012.
© Institute for Computer Sciences, Social Informatics and Telecommunications Engineering 2012

interest of mobile providers for machine-to-machine services (M2M) open possibilities for continuous monitoring employing wireless sensor networks on a patient, and/or in the environment. However, key challenges to wider application are:

1. Technical - accuracy, energy efficient long-term operation, reliability;
2. Clinical testing, certification, proof of effectiveness and
   patient adherence [3];
3. Adoption of adequate business models for deployment of the system.

In this paper we broach the topic of technical issues, first by providing a short overview of current state of the art in asthma telemonitoring. Further, we propose an architecture of the asthma monitoring system. In the second part of the article we focus on the wireless body sensor node for continuous 24-hour telemonitoring and present our current work on signal processing for wheeze detection.

2   Recent Advances in Asthma Monitoring

With the advent of wireless sensor networks (WSN) in the last decade, a number of environmental sensing applications emerged, including those dealing with asthma. Shanoy [5] described the concept of WSN with stationary nodes monitoring levels of gasses potentially triggering asthma. Seto [6] proposes mobile, body-worn nodes for sensing environmental asthma triggers (pollen) in the immediate surroundings of patient and his physical activity. Fu [7] shows the concept of a body-worn sensor network for simultaneous monitoring of triggers in patient's surrounding and respiratory parameters (breath rate). Wisniewski [8] also emphasizes the need of combined monitoring of the environment and the patient and proposes the continuous wheeze monitoring by smartphone. Nanyang University group [9] proposes wearable node with local signal processing for continuous detection of wheezes.

Among mature products offering wheeze monitoring, most recent are the certified products for short-term ambulatory use [10]. Also, much public attention was attracted to the project Asthmapolis featuring indirect collaborative sensing of the potentially dangerous zones by mean of sensorizing asthma-inhaler pumps, and equipping them with GPS/GPRS modules and combining with ubiquitous mobile phones [11]. Also a multitude of iPhone/Android asthma diary applications exist.

3   Architecture of Asthma Monitoring System

Following the review in the previous section, we agree on the significance of both monitoring of respiratory function and monitoring of triggers in the environment and propose system architecture as shown in Fig. 1.

We propose a continuously worn body sensor node as a core of our monitoring system. Continuing our research of power management of wireless sensor nodes [4], we aspire towards the battery operated device with locally implemented signal processing for multiple weeks of autonomous on-board classification of respiratory sounds in real-time. The accent is put on wearability and minimum of intrusiveness and maintenance by user.

Upon detection of abnormal sounds (eg. wheezing), data is sent to the patient's smart phone, used as a presentation layer and data gateway to medical database. Bluetooth peak-flow-meters may be used as an optional device for control. Emphasis is put on self-management of the disease, rather than on direct communication with medical staff (in times of exacerbations). Data is typically accessed by medical staff during patient's scheduled checks, with content and form of presentation also being key elements.

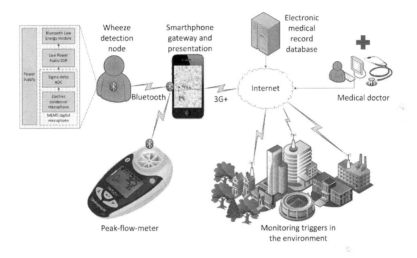

Fig. 1. Architecture of the asthma monitoring system

Asthma triggers are sensed from the environment by stationary nodes. Our preliminary internal study of the off-the-shelf electronic components currently marketed has shown little advance in the field of air quality sensing, with high consumption, low precision MOX gas sensors and bulky optical particulate matter sensors still being state of the art. This restricts the deployment of a feasible body worn network for sensing triggers. Meteorological data, such as air temperature, humidity or pressure, can be monitored by low power digital sensors. Thus, in the rest of this study we focus on the body sensor node for continuous monitoring of wheezes.

4   BSN Node Architecture

Architecture of the BSN node is presented in Fig. 1. It consists of several functional blocks: audio signal acquisition, DSP for local wheeze detection and Bluetooth Low Energy communication.

Major guidelines on acquisition and preprocessing of respiratory signals are defined by [12] and [13]. Both microphone and accelerometer devices are recommended for acoustical signal acquisition. To shorten the analog signal path, we propose an evaluation of the analog capacitive MEMS with integrated amplifiers or digital MEMS microphones with integrated sigma-delta ADC-s. Except anti-alias filtering, a high pass filter of cutoff frequency around 100 Hz is needed for attenuation of heart sounds.

Bluetooth is proposed for communication between the body sensor node and a smartphone, because of its widespread compatibility and its acceptance among interoperability and standardization bodies. Its main disadvantages are high power consumption and long and complicated pairing/connection protocols. In our scenario, data is transferred to a smartphone only upon the detection of a wheeze. Due to the low data rate and power consumption, we strive to implement Bluetooth Low Energy(Bluetooth 4.0) as a next generation solution.

5   LPC Wheeze Detection Algorithm

Signal processing for asthma management includes wheeze detection, respiratory cycle detection and noise cancellation techniques. In the rest of the text we concentrate on one wheeze detection technique.

Wheezes are continuous adventitious respiratory sounds consisting of a single or a small set of discreet frequencies, of defined duration, superposed on normal respiratory sounds [13]. Thus, wheeze detection converges to the problem of instantaneous frequency estimation. A multitude of short-term Fourier transform (STFT) based algorithms already exist which follow the straight-forward approach of calculating the fast Fourier transform (FFT) of the signal block, calculating power spectrum and applying the set of rules regarding energy distribution throughout the spectrum and duration of spectral peaks [14].

In perspective of implementing the algorithm on a low-power, wireless body-worn sensor node eventually not featuring dedicated FFT coprocessor, one of the ideas is to avoid common FFT based algorithms, and to search for simpler algorithms operating on a minimum feature set extracted directly from the time domain. In this light, we evaluate a variant of the linear predictive coding (LPC).

The method originates from speech coding. It has already been used by several authors in respiratory sounds analysis, mostly for crackle detection, as LPC coefficients exhibit changes correlated to changes in signal waveform [15].

LPC is an autoregressive time domain estimator $\hat{s}[n]$ of current signal sample $s[n]$ based on linear combination of previous p samples and LPC coefficients a_k, $k = 1...p$:

$$\hat{s}[n] = \sum_{k=1}^{p} a_k s[n - k]. \tag{1}$$

Energy of error of the predictor E is taken as a measure of quality of prediction:

$$E = \sum_{n}(s[n] - \hat{s}[n])^2 = \sum_{n}(s[n] - \sum_{k=1}^{p} a_k s[n-k])^2. \tag{2}$$

Prediction coefficients $a_k, k = 1...p$, are found by minimizing E, $\frac{\partial E}{\partial a_k}$ which converges to solving a linear equation with p variables. Here, a variant of LPC solver using short-term autocorrelation and Durbin's algorithm is used. Each pass through the Durbin's recursion generates the prediction error energy of that order k, $E^{(k)}$:

$$E^{(k)} = (1 - (a_k^{(k)})^2)E^{(k-1)}, \tag{3}$$

where $a_k^{(k)}$ is k-th prediction coefficient of k-th order. Thus, during process of calculation of LPC coefficients of order p, $E^{(0)}...E^{(p)}$ are produced. $E^{(0)}$ is energy of the analysed signal block (in fact autocorrelation for offset 0). Generally, it can be observed that with each subsequent recursion pass, prediction error energy $E^{(k)}$, $k < p$ exhibit the fall in value. Fall is mostly pronounced for lower orders and is dependent on the analysed signal. Correlated signals with small number of discreet frequency components such as wheezes exhibit higher fall than normal respiratory sounds (wide-band signal). Fig. 2 shows this in form of $E^{(0)}/E^{(k)}$ ratio. Given ratio was used as a feature to classify signals into normal and wheezing class.

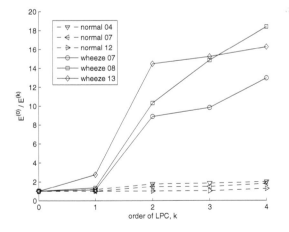

Fig. 2. E(0)/E(k) ratio for wheeze and non-wheeze (normal) signals as a function of the LPC order

6   Materials and Methods

The implementation of the algorithm is depicted in Fig. 3. The algorithm operates on blocks of 256 successive samples sampled by ADC at 2 kHz. FIR filter is used to attenuate hearth sounds. Hamming window is used. Low order of 4 of predictor is used to prevent accumulation of numerical error during implementation of fixed-point Durbin's recursion. For the same reason, dithering noise was added. Special care was taken to detect and avoid overflows. $E^{(0)}/E^{(4)}$ ratio was used as a feature for the classification of respiratory signal. The simplest fixed threshold classification of $E^{(0)}/E^{(4)} > \theta$ was used, with θ empirically determined to be 6.

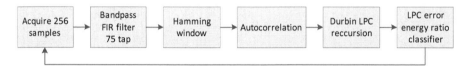

Fig. 3. LPC wheeze detection algorithm

The LPC algorithm was implemented on a 16-bit fixed point DSP TI TMS320-VC5505 evaluation board. The test was conducted by feeding the test signals through evaluation board's analog audio-in interface.

Test signals were acquired from various open-access Internet sources. A total of 13 wheezing signals (W01...W13) and 13 normal signals (N01...N13) of various duration, sampling frequency, SNR, body-locations and of patients of various age and gender were used. Generally, a lack of standard databases of pre-recorded respiratory sounds analysis, for evaluation of algorithms is identified.

Evaluation was conducted by auditory and visual inspection of the spectrogram of all sound recordings. Spectrograms were segmented into inspiratory and expiratory phases. For each segment (respiratory phase), occurrence of wheeze on spectrogram was compared with resulting DSP output stream in order to identify true positive (TP), true negative (TN), false positive (FP) and false negative (FN) regions and calculate sensitivity $SE = TP/(TP + FN)$, specificity $SP = TN/(TN + FP)$ and accuracy $ACC = (TP + TN)/(TP + FP + FN + TN)$. The results were averaged.

7   Results

Results of the testing are given in Table 1. The mean sensitivity of the algorithm was found 70.9%, specificity 98.6% and accuracy 90.29%.

A posteriori analysis of the algorithm performance has been conducted in order to determine hardware constraints for proposed solution. At the operating frequency of DSP of 60 MHz, and audio sampling frequency of 2000 Hz, execution time of one cycle of the algorithm (calculation of one set of 4-th order

Table 1. Performance of the LPC algorithm

File	TP	FN	TN	FP	**SE** [%]	**SP** [%]	**ACC** [%]
W01, N01	4	2	11	2	66.7	84.6	78.98
W02, N02	2	3	13	0	40	100	83.33
W03, N03	2	5	13	0	28.6	100	75
W04, N04	3	0	5	0	100	100	100
W05, N05	2	1	4	0	66.7	100	85.71
W06, N06	2	2	8	0	50	100	83.33
W07, N07	6	0	10	0	100	100	100
W08, N08	3	0	7	0	100	100	100
W09, N09	0	3	9	0	0	100	75
W10, N10	4	0	14	0	100	100	100
W11, N11	13	2	25	0	86.7	100	95
W12, N12	1	0	7	0	100	100	100
W13, N13	2	0	12	0	100	100	100
Total	44	18	138	2	**70.9**	**98.6**	**90.29**

coefficients on 256 samples) is 554,872 DSP cycles. This results in occupying of 7.23% of the DSP time and yields theoretically achievable average consumption of (only) signal processing of 0.85 mA. Duration of the execution is invariant of the outcome of the detection.

8   Conclusion

We have proposed a feasible system architecture of a wireless sensor network for monitoring of asthma, using the state of the art off-the-shelf components.

The proposed LPC wheeze detection algorithm proved to be capable for real-time operation. Due to low workload, the DSP can be theoretically put to standby for more than 90 % of the duty-cycle or algorithm can be implemented on the platforms with significantly lower power consumption (where FFT may not be applicable).

Main drawback of the algorithm is its inherently low sensitivity due to the simple feature set capable of detecting only the degree of correlation of the signal. Thus, method can not be implemented as a trigger for another high sensitivity method as relatively great amount of wheezing blocks are interpreted as normal (non-wheezing).

In the future we plan to investigate the improvement of sensitivity by extending the feature set and employing more sophisticated classifiers. Also, extensive testing of the operation of the algorithm in conditions of low SNR, comparison of the performance with the STFT based algorithms and detailed evaluation of power consumption are required.

Acknowledgment. This study is supported by RECRO-NET through project *Biomonitoring of physiological functions and environment in management of asthma* conducted at University of Zagreb, Faculty of Electrical Engineering and Computing. We would like to thank prof.dr.sc. Davor Petrinovic for valuable advices.

References

1. Ostojic, V., Cvoriscec, B., Ostojic, S.B., Reznikoff, D., Stipic-Markovic, A., Tudjman, Z.: Improving asthma control through telemedicine: a study of short-message service. Telemed. J. E Health 11, 28–35 (2005)
2. Jaana, M., Paré, G., Sicotte, C.: Home telemonitoring for respiratory conditions: a systematic review. Am. J. Manag. Care 15, 313–320 (2009)
3. Bonfiglio, A., Rossi, D.D.: Wearable Monitoring Systems. Springer, Heidelberg (2011)
4. Oletic, D., Razov, T., Bilas, V.: Extending lifetime of battery operated wireless sensor node with DC-DC switching converter. In: IEEE Instrument. and Measurement Tech. Conf. I2MTC, Hanzhou (2011)
5. Shenoy, N., Nazeran, H.: A PDA-based Network for Telemonitoring Asthma Triggering Gases in the El Paso School Districts of the US - Mexico Border Region. In: Proc. 27th Annual Int. Conf. of the Engineering in Medicine and Biology Society IEEE-EMBS, pp. 5186–5189 (2005)
6. Seto, E.Y.W., Giani, A., Shia, V., Wang, C., Yan, P., Yang, A.Y., Jerrett, M., Bajcsy, R.: A wireless body sensor network for the prevention and management of asthma. In: Proc. IEEE Int. Symp. Industrial Embedded Systems SIES 2009, Lausanne, pp. 120–123 (2009)
7. Fu, Y., Ayyagari, D., Colquitt, N.: Pulmonary disease management system with distributed wearable sensors. In: Proc. Annual Int. Conf. of the IEEE Engineering in Medicine and Biology Society EMBC, Minneaplolis, pp. 773–776 (2009)
8. Wisniewski, M., Zielinski, T.: Digital analysis methods of wheezes in asthma. In: Proc. Int. Signals and Electronic Systems (ICSES) Conf., Gliwice, pp. 69–72 (2010)
9. Ser, W., Yu, Z.-L., Zhang, J., Yu, J.: Wearable system design with wheeze signal detection. In: Proc. 5th Int. Summer School and Symp. Medical Devices and Biosensors ISSS-MDBS, Hong-Kong, pp. 260–263 (2008)
10. Karmelsonix Website, http://www.karmelsonix.com/
11. Asthmapolis - asthma inhaler tracking, http://asthmapolis.com/
12. Pasterkamp, H., Kraman, S.S., Wodicka, G.R.: Respiratory sounds. Advances beyond the stethoscope. Am. J. Respir. Crit. Care Med. 156, 974–987 (1997)
13. European Respiratory Review. European Respiratory Journal (2000)
14. Alic, A., Lackovic, I., Bilas, V., Sersic, D., Magjarevic, R.: A Novel Approach to Wheeze Detection. In: Proc. World Congress IFMBE, pp. 963–966 (2007)
15. Bahoura, M.: Pattern recognition methods applied to respiratory sounds classification into normal and wheeze classes. Computers in Biology and Medicine 39, 824–843 (2009)

Reduced Power Consumption for 3GPP-Compliant Continua Health Devices by Deployment of Femtocells in the Home Environment

Edward Mutafungwa, Zhong Zheng, and Jyri Hämäläinen

Department of Communications and Networking, Aalto University
P.O. Box 13000, 00076 Aalto, Espoo, Finland
edward.mutafungwa@tkk.fi,
{zhong.zheng,jyri.hamalainen}@aalto.fi

Abstract. The Continua reference architecture is now widely considered to be the *de facto* framework for implementation of personal telehealth systems. Existing Continua guidelines specify the use of personal or local area network technologies (Bluetooth, USB, ZigBee) for personal health device connectivity within the monitored person's local environment. However, there are now efforts to update the Continua reference architecture to include personal health devices with integrated 3GPP radio interfaces (GPRS, HSPA, LTE etc.). To that end, indoor femtocell home gateways provide several benefits for personal telehealth systems with such 3GPP-compliant Continua health devices. A notable benefit is the reduced power consumption by the devices compared to case where connectivity is via outdoor macro base stations. This paper supports this premise with some results from comparative simulations between femto- and macro-based implementations of the Continua personal telehealth systems.

Keywords: Personal Teleheath, Femtocells, Power Consumption, Energy Efficiency, Machine-Type Communications, Continua.

1   Introduction

The Continua Health Alliance[1] (henceforth, referred to simply as "Continua") is arguably the leading open industry group providing interoperability guidelines and certification programs for the implementation of personal telehealth systems. The guidelines specify an end-to-end harmonized Continua reference architecture to represent the high-level structure of personal telehealth systems and provide commonly-agreed terminology for different health device classes and system interfaces [1]. Legacy Continua devices in the patient's domain currently use Bluetooth, Universal Serial Bus (USB) or Zigbee standards for data transport[1], and the baseline IEEE 11073 Personal Health Data (PHD) standards for interoperable data formatting, session control and so on [2]. The next generation of Continua health

---

[1] Continua Health Alliance: http://www.continuaalliance.org/

K.S. Nikita et al. (Eds.): MobiHealth 2011, LNICST 83, pp. 173–180, 2012.

devices may also include Third Generation Partnership Project (3GPP) standardized radio interfaces (GSM/GPRS, HSPA, LTE, etc.).[2]

The 3GPP-compliant Continua health devices extend direct mobile network connectivity to personal telehealth systems allowing for remote device management (e.g. fault management). Furthermore, it exposes mobile operator service APIs (e.g. location based services), subscriber-related data repositories (e.g. Home Subscriber Server) and billing platforms, thus enabling development of innovative personal telehealth services and business models [4]. However, the radio access links from indoor located 3GPP-compliant Continua health devices would typically terminate at an outdoor base station located kilometers away. This means in order to achieve radio performance targets (e.g. Signal to Interference and Noise Ratio) higher device transmit powers are needed to compensate for the distant-dependent path losses, attenuation through building walls and co-channel interference [5]. This will in-turn necessitate more frequent battery recharging or replacement for portable personal health devices, a scenario that should be avoided, particularly for elderly patients living independently with frail physical and/or cognitive capabilities, or for cases where health device downtime may have serious ramifications.

Subscriber-deployed miniature indoor base stations or femtocells [3] may provide several benefits (one being reduction in required device transmit powers) when implemented as gateways for personal telehealth systems [4]. In this paper, we present comparative simulations between the femto- and macrocellular scenarios that quantify the possible power savings (prolonged battery lifetimes) for 3GPP-compliant devices in Continua personal telehealth systems. The remainder of the paper is organized as follows. Section 2 provides a background on femtocells and potential benefits in personal telehealth systems, while Section 3 outlines the system model that is adopted for the study. Section 4 describes the simulation methodology and results, while concluding remarks are presented in Section 5.

2   Usage of Femtocell for Personal Telehealth Systems

As more information gathering and processing functionality is increasingly embedded in everyday objects, the need for converged home gateways for horizontal integration and management of disparate networked home systems (for telehealth, smart metering, home security, etc.) becomes more acute. To that end, femtocell technologies provide an attractive way for implementing the converged gateway for the systems, such as, the personal telehealth system discussed in this paper. Access to femtocellular services (voice, mobile data etc.) is usually restricted to a closed subscriber group (CSG), such as, household members. Femto base stations that operate in the 3GPP Universal Mobile Telecommunication System (UMTS) and Long-Term Evolution (LTE) environments are known as, Home Node Bs (HNBs) and Home eNodeBs (HeNBs), respectively [3]. These terminologies provide distinction from the NodeB and eNodeB terms used for UMTS and LTE macro base stations, respectively.

---

[2] The Global Certification Forum (GCF) certifies 3GPP-compliant mobile devices and has announced in early 2011 a partnership with Continua on joint certification of 3GPP-compliant devices: http://www.globalcertificationforum.org/WebSite/public/continua.aspx

Figure 1 illustrates a simplified end-to-end high-level view of heterogeneous macrocell and femtocell deployments to support a Continua personal telehealth system. Mobile network connectivity for 3GPP-compliant Continua health devices is possible via a macro nodeB (or eNodeB) located outdoors, or alternatively via a femto HNB (HeNB) deployed within the home environment. On the other hand, mobile network connectivity for legacy Continua Personal Area and Local Area Network (PAN/LAN) health devices is enabled on the first hop via a Continua Application Hosting Device (AHD), such as, a Smartphone, which provides Continua PAN/LAN interfaces towards the health devices and 3GPP interfaces towards the macro or femtocullar mobile network. From the mobile core network perspective, all 3GPP-compliant Continua personal health devices and AHDs devices are viewed as 3GPP user equipment (UE). Henceforth, the term UE will be used to refer to any of those aforementioned 3GPP-compliant Continua devices.

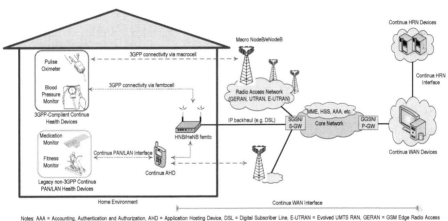

Notes: AAA = Accounting, Authentication and Authorization, AHD = Application Hosting Device, DSL = Digital Subscriber Line, E-UTRAN = Evolved UMTS RAN, GERAN = GSM Edge Radio Access Network, GGSN = Gateway GPRS Support Node, GPRS = General Packet Radio Service, GSM = Global System for Mobile, HRN = Health Reporting Network, HSS = Home Subscriber Server, IP = Internet Protocol, LAN = Local Area Network, LTE = Long Term Evolution, MME = Mobility Management Entity, PAN = Personal Area Network, P-GW = Packet Data Network Gateway, RAN = Radio Access Network, SGSN = Serving GPRS Support Node, S-GW = Serving Gateway, UMTS = Universal Mobile Telecommunications System, UTRAN = UMTS RAN, WAN = Wide Area Network, WCDMA = Wideband Code Division Multiple Access, 3GPP = Third Generation Partnership Project.

Fig. 1. High-level view of MTC and femtocell utilization for implementing the Continua personal telehealth system

The use of femtocells, as illustrated in Figure 1, provides additional benefits for 3GPP-compliant Continua personal telehealth systems, such as [4]:

- Enhanced access control strategies by the management of the femtocell CSG restrictions for personal health devices, temporary care givers etc.;
- Improved indoor coverage and capacity, ensuring service availability even usual macro coverage dead spots (e.g. attic) and scalability in terms of achievable throughput;
- Exposure of APIs for value-added services (e.g. virtual fridge notes for medicine reminders) that exploit femtocell awareness information of devices and subscribers within their coverage areas [6];

- Enhanced local area internetworking among home devices by implementation of multi-radio femtocell designs[3] and Local IP Access (LIPA) traffic breakout to alleviates congestion of macrocellular networks [7];
- Reduced device power consumption (less device uplink transmit power requirements) compared to macrocellular case.

This paper provides a more detailed simulation study of the reduced power consumption benefit enabled by femtocell connectivity.

3   System Model

The radio frequency (RF) uplink models are useful for the study of transmission power requirements (hence power consumption) of wireless health devices. The typical building blocks for a 3GPP-compliant mobile wireless device include integrated circuits for RF, baseband and mixed-signal processing functions [9, 10]. Each of the functional blocks contributes to the overall budget for power consumption of the device. To that end, the majority of power consumption budget is attributed to the RF transceiver and modem circuitry [10, 11]. The fraction of RF components power consumption is even more significant in wireless embedded devices that would usually have relatively less as or no user interface components (microphones, speakers, backlit displays, etc.) and reduced set of integrated secondary radio interfaces (Bluetooth, GPS, NFC, etc.).

An LTE radio access environment is assumed in this study, as there are ongoing efforts within 3GPP to modify features (e.g., signaling) in LTE, and evolutions beyond, for the impending proliferation of a large number of 3GPP embedded wireless devices [8]. In LTE networks the UE that are ON may be in either Radio Resource Control (RRC) connected or RRC idle state depending on whether the UE context is registered at a serving (H)eNB or not [5]. In the RRC_Connected state, the UE transfers (or receives) data packets, monitors the shared signaling channels for any scheduled resource allocation and provides measurement report feedback to the (H)eNB. By contrast, no transmission (or reception) of user data occurs in the RRC_Idle state except for periodic monitoring of common signaling and paging channels,. This inactivity in RRC_Idle state allows for most of the device circuitry to be powered down (sleep mode) so as to prolong battery life. Battery conserving opportunities can also be obtained by exploiting inactive periods in the RRC_Connected state (e.g. due to bursty traffic) [12]. Reduction of transmit power by reducing radio link distance provides even higher savings in power consumption in RRC_Connected state [13] and is the focus of this paper.

The LTE uplink adopts Single-Carrier Frequency-Division Multiple Access (SC-FDMA) as the radio interface. The SC-FDMA scheme maintains a low Peak-to-Average Power Ratio (PAPR) thus reducing UE power consumption, and also provides the benefit of high frequency efficiency due to compact subcarrier spacing [5].

---

[3] Example: Argela Femtocell and Home Gateway integrates both 3GPP (WCDMA/HSPA) and non-3GPP (WiFi, ZigBee) interfaces: http://www.argela.com/solutions.php?cid=femtocell

The total system bandwidth W is divided into subcarriers with an inter-carrier spacing 15kHz. Twelve subcarriers with 180 kHz bandwidth in total are grouped into a physical resource block (PRB) with 0.5 ms temporal duration, which is the basic radio resource unit allocated to UE. In this study, we assume the base station scheduler operates in a round-robin manner. The radio propagation between UE and (H)eNB experiences path losses including distance dependent path loss, shadowing and fast fading. The penetration loss due to walls separating the apartments is explicitly modelled, while penetration loss due to internal walls within the apartment is taken into account by using a log-linear model that depends on the separation distance. In this paper, we adopt the 3GPP TR 36.814 channel models [14], to evaluate the path loss PL for all possible radio propagation scenarios, namely: indoor UE to HeNB, outdoor UE to eNB, indoor UE to eNB, and outdoor UE to HeNB propagation.

The UE uplink transmission power is determined by the 3GPP fractional power control method for the Physical Uplink Shared Channel (PUSCH) that bears the LTE uplink user data. We use a similar approach to that proposed in [15], by ignoring the closed-loop corrections which results in the following open-loop power control scheme where the UE transmit power P_t is expressed as:

$$P_t = \min\{P_{max}, P_0 + 10\log_{10}M + \alpha PL\} \qquad (1)$$

where P_{max} is the maximum UE transmit power, P_0 is a UE- or cell-specific parameter indicating the possible minimum UE transmit power, M is the number of PRBs assigned for a certain UE, α is the cell-specific path-loss compensation factor, and PL is downlink path loss estimated by the UE. The uplink performance is measured by investigating the UE uplink throughput by summing the data throughput over all the PRBs allocated to the UE. The throughput over PRB j is calculated by mapping the signal-to-interference plus noise ratio (SINR) on PRB j with the equations,

$$S_j = \begin{cases} 0 & SINR_j \leq SINR_{min} \\ BW_{PRB} \cdot BW_{eff} \log_2\left(1 + SINR_j / SINR_{eff}\right) & SINR_{min} \leq SINR_j \leq SINR_{max} \\ S_{max} & SINR_j \geq SINR_{max} \end{cases} \qquad (2)$$

where BW_{PRB} is the bandwidth of a single PRB, BW_{eff} and $SINR_{eff}$ are bandwidth and SINR efficiencies that represent capacity loss due to system implementation and signal processing procedures. The link-level throughput mapping (2) was initially proposed in [16] and the parameters BW_{eff} and $SINR_{eff}$ are aligned with 3GPP TR 36.942 [17] and are also listed in simulation parameters Table 1 of next Section.

4   Simulation Methodology and Results

The simulated environment was covered by an overlaying macrocellular network composed of seven uniformly placed eNB sites as shown in Figure 2. Each site covers three hexagonal sectors using sectorized antennas and maintains separate radio resource management procedures. The eNBs provide universal radio access services to all registered subscribers. In the central cell, one large residence building is modelled either at cell centre or cell edge (see Figure 2, inset). The building has 100m^2 apartments arranged in a 5x5 grid layout that is commonly used in 3GPP simulation studies [18]. Furthermore, each apartment may have either one or no HeNB deployed, and the HeNBs operate in same spectrum band as the macro eNBs.

The system simulation parameters are listed in Table 1 and are in line with 3GPP simulation guidelines outlined in [14]. Comparative studies of the macro and femto systems were performed using a Matlab static simulator over a large number (10^4) of random snapshots (Figure 2 being an example snapshot). For each snapshot, the positions of the UE (both indoor and outdoor) and HeNBs (indoor) were generated randomly and the performance metrics evaluated.

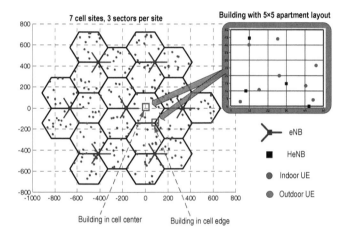

Fig. 2. Macro- and femtocellular system layout

Table 1. Simulation parameters

Parameter	Value
System Parameters	
Inter-site Distance	3GPP Macro Case 1: 500m
Carrier Frequency	2 GHz
Bandwidth	10 MHz, 48 PRBs for data+2 PRBs for signalling
Bandwidth Efficiency	0.4
SINR Efficiency	1
Thermal Noise PSD	-174 dBm/Hz
(H)eNB Parameters	
eNB Antenna Gain	14 dBi
Received Noise Figure	7 dB
Antenna Pattern	eNB: $A(\theta) = -\min[12\,(\theta/\,\theta_{3dB})^2, A_m]$ $\theta_{3dB} = 70^\circ$ and $A_m = 20$ dB
	HeNB: Omni-directional
UE Parameters	
Maximum Transmit Power	23 dBm
Maximum Antenna Gain	0 dBi

The generated cumulative distribution function (CDF) plots of the average UE transmission power for the scenarios where the indoor UEs (Continua health devices) are connected via the eNB and the HeNB are shown in Figure 3 (left). For the HeNB case, simulations are run for different values of power control parameter P_0 so as to observe transmission power requirements for different power control scenarios. For the eNB, simulations are repeated for the scenario where the building is in cell edge and cell center. Considering same combination of power control parameters ($P_0 = -73$ dBm and $\alpha = 0.8$) for eNB and HeNB in Figure 4, the UE connection via the HeNB can provide 50%-ile UE transmission power saving of 39 dB. To put that power saving into perspective using a simple example, an alkaline AAA-size cell (1.41 Wh capacity) could potentially increase its lifetime by 2-3 orders of magnitude.

Figure 4 (right) shows the CDFs for the average achievable UE throughputs for the same set of simulation runs. The results demonstrate ten-fold 50%-ile throughput increase even for the case of UE with lower power ($P_0 = -83$ dBm and -93 dBm, with $\alpha = 0.8$). For typical periodic and bursty telemetry traffic from monitoring devices, a higher throughput reduces the amount of time the UE remains in connected state, which in turn reduces power consumption [19]. This benefit becomes more significant for cases with increased volumes of telemetric data (e.g. video clips).

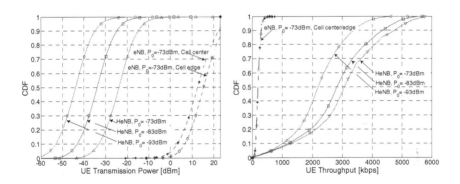

Fig. 3. CDFs of average UE transmission power (left) and throughput (right)

5   Conclusions

This paper studied the potential device power consumption reduction benefit enabled by femtocells in the implementation of Continua personal telehealth system. To that end, we obtained simulation results for a given case study that indicated significant device transmit power savings for a femto-based system compared to an equivalent macro-based personal telehealth system. We plan to continue study with further elaborate simulations that for instance take into account real traffic patterns of commercial off-the-shelf personal health devices. Furthermore, we intend to reproduce the teleheath system setup in a femtocell testbed to obtain even more realistic evidence of the power saving benefits and demonstrate other femtocell features that add value to personal telehealth use cases.

Acknowledgments. This work was prepared in MOTIVE UBI-SERV project supported by the Academy of Finland (grant number 129446); and the CELTIC HOMESNET (CP6-009) project supported in part by the TEKES, NSN and ECE.

References

1. Wartena, F., Muskens, J., Schmitt, L., Petković, M.: Continua: The Reference Architecture of a Personal Telehealth Ecosystem. In: Proceedings of the 12th IEEE International Conference on e-Health Networking Applications and Services (Healthcom), Lyon, p. 6 (2010)
2. Martínez-Espronceda, M., Martínez, I., Escayola, J., Serrano, L., Trigo, J., Led, S., García, J.: Standard-Based Homecare Challenge. In: Yogesan, K., Bos, L., Brett, P., Gibbons, M.C. (eds.) Handbook of Digital Homecare. BIOMED, vol. 2, pp. 179–202. Springer, Heidelberg (2009)
3. Zhang, J., de la Roche, G.: Femtocells: Technologies and Deployment, 1st edn. John Wiley & Sons Ltd., Chichester (2009)
4. Mutafungwa, E.: Applying MTC and Femtocell Technologies to the Continua Health Reference Architecture. In: Proceedings of the International Workshop on Health and Well-being Technologies and Services for Elderly (HWTS 2011), Oulu, p. 10 (2010)
5. Holma, H., Toskala, A.: WCDMA for UMTS: HSPA Evolution and LTE, 5th edn. Wiley & Sons Ltd., Chichester (2009)
6. Femto Forum Services SIG: Femto Services Version 1 API (2011)
7. 3GPP TR 22.368: Local IP Access & Selected IP Traffic Offload (LIPA-SIPTO) (2011)
8. 3GPP TR 22.888: System Improvements for Machine-Type Communications (2010)
9. Shearer, F.: Power Management in Mobile Devices, 1st edn. Newnes, Burlington (2008)
10. Silven, O., Jyrkkä, K.: Observations on Power-Efficiency Trends in Mobile Communication Devices. EURASIP J. Embed. Sys., 1–9 (2007)
11. Kim, H., de Veciana, G.: Leveraging Dynamic Spare Capacity in Wireless Systems to Conserve Mobile Terminals' Energy. IEEE/ACM Trans. Netw., 802–815 (2010)
12. Bontu, C., Illidge, E.: DRX Mechanism for Power Saving in LTE. IEEE Comm. Mag. 47, 48–55 (2009)
13. Haq Abbas, Z., Li, F.: Distance-Related Energy Consumption Analysis for Mobile/Relay Stations in Heterogeneous Wireless Networks. In: Proceedings 7th International Symposium on Wireless Communication Systems (ISWCS), York (2010)
14. 3GPP TR 36.814: Further advancements for E-UTRA Physical Layer Aspects (2010)
15. Castellanos, C.U., Villa, D.L., Rosa, C., Pedersen, K.I., Calabrese, F.D., Michaelsen, P., Michel, J.: Performance of Uplink Fractional Power Control in UTRAN LTE. In: Proceedings of IEEE VTC Spring, Singapore, pp. 2517–2521 (2008)
16. Mogensen, P., Wei, N., Kovacs, I.Z., Frederiksen, F., Pokhariyal, A., Pedersen, K.I., Kolding, T., Hugl, K., Kuusela, M.: LTE Capacity Compared to the Shannon Bound. In: Proceedings of IEEE VTC Spring, Dublin, pp. 1234–1238 (2007)
17. 3GPP TR 36.942: Evolved Universal Terrestrial Radio Access (E-UTRA); Radio Frequency (RF) System Scenarios (Release 10) (2010)
18. 3GPP R4-092042: Simulation assumptions and parameters for FDD HeNB RF requirements. WG4, Meeting 51 (2009)
19. Wang, L., Manner, J.: Energy Consumption Analysis of WLAN, 2G and 3G Interfaces. In: Proceedings of 2010 IEEE/ACM International Conference on Green Computing and Communications, Hangzhou, pp. 300–307 (2010)

A New Platform for Delivery Interoperable Telemedicine Services

Foteini Andriopoulou and Dimitrios Lymberopoulos

Wire Communications Laboratory, Electrical and Computer Engineering Department,
University of Patras,
University Campus, 265 04 Rio Patras, Greece
{fandriop,dlympero}@upatras.gr

Abstract. This paper represents a new concept for Telemedicine service delivery. It is based over the existed Service Delivery Platforms (SDPs) with the mentality of SOA and Enterprise Service Bus (ESB) architecture to provide integration and interoperability in telemedicine aspects over the Next Generation Network (NGN). Telemedicine SDP (T-SDP), which is proposed, is a lighter middleware to provide flexibility and integration in emergency cases. Lighter middleware supports the benefits of SDP, real time monitoring and communicating in responses. T-SDP is able to leverage contexts, semantics and events via a service engine and Parlays. In the same platform events from the biomedicine devices, context from user profiles and messages or ontologies and semantics can interoperate with each other and create complex services useful for the healthcare agents.

Keywords: telemedicine, healthcare, interoperability, integration, SDP, ESB, NGN.

1 Introduction

According to recent statistic researches, world's ageing population and average of chronic diseases are increasing rapidly, leading to high demand of healthcare services. Telemedicine involves the delivery of healthcare and related information over long distances, reducing the response time and cost [10].

During the last years many efforts have accomplished to create secure, interoperable, flexible and user-friendly telemedicine systems. Mobile telemedicine has enhanced, smart homes and smart cars have created but the aspect of interoperability, integration and real timing response is always a sensitive and significant issue.

Nowadays, there are different types of service buses that are depending on what a device, a web service, etc use as an output e.g. events, contexts or semantics. It is useless, expensive and doesn't support interoperability to establish different service platforms for all these kinds of ESBs or create a single SDP, which contains all the different types of service buses in order to create an interoperable common service delivery platform that could be standardized and work properly with all enterprises. In this paper we introduce a telemedicine ESB (TSB) in a SDP to solve integrated and

K.S. Nikita et al. (Eds.): MobiHealth 2011, LNICST 83, pp. 181–188, 2012.

interoperable problems in large scale enterprises such as hospitals, laboratory institutes, etc and provide flexibility and maintainability. The proposed SDP for telemedicine purposes (T-SDP) is used to delivery, compose, deploy services to create reusability and make different context, event, semantic components to become interoperable with each other.

The innovation of our proposition is, a service delivery platform based on ESB and SOA which provides a *lighter service execution environment*. The lighter the SDP is, the more efficient services are used, reused, created, bund and invoked dynamically in a real world period of time. This aspect is significant in telemedicine field where we have thousand of events only from one sensor and many agents in a common scenario to avoid "spaghetti" communication problems. Not only it is important to have the best architecture but also to be interoperable and flexible in real time responses when the patient is moving, running or when he changes connecting devices and destinations which are not according to his daily program. In our proposition the T-SDP isn't consisted of enablers that are used in classical architectures. This makes T-SDP lighter and flexible, moreover according to the enablers from the network can support properly through APIs and JAINSLEE all different type of output which referred above.

These characteristics make our platform valuable in the telemedicine and enterprise fields and are expected to play an active role in standardizing SDPs.

2   Basic Approach

As a result of the increased life expectancy and chronic diseases, the growing financial costs in healthcare leads to the need of organizing healthcare away from hospitals and clinics and providing a patient- centric, more user-friendly system where patients would have a significant and centric role in their own treatment and care of their illness. Moreover, connects patients with their healthcare professional team and helps them to be socialized without changing their preferences and daily habits.

Figure 1 represents a common telemedicine cooperative working group of subjects, hereafter referred as agents. Each one of the cooperative agents can be simultaneously either a provider or a consumer (such as client - server).

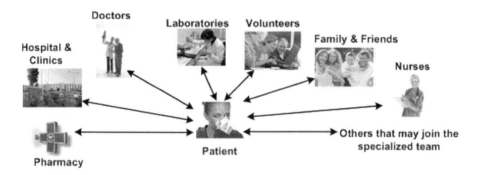

Fig. 1. Cooperative agents in healthcare based on patient-centric philosophy

Due to the fact that healthcare is transferred in the environment of patients (home, car, office, etc) it is important to support patient mobility (monitoring of remote vital signs, mobile medical records etc). S.Van Hoecke, K.Steurbaut, et al, have designed and implemented a secure and user-friendly broker platform supporting the end to end provisioning of e-homecare services to solve the problem of patient mobility and provide interoperability via an open ESB [1]. Although, the solution for healthcare interoperability provides many advantages for the use of ESB and eliminates the startup time since it is required only a login to be authenticated and accessed to the e-environment services, it is premised all the e-environment services to be integrated in a single client application and doesn't support different type of contexts, events etc. This paper proposes not only a middleware architecture like ESB, but a SDP middleware based on the benefits of ESB. To achieve authentication with only one sign in and save time we have lighten the SDP platform removing the service components, so that e-environment services can be anywhere, integrated and transferred via APIs, Parlay, OSA etc in the SDP via the JAINSLEE and ESB (Fig. 2). The authentication will take place in SDP and since the validation is controlled and guaranteed is the first and last time the system requires the electronic health pass card. Moreover, due to the fact that the application, services are integrated and connected to TSB can support different contexts.

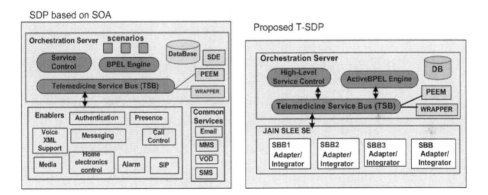

Fig. 2. Traditional architecture of SDP based on SOA and the proposed T-SDP

Jaime Martin, Mario Ibanez et al have published an e-Health system for a complete home assistance [2]. E-health system tried to adjust integration and openness to telemedicine services and leave for future work a manner to generate medical alerts based on the patient health data and usability of applications. In our approach ESBs are established in the environment of patient and biomedicine vital signals are sent in SDP. The ESB provides management mechanisms for troubling states on the NGN, that's the reason ESB is expanded in the business support system of NGN as shown in figure 3. Moreover, according to a survey on web services in Telecommunications [3], JAINSLEE's framework contains Alarms, Timers, profile information and emergency states which are mediated via ESB to inform agents. The reusability of all the services and applications is guaranteed in SOA and ESB architecture.

3   Architecture of the Proposed T-SDP

T-SDP is a middleware software architecture which is established between the service stratum of NGN and the applications provided by 3rd party providers. T-SDP communicates through Telemedicine Service Bus with NGN Business Support System (BSS) for the purpose of management in the service stratum and Operation Support System (OSS) in transport stratum. Figure 3 depicts the proposed T-SDP structure, which is aligned with the majority of SDP architectures.

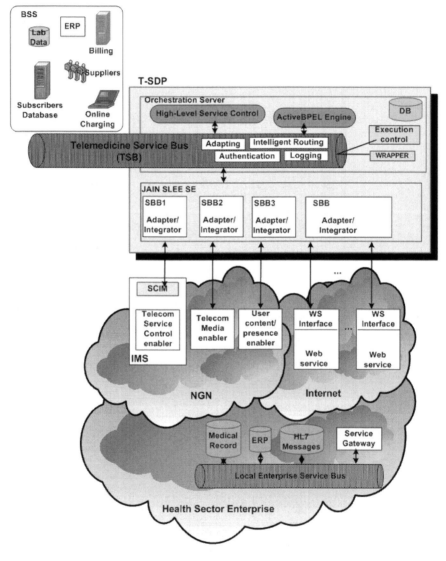

Fig. 3. Proposed architecture of T-SDP

In our approach, enablers are provided from the network, internet and health sector enterprises. By the term health sector enterprises we depict smart e-homes, smart cars, e-ambulance, hospitals, etc. T-SDP leverages adopters and integrators of telecom Media enablers (voice, video, audio etc), Service Control enablers (signaling, calling, controlling etc), User content enablers of networks like GSM, 3G, NGN, web services, and applications and services from the health sector enterprises. The Service Capability Interaction Manager (SCIM) function block is necessary to control SIP messages between the network and enablers. The SCIM distributes messages to appropriate destinations based on customer profiles or SIP parameters, blocks inconsistent messages, relays messages between Application Servers (ASs) and enablers, and controls service contention [5]. Then, T-SDP authenticates patients via the PEEM, mediates intelligent route requests and composes more complex services and application from these fine- grained components to create a new required service for a healthcare agent or patient.

Since T-SDP is a light middleware software architecture, it doesn't contain enablers but points of integrations supported by JAINSLEE service engine. Integration of service components and APIs takes place away from T-SDP, and in the database of T-SDP there are only copies of required services and not all the information of applications servers. These reasons provide flexibility in the T-SDP and reusability in real time action, so make T-SDP platform challenging for emergency occasions.

The T-SDP is composed of:

The *Orchestration server* is provided by a service bus in the role of service broker which connects and controls all workflows and manages processes and components such as enablers and BPEL engines. Orchestration server controls the business process, implements the SOA paradigm (find, publish, bind and invoke) and involves the intelligent routing and capabilities of Service Bus. Orchestration server is composed of:

Active BPEL (Business Process Execution Language) engine. BPEL is a process - oriented language which is used in web services for describing workflow composition of services, invoking purposes, handling asynchronous communication and event subscription. BPEL engines are stateful and able to interpret the BPEL language. Moreover support different context awareness scenarios for service coordination depending on patient's or agent's needs and preferences. Active BPEL is an open source BPEL engine written in Java which can read BPEL code and related standards, and create its own representation of BPEL processes. When a client invokes a BPEL process, the engine creates a new instance of it by translating patient's or agent's request in each scenario. The appropriate services are found according to policy, bound and invoked dynamically. Because Active BPEL is an open source is the appropriate BPEL engine for our approach and open environment.

Service Database (DB) is where the data are stored, manages registration, advertisement and discovery in electronic health records - EHR, agent's profiles and preferences, laboratories tests etc.

Execution control provides Policy Evaluation, Enforcement and Management (PEEM) capabilities. It provides Single Sing-On SSO authentication and guarantees the SLA by observing the response time.

Telemedicine Service Bus is a middleware software architecture and a standard-based messaging engine, actually a message broker, which is driven by events and provides fundamental services for more complex software systems. TSB is used in medical care environment and provides connection and integration to distributed health services and agents with a secure and reliable way, regardless of the type of software or hardware which they use. Usually service buses are realized as distributed service containers and connected to the services via endpoints such as routers, application adapters, message-oriented middleware bridges and other communication facilities. The TSB is also used to provide mechanisms to bind the federated required services in the composed service solution and provides infrastructure services such as mediation services, security, logging, service discovery, service description transformation services.

Wrapper is a program that sets the rules for other applications and services which are more complex and are composed of services and interfaces. Leverages application interfaces to web services format that can be used for further composition and reusability of other services to make services more reliable.

The *JAIN SLEE SE* is a high throughput, low latency event processing application environment Java standard for a Service Logic Execution Environment (SLEE). The JAIN SLEE SE is used to make sure that SLEE ruins in the TSB environment as a service engine. Is composed of messaging processing, lifecycle, deployment unit, registration management module and is designed to achieve scalability, interoperability and availability through federating services and service oriented architectures. Finally, JAIN SLEE is the point of integration for multiple network resources and protocols. The integration between the enablers, web services and other health sectors is achieved through adapter/ integrator service building blocks. [3], [8]

4   Creation of High Level Telemedicine Service

In this section, we use T-SDP in order to represent the creation of a complex high level telemedicine service, which is composed of fine grained services such as multimedia conferencing, media records, vital examination and user content services (Fig. 4). Moreover we represent the implementation of a multimedia conference scenario between a patient and an agent (Fig. 5).

As depicted in fig.4, each one of the granularity services is created from a set of reusable service building blocks (SBBs), which are usually wrapped around a set of applications and services in interaction with databases, relying on a system infrastructure and delivered to the customer via an access network, NGN enablers, IMS (IP Multimedia Subsystem) in coordination with Service Capability Interaction Manager (SCIM) and health sector enterprises networks. For example, multimedia conferencing service is composed of voice, video, audio and data SBBs (Fig.5). SBBs from NGN, Internet and Health sector enterprises are integrated and driven in JAIN SLEE service engine. JAINSLEE service engine can be easily connected to different application platforms or enablers (network elements) and store temporary the appropriate enablers and services for each purpose. Then JAINSLEE service engine and Active BPEL engine are integrated to TSB which transforms, routes, mediates, etc, the SBBs and common services and creates the required application. The created service is registered in the service repository and is ready for the next time someone will need it. With the same way and according to the preferences of the consumer (Active BPEL engine), the policy enforcement, QoS, SLA agreement etc, the user content, medical records and vital examinations are created from all the other fine grained components and services. Moreover, Active BPEL engine provides the capability to manage the conference process by orchestrating conference services. [7]

The high level Telemedicine service is controlled by the service control engine and is provided to the consumer.

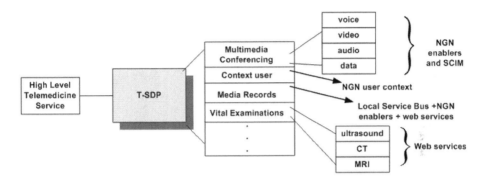

Fig. 4. Creation of a High level Telemedicine service

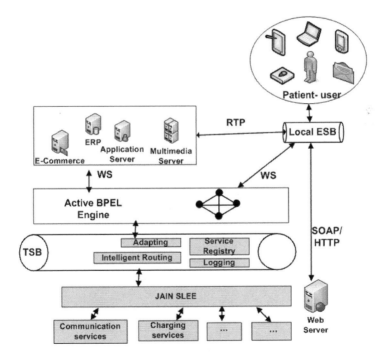

Fig. 5. Implementation of Multimedia Conferencing service

5   Conclusion

This paper has presented an advanced T-SDP and TSB approach in telemedicine field based on agent and patient oriented demands. The proposed system would provide a flexible, interoperable and integrated way of delivering and composing healthcare services in patient's and cooperative's agents than is currently available in a number of healthcare scenarios in remote, emergency, rural cases.

Lighter middleware of T-SDP could be significant in ubiquitous telemedicine environments because it has the ability to provide secure services based on private policies, preferences, profiles and integrates agents and patients dynamically in a short period of time so that is efficient in emergency events. Moreover, a standardized TSB appropriate for events, semantics, and contexts is expected to solve interoperable problems and establish more flexible and efficient telemedicine systems.

References

1. Van Hoecke, S., Steurbaut, K., et al.: Design and implementation of a secure and user-friendly broker platform supporting the end to end provisioning of e-homecare services. Journal of Telemedicine and Telecare 16, 42–47 (2010)
2. Martín, J., Ibañez, M., Martínez Madrid, N., Seepold, R.: An eHealth System for a Complete Home Assistance. In: García-Pedrajas, N., Herrera, F., Fyfe, C., Benítez, J.M., Ali, M. (eds.) IEA/AIE 2010, Part II. LNCS, vol. 6097, pp. 460–469. Springer, Heidelberg (2010)
3. Grifin, D., Pesch, D.: A survey on Web Services in Telecommunications. IEEE Communication Magazine (July 2007)
4. Sorwar, G., Ali, A.: Advanced Telemedicine System Using 3G Cellular Networks and Agent Technology. In: Takeda, H. (ed.) E-Health 2010. IFIP AICT, vol. 335, pp. 187–197. Springer, Heidelberg (2010)
5. Maes, S.H.: Service Delivery Platforms as IT Realization of OMA Service Environment: Service Oriented Architectures for Telecommunications. In: IEEE, WCNC 2007, pp. 2885–2890 (2007)
6. Sunaga, H., Yamato, Y., Ohnishi, H., Kaneko, M., Iio, M., Hirano, M.: Service Delivery Platform Architecture for the Next-Generation Network
7. Zhu, D., Zhang, Y., Chen, J., Cheng, B.: Enhancing ESB based Execution Platform to Support Flexible Communication Web Services over Heterogeneous Networks. In: IEEE, ICC 2010 (2010)
8. JAIN SLEE, http://www.jainslee.org/slee/slee.html
9. Menge, F.: Enterprise Service Bus. In: Free and Open Source Software Conference (2007)
10. Lin, C.-F.: Mobile Telemedicine: A survey Study. Journal of Medical Systems, Online First^{TM} (April 27, 2010)

Social Media in Healthcare - User Research Findings and Site Benchmarking

Timo O. Korhonen, Maija Pekkola, and Christos Karaiskos

Aalto-University,
P.O. Box 3000, FIN-02015 TKK, Finland
{first name.last name}@aalto.fi

Abstract. Still nowadays social media is widely spread, it is recognized that general purpose sites support health issues sporadically only. In this paper, we report a user experience study of a Finnish health-oriented online social network service, Hoitonetti.fi and benchmark it to some other, comparable sites to qualitatively compare service palettes and to scope future pathways. In our research we have found out that social media can indeed efficiently support accessing and sharing healthcare information. However, our results indicate that the health portals could anyhow better engage the users. Medical professionals should author blogs, participate to discussions, and provide safety support. If sites could more actively support professionals to develop their diagnosing and care taking practices their commitment and dedication to site up keeping could be improved. Also, anonymized data gathered could be available for research. Even respective statistical tools could be incorporated to the sites by using some web 2.0 technology as Ajax. Efficient linking of external health records could be supported. Patient-secure links to general health portals as Google Health or Microsoft Health Vault could be created. All this could enable new service ecosystems to be created too.

Keywords: online social media, social websites, online social networks, healthcare, well-being.

1 Introduction

Worldwide health statistics of Continua Health Alliance reveal that there are 1 billion adults that are overweight, and 860 million individuals with chronic conditions. Also, there are 600 million individuals age of 60 or older. 75-85% of healthcare costs are due to chronic condition management. On the other hand, there is increasing loneliness, and need for social activities especially with elderly. Clearly we should rise up awareness of our own role in up keeping physical, social and mental health. Important tool for this is the developing social media that allows us to connect, communicate, socialize and interact virtually and to generate, publish and share all kind of content. Social media includes multiple services, such as online social networking, multimedia sharing (e.g. video, audio and photography), content publishing (e.g. blogs and micro-blogs), information mining and online collaboration

K.S. Nikita et al. (Eds.): MobiHealth 2011, LNICST 83, pp. 189–196, 2012.

(e.g. wikis). Health related social websites are now quickly appearing worldwide (e.g. Google Health, Microsoft Health Vault, MedHelp.org, DailyStrength.org, and PatientsLikeMe.com). Also, there exist sites that are tailored especially for elderly (growingbolder.com, EONS.com, ELDR.com, and Seniornet.org). Important aspect in these sites is that issues relating to privacy are greatly circumvented by allowing anonymous participation. Sites apply, as a rule, internal privacy policies that are independent of governmental policies. This enables sites to be developed more freely and it gives also a greater freedom to develop new services that are based on (anonymous) user data. These include for instance tailored user interfaces tuned to follow users' personal health facts and/or on-site discussions. Also, when the site becomes bigger, user data can be used for statistical scientific research as for instance for investigation of diagnosing techniques as experienced by the site users. Also, efficiency of health care practicalities and medications for particular sicknesses can be analyzed.

In this paper, we report and discuss our user experience research of a Finnish health-related online social networking service, Hoitonetti.fi. Also, we benchmark Hoitonetti.fi with some similar sites in order to create a critical view of current state-of-the-art, and to evaluate how the medical social media could be seen to develop in the future. Findings of the user research and site benchmarking are reported in sections 3 and 4 respectively. We begin by considering the general role and potentials of social media for well-being and healthcare in section 2.

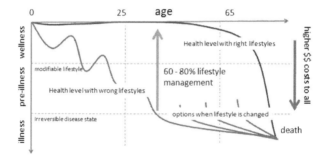

Fig. 1. Well-being vision of Continua Health Alliance (fig. is modified from [1].). We cause major part of our sicknesses; therefore we can also avoid them by fostering well-being. This can result longer life, higher-quality life in general, and cut down health care costs.

2   Social Media and Health Care

In general it is known that in an individual's life there is an association between social support and psychological well-being [2] and the importance and benefit of social support for recovery, rehabilitation and adaptation to illness are acknowledged [3]. The importance of introducing online and computer-mediated social support is thus well recognized. For example the use and contents of health-related online support groups have previously been analyzed for example in the case of an Alzheimer caregivers' e-mail group [4] and an online bulletin board for breast cancer patients [5]. Also the use of a discussion board on a health website for older adults has been documented already few years ago [6]. The techniques of Web 2.0 essentially deployed for online social media enrich the patient-to-patient communication by providing diversified, more socially oriented ways for the patients and care givers to connect and share experiences and advices with their peers. It is generally recognized, that improved health and reduction of health care costs should be based on understanding, marketing and implementing well-being practices as early as possible. Obesity, diabetes, and most heart diseases, for instance, are nowadays known to have their roots in our ways of living. They are causing the population to fall away from their peak health potential that degrades our life's quality and creates a substantial financial burden too. In Continua Health Alliance vision [1] realization of peak health potential (following right lifestyles) is in great deal, our responsibility (Fig. 1). An important challenge is how to motivate people to follow or make the change to follow better lifestyles. Also, how the relating personal objectives should be defined and how to obtain sustainable improvements? Online social media can indeed be seen offering tools to empowerment of patients and care takers for improved care in many dimensions [7] that we investigate in this paper by a site survey and benchmarking study.

2.1   Defining Well-Being

Research of social and psychological impacts of social websites and engaging by social media sheds light to the characterization of these services. Dynamic and sustainable states of physical, environmental, economical, social, emotional, intellectual, and ethical well-being are important components of the multi-dimensional health (Table I). Therefore site service palette should address them wisely while targeting the overall health, well-being and communication.

Table 1. Some dimensions of well-being, modified from [8]

Physical	Environ.	Econom.	Social	Emotional	Intellect.	Ethical
Structures and functions of the body	The environment where we live and work	Our disposable income and financial health	Make and maintain relationships in various social contexts	Recognize and express needs and feelings adequately and appropriately	Learning to think clearly and coherently to enable rational decisions	The art of understanding and applying concepts of right and wrong decisions

3   Characterizing Hoitonetti.fi and the Site Survey

Hoitonetti.fi is a Finnish social media based health care site. The service has approximately 2300 registered users. The number of unregistered visitors highly exceeds the number of registered users. The age distribution of services spans from 10 to 80+ years most users being 50-59 years (580 users). The next largest user groups are 40-49 years (450 users), 60-69 years (380 users), and 30-39 years (320 users).

The idea behind the service is in aiming to create an interactive collection of health information, where experts' knowledge and user-originating content are combined. The site consists of registered users' profile pages and illness-specific pages (e.g. asthma) for symptoms (e.g. dizziness), treatments (e.g. antihistamines) and health measurements (e.g. haemoglobin concentration). The information pages are written by medical experts and controlled by Hoitonetti.fi maintenance staff. Users' profile pages are updated by the users themselves. Users can add illnesses, symptoms, treatments and health statistics to their profiles together with personal experiences interlinking the users' profile pages to information pages. Each information page contains also a discussion section where both registered and unregistered users can participate, share thoughts and experiences and ask questions. A key feature of Hoitonetti.fi is that its maintenance team includes responsible doctors actively monitoring and taking part to discussions. Registered users can create micro-blogs, send status messages, become friends with other registered users and join communities. Besides discussions on information and community pages, the forms of user-to-user interaction include commenting each other's updates and sending private messages and virtual hugs. While registering to the website, user can choose own screen name. Therefore, it is not required to interact by real name. The service can be used totally anonymously.

3.1   Site Survey

An online questionnaire consisting of eight open-ended questions was created in order to collect information on the users' experiences, perceptions and views of using Hoitonetti.fi. A total of 31 responses were received none of which were excluded from the analysis. Some results are gathered to Table 2. The most prevailing themes found in the analysis were 1) community, support and sharing and 2) information sourcing, medical professionals' participation and safety.

1) Community, Support and Sharing. A strikingly strong emphasis in the respondents' descriptions was on finding peer support and sharing experiences via the online social network service. They were highly valued and one of the key factors in the respondents' engagement in the service. Peer support and sharing of experiences hold an emotional aspect as well as a practical aspect.

2) Information Sourcing, Medical Professionals' Participation and Safety. Finding health related information was the single most often mentioned factor in the responses. Having a doctor answering questions was considered important and desirable service feature, but it was also considered to be an essential part in making the virtual environment feeling safe. Actually, for some of respondents, this was the reason why they initially decided to use Hoitonetti.fi. Therefore, based on our research, safety and trust are very important aspects in service experience. Note that this includes also moderation of discussions where any inappropriate communication is removed.

Table 2. Outlining survey results

Question related to	Most common codes (frequency of occurrence)	
Reasons for registering to the service	1.	interest in the topics (2), presence of medical professionals (2), sharing own experiences and reading about other people's experiences (2), usefulness of the service (2)
	2.	positive atmosphere (1), finding people to talk to (1), passing the time (1), receiving support (1)
Reasons and purposes for using the service	1.	getting information (18)
	2.	sharing own experiences and reading about other people's experiences (8)
	3.	giving and/or receiving support and encouragement (7)
Best about the service	1.	giving and/or receiving support and encouragement (9)
	2.	friends and other users (5), positive and encouraging atmosphere (5)
	3.	clear and easy to use (4), hugs (4), presence of medical professionals (4)
Bad about the service	1.	too little or lack of information on some specific medical issues (3), too much of virtual hugging (3), usability problems (3)
	2.	lack of trust in the doctor's answers (2)
	3.	lack of information about the maintenance staff (1)
Reasons to keep using the service	1.	friends and other users (10)
	2.	giving and/or receiving support and encouragement (6), positive and encouraging atmosphere (6),
	3.	getting information (5)
Most useful features of the service	1.	information (11)
	2.	possibilities for discussion (8)
	3.	doctor's answers (4)
Problems in using the service	1.	usage was difficult in the beginning (7)
	2.	varying usability problems (3)
Requests for development	1.	more of medical professionals participation (4)
	2.	varying requests for certain medical information (3)
	3.	making things more clear (2)

4   Benchmarking

In the benchmarking four sites were compared of which summarizing results are extracted in Table 3. We focused on in getting overall impression of user experience in medical substance and in support of communication. Generally, all sites offer relatively high quality patient-to-patient peer support. We considered patient-to-patient interaction to include services such as adding friends, sharing messages, participating in health discussions in forums, sharing personality traits etc.

As far as expert advice was concerned, the widest variety of features was offered by MedHelp.org, where there are specialized forums and chat rooms for immediate patient-to-doctor interaction. Physicians too have own profiles revealing areas of specialization and some details of their medical careers too. Also, pictures, blog entries and previous posts of physicians were easy to access. This enabled patients to truly rely on the validity of answers that is, as we noticed in Hoitonetti.fi survey, generally highly appreciated.

Website environments differ quite a lot between the four sites. Medhelp.org offers the most complete environment and could easily compete with any other social networking website due to its seemingly high usability and variety of services offered. Patientslikeme.com follows Facebook-inspired interface with multiple services and its usage is quite straightforward especially for Facebook-users.

Table 3. Benchmarking of well-being sites

	Hoitonetti .fi	MedHelp .org	DailyStrength .org	Patientslikeme .com
Target Age Group	*elderly*	*any*	*any*	*any*
Patient-to-Patient Interaction				
-Discussion forums	*Yes*	*Yes*	*Yes*	*Yes*
-Private/public messages	*Yes*	*Yes*	*Yes*	*Yes*
-Add friends	*Yes*	*Yes*	*Yes*	*Yes*
-Share health statistics	*Yes*	*Yes*	*Yes*	*Yes*
-Other	*-Virtual Hugs*	*-iPhone Apps, -Log In with Facebook*	*-Virtual Hugs*	*-"Say Thanks" Option, -Simple Surveys to share personality traits*
Direct Expert Advice				
-Own space for Dr. consultations	*No*	*Yes*	*Yes*	*No*
-Proof of expertise	-	*Yes*	*Yes*	-
Anonymity				
-Nicknames	*Yes*	*Yes*	*Yes*	*Yes*
-Real credentials	*No*	*Optional (Facebook login)*	*No*	*Optional*
Website Environment				
-Intuitive	*Fair*	*Yes*	*Fair*	*Yes*
-Friendly atmosphere	*Yes*	*Yes*	*Fair*	*Yes*
-Easy site-help	*No*	*Yes*	*No*	*Yes*
-Variety of services	*Fair*	*Great*	*Fair*	*Good*
Info services, education of patients	*-Information about symptoms, treatments, health values*	*-Latest Health News, - Research, -Doctor Blogs*	*-Health Blogs by Health Advisors*	*-User Data Statistics in Graph Form, -Disease Descriptions*

Hoitonetti.fi and Dailystrength.org rely on simpler interfaces and offer basic services. This might actually be quite good for some users groups, as elderly, because restricted palette of services can lower the barrier to start using the service and also to keep oneself on site because the required services are easy to access. Finally, some educational/learning environmental aspects are present in all the four sites. They support learning and information sharing by having blog entries authored by health professionals. There are also health news, disease descriptions and many kinds of information about treatments and medications. These components are especially import for medical industry to note, because these sites offer vivid customer interface to market and develop products, services and to foster new customer policies.

5   Conclusions

According to an American survey, online social networking sites like MySpace and Facebook are only rarely used for health queries and updates [9]. One of the reasons behind this could be that these sites have been developed to serve very general social networking purposes. Based on our research, we suggest that sites specifically dedicated to health issues really do serve the needs of the online patients and health care professionals much better. Tailored social media can work as a very informative tool in health-related issues and to provide group support.

We found out in our site survey that positive, open atmosphere is very important and anonymity is likely to have a major role in this. Anonymity is not only a factor in the face-to-face versus virtual discussions, but also a factor between a health specific versus general purpose online social networking service. Facebook, for instance, is a "nonymous" (as opposed to anonymous) online environment where online relationships are typically anchored to offline life since online partners can be identified and locate offline in many ways. Therefore users of Facebook tend to produce and maintain some online image of themselves that is socially desirable [10]. Users are therefore naturally very careful in managing their online identities. For example, discussing own illnesses may not be appropriate in Facebook. In the case of peer support and sharing personal experiences in health and wellbeing related issues, especially in the delicate ones, our research confirms that it is important to be able to openly expose own thoughts and feelings.

Secondly, a service dedicated specifically to health issues enhances ways of medical professionals to interact with patients and other professionals. Our findings support the notion that medical professionals need to participate to online social media and that there is a customer demand for their online presence. Their participation provides current medical information, for example in the form of answering questions and moderating online discussions. Doctors', nurses' and other medical staff's online role should be clearly distinguished from the regular users. There needs to be trust between the online patients and the online medical professionals. This can be achieved by assigning dedicated professional online profiles to the medical staff and by clearly stating their education, profession, specialties and affiliations. General purpose social networking sites do not necessarily enable maintaining specific professional online identities. On the same note, the medical staff's online participation needs to be consistent. This can be achieved for example by creating general guidelines of best practices for ways to work and interact in the online world. Health portals could also better engage the users thus increasing their activity. Our results indicate that more activity is expected from site professionals that can be a factor in regular user site engagement too. One reason for loose professional engagement can be in lack of motivation. Therefore site should serve professionals better too. It should be possible for them to use the user data for developing their clinical practices for instance by summarizing site discussions or even enabling some data mining. Also, new earning models could be more actively developed using social media associated services. If links to medical records of Google Health would be provided, for example, potentials for new service mash-ups would radically increase, that would certainly benefit both regular users as well as professionals.

In many cases, doctors indeed need to have a face-to-face contact with the patients in order to be able to properly diagnose and advice them. Online social media can provide tools for this as well. For example, on HelloHealth.com instant messaging and video chats are utilized for online consultations. In the case where doctors are giving actual online consultations to the patients there is the need to identify both the doctor and the patient. This is in conflict with the importance of anonymity as discussed previously. Also in the case of utilizing online patient records the information is anchored outside of the site. Therefore, there is some need for identity / security rankings. They can then be agreed in site sign in or later when special communication / treatment requirements appear. Detailing this issue is among the important topics of future investigations.

References

1. http://www.continuaalliance.org/static/cms_workspace/
   Continua_Overview_Presentation_4.22.2011.pdf
2. Turner, R.J.: Social support as a contingency in psychological well-being. Journal of Health and Social Behavior 22, 357–367 (1981)
3. Wallston, B.S., Alagna, S.W., DeVellis, B.M., DeVellis, R.F.: Social Support and Physical Health. Health Psychology 2, 367–391 (1983)
4. White, M.H., Dorman, S.M.: Online Support for Caregivers: Analysis of an Internet Alzheimer Mailgroup. Computers in Nursing 18, 168–179 (2000)
5. Weinberg, N., Schmale, J., Uken, J., Wessel, K.: Online Help: Cancer Patients in a Computer-Mediated Support Group. Health & Social Work 21, 24–29 (1996)
6. Nahm, E., Resnick, B., DeGrezia, M., Brotemarkle, R.: Use of Discussion Boards in a Theory-Based Health Web Site for Older Adults. Nursing Research 58, 419–426 (2009)
7. Hawn, C.: Take Two Aspirin and Tweet Me in the Morning: How Twitter, Facebook, and Other Social Media are Reshaping Health Care. Health Affairs 28, 361–368 (2009)
8. http://www.environment.gov.au/education/publications/tsw/
   modules/module11.html; http://www.hwbuk.com/
9. Fox, S., Jones, S.: The Social Life of Health Information. Pew Internet & American Life Project (2009), http://www.pewinternet.org/~media//Files/Reports/
   2009/PIP_Health_2009.pdf
10. Zhao, S., Grasmuck, S., Martin, J.: Identity construction on Facebook: Digital empowerment in anchored relationships. Computers in Human Behavior 24, 1816–1836 (2008)

Diabetes Management:
Devices, ICT Technologies and Future Perspectives

Emmanouil G. Spanakis and Franco Chiarugi

Computational Medicine Laboratory (CML)
Institute of Computer Science (ICS)
Foundation for Research and Technology – Hellas (FORTH)
{spanakis,chiarugi}@ics.forth.gr

Abstract. Diabetes, a metabolic disorder, has reached epidemic proportions in western countries. Contextualized continuous glucose monitoring measurements to control directly the insulin delivery can provide optimum management of the disease and could provide diabetes patients with insulin profiles close to that of the normal patient, however problems related to delivery systems, control algorithms and safety of the system have to be properly solved for realizing a long term accurate continuous monitoring systems. In this paper we describe the current status of diabetes management and insulin administration and the future perspectives and aspects to support long term management of diabetes based on wearable continuous blood glucose monitoring sensors.

Keywords: Diabetes management, Blood glucose measurement, Insulin delivery devices, Continues glucose monitoring, Close loop control.

1 Introduction

Diabetes mellitus is a metabolic disorder characterized by hyperglycemia (high blood sugar) resulting from defects in the production or response to insulin [1]. The disease has two main forms: type 1 and type 2. Type 1 disease is characterized by diminished insulin production resulting from the loss of beta cells in the pancreatic islets of Langerhans, in most cases caused by immune-mediated cell destruction. Disease management entails administration of insulin in combination with careful blood glucose monitoring. Type 2 diabetes patients, usually over 50 years old with additional health problems, especially cardiovascular disease (CVD), exhibit reduced insulin production and resistance or reduced sensitivity to insulin. Earlier stages of type 2 diabetes are characterized by high plasmatic insulin concentration due to the insulin resistance and the capability of beta-cells to secrete insulin. In later stages of type 2 diabetes, beta-cells are unable to produce enough insulin and in such cases type 2 becomes more similar to type 1 diabetes. Management principally involves the adjustment of diet and exercise level and the use of oral anti-diabetic drugs (OADs) and, eventually, insulin to control blood sugar. Diabetes type 2 is one of the faster growing chronic conditions in the developed world and it is closely linked to the emerging epidemy of obesity and bad life style, which is now a major cause of preventable health problems.

K.S. Nikita et al. (Eds.): MobiHealth 2011, LNICST 83, pp. 197–202, 2012.
© Institute for Computer Sciences, Social Informatics and Telecommunications Engineering 2012

2   Glucose Control in Diabetes Therapy for Insulin Dependent Patients

Blood glucose is typically measured in a drop of capillary blood using a disposable dry chemical strip and reader device, an uncomfortable and slow process. Tight Glucose Control (TGC) requires almost continuous measurements and different sensors for continuous blood glucose measurement have been under development in the last two decades. Minimally invasive sensors able to measure glucose in interstitial fluid, more suitable for self-monitoring, have also been developed. To date, however, none of these has delivered a level of performance sufficient for use in routine glucose monitoring. Robust, clinically acceptable devices are however widely expected to become available in the near term.

Although several guidelines for treatment regimen for management of diabetes have been defined [2, 3, 4], no clear definitions of treatment regimen have been found for the establishment of glycaemic control of patients [5, 6]. Hyperglycaemia and insulin resistance are common in severe illness and are often associated with physical and mental stress. Studies have shown that frequency of hyperglycaemia in surgical ICU's (Intensive Care Units) can amount to –50-70 % of all admitted patients [7].

Management of diabetes has to be performed at 360° in hospital and outside healthcare premises. Based on the emerging clinical evidence from several clinical studies, there are increasing efforts world-wide to establish tight glycaemic control in critically ill and hospitalized patients [8, 9, 10, 11]. One of the major differences between inpatient and outpatient control of glycaemic levels is the fact that tight glycaemic control in hospitalized patients has to be provided by healthcare physicians and/or nurses. Achieving the goal of tight glycaemic targets requires extensive nursing efforts, including frequent bedside glucose monitoring, training to handle control algorithms or guidelines with intuitive decision taking and most importantly additional responsibility to prevent hypoglycaemic episodes. In summary, there is an urgent need for a safe method to establish tight glycaemic control without any risk of hypoglycaemia for patients in intensive care units and the general ward.

Traditional diabetes therapies for insulin dependent patients try to achieve normal glycaemia by administrating synthetic insulin to control patient's blood sugar level. Given that insulin cannot be taken orally patients must turn to special type of devices to administer insulin. The overall insulin the body needs is covered with the administration of "basal insulin" and "bolus insulin". Using a continuous delivery device it is possible to deliver continuously the "basal insulin" at a "basal rate".

3   Insulin Delivery Devices

During the last years a number of new insulin delivery systems have become available making insulin administration much more easily. Many factors influence the choice of the appropriate device for each patient including patient conformance and self-care capacity. Most commonly, insulin is delivered, into the fat under the skin (subcutaneous) or into the blood (intra-venous), using a needle injection or a syringe, an insulin pen, an insulin pump.

A syringe is the simplest device used for the injection of insulin. To administer insulin, the patients typically use disposable units to prevent contamination and infection. An insulin pen is composed of an insulin cartridge and a dial to measure the dose. There are two different types of insulin pens: a durable pen using a replaceable insulin cartridge allowing the disposal and replacement of an empty insulin cartridge with a new one; and an entirely disposable prefilled pen, where the pen comes pre-filled with insulin, and when the insulin cartridge or reservoir is empty the entire unit is discarded. Insulin pens are convenient and may cause less pain than syringes for the injection of insulin.

The advantage of insulin pens over insulin syringes is that they are much more convenient and easier to transport and use than traditional vial and syringe. They can repeatedly administrate more accurate dosages (especially for patients with visual or motor skills impairments). Insulin pens usually cause less pain to the patient. One of the disadvantages of insulin pens over insulin syringes is that unlike traditional syringe two different insulin solutions cannot be mixed in an insulin pen. Also using pens needles is usually more expensive than using the traditional vial and syringe method.

A major disadvantage of insulin injection devices is that they are only designed to administer insulin in large boluses, which can cause peaks and valleys in the blood sugar levels in patients. A solution to that problem (insulin shots) is provided when using insulin pumps. These devices, which are worn outside of the patient's body, can be programmed to deliver a steady supply of insulin throughout the day and/or programmed to deliver larger boluses of insulin after meals. An insulin pump delivers tiny selected amounts of insulin continuously throughout the day (the basal rate) and is able to provide additional doses if necessary (bolus doses). Each basal rate can be variable since the pump can be programmed to deliver different basal rates throughout the day. The main advantage of insulin pumps is that the individuals do not need to take multiple injections or shots every day allowing them to continue with their daily activities without any problem. With a pump, continuous doses of background insulin are delivered to support the body's needs between meals and with a button press it is possible to obtain an "on demand" dose of insulin (a "bolus") to cover instant needs.

The main disadvantage of insulin pumps, apart from the obvious discomfort that a person might feel wearing it, is the cost of the pump and the cost of maintenance that can get very high. More than that, patient's activity could force the infusion set of the pump to slide off and stop the necessary delivery of insulin to the patient's body. It is thus very important for patients to monitor their blood glucose levels much more frequent making sure that the pump is working correctly and, thus, also avoiding risks of ketoacidosis.

Other approaches include insulin inhalers that deliver insulin by spraying a blast of insulin powder into the patient's respiratory system. They operate similarly to other kinds of inhalers but are not as efficient at delivering insulin into the body which means patients must use larger quantities of insulin. Pulsatile insulin uses micro-jets to pulse (transdermal) insulin into the patient, mimicking the physiological secretions of insulin by the pancreas. The transdermal insulin administration aspect remains even today experimental.

4   ICT Technologies in Diabetes Management

4.1   Close-Loop and Self-management of Diabetes with an ICT Platform

A possible solution would be envisioned with a state-of-art technological solution that will assist healthcare professional, patients and informal carers, to better manage diabetes insulin therapy in a variety of settings, help patients understand their disease, support self-management and provide a safe environment by monitoring adverse and potentially life-threatening situations with appropriate crisis management.

Within the REACTION project [12] we aim to support long term management of diabetes based on wearable, continuous blood glucose monitoring sensors, and automated closed-loop delivery of insulin. The platform will provide integrated, professional, management and therapy services to diabetes patients in different healthcare regimes, including professional decision support for in-hospital environments, safety monitoring for dosage and compliance.

At such purpose, the availability of glucose sensors with proper accuracy is requested for a proper functioning of the overall system.

4.2   Glucose Sensor Technology

The standard-of-care for measuring glucose levels is by "finger-stick" blood glucose meters. For these a drop of blood, usually drawn by piercing the skin of a finger, is brought in contact with a test strip. A chemical reaction, commonly mediated by glucose oxidase, glucose dehydrogenase or hexokinase enzymes, triggers an electrochemical sensor or a color reaction that is detected in a reader. The drawback of this method is that only a few measurements can be performed in the course of a day.

The development of a control system that infuses insulin on the basis of glucose measurements could permit tighter glycaemic control and improve clinical outcome without increasing workload of the health care professionals [13]. Only a few commercially available sensors, that allow continuously monitoring the blood glucose level (CGM), have been approved for use. These sensors rely on electrochemical detection of an enzymatic reaction and are minimally invasive.

A range of other sensor technologies are currently being tested for their suitability for glucose monitoring. The most promising technologies for continuous glucose monitoring can broadly be classified as follows: Enzymatic (electrochemical), Impedance Spectroscopy / Dielectric spectroscopy, Optical (IR-/NIR-Absorption, mid-IR emission, Polarimetric (e.g. anterior chamber of the eye), Refractive index, Raman spectroscopy (inelastic photon scattering), Photoacustic (pulsed light absorption dependent on glucose concentration), and other.

Moreover, alternatives for invasive sampling are being investigated for electrochemical detection, for example samples may also be collected by iontophoresis or suction blister extraction through the skin. Despite significant efforts these technologies are still in a development or evaluation phase and yet have to prove their reliability and accuracy.

Wearability for sensors for some time can help in the collection of the required measurements. For this purpose the use of the ePatch technology has been considered (see Fig. 1).

Fig. 1. Wearable health monitoring ePatch system from DELTA

The ePatch sensor is a small body sensor, which senses physiological signals and is embedded in a skin-friendly adhesive. It can contain various types of miniaturised body sensors to measure physiological parameters, microelectronics for data analysis, a wireless radio module for communication and a battery power source. The skin adhesive of the ePatch ensures optimised for wearability and bio compatibility. The basis for the adhesive will be hydrocolloid pressure sensitive adhesives. This category of adhesives is extensively used in a number of medical devices like ostomy products, blister patches, wound dressings and for other skin applications. Wireless communication between sensors and the central node in the Body Area Network BAN: off-the-shelf radio chips will be benchmarked to identify components that optimize the trade-off between bandwidth, reliability and low-power performance.

The loop has to be closed to health professionals inside hospitals or to the patients when outside the healthcare premises. Then, they will deliver the insulin using the appropriate insulin device. In case of automatic glucose management the loop will be automatically closed on an insulin pump.

5   Conclusion

There is abundant evidence, for the future diabetes management and therapy, that tight control of the blood glucose level is vital for good diabetes management and insulin therapy. Good glucose control requires frequent measurement of blood glucose levels and complicated algorithms for assessing the insulin dose needed to adjust for short term variations in activity, diet and stress. On the other hand, good control of diabetes, as well as increased emphasis on blood pressure control and lifestyle factors, may improve the risk profile of most complications and attain future good health.

Acknowledgment. This work was performed in the framework of FP7 Integrated Project Reaction (Remote Accessibility to Diabetes Management and Therapy in Operational Healthcare Networks) partially funded by the European Commission under Grant Agreement 248590.

References

1. Zimmet, P., Alberti, K.G., Shaw, J.: Global and societal implications of the diabetes epidemic. Nature 414, 782–787 (2001)
2. Sakharova, O.V., Inzucchi, S.E.: Treatment of diabetes in the elderly. Addressing its complexities in this high-risk group. Postgrad. Med. 118, 19–26, 29 (2005)
3. Das, S.K., Chakrabarti, R.: Non-insulin dependent diabetes mellitus: present therapies and new drug targets. Mini. Rev. Med. Chem. 5, 1019–1034 (2005)
4. Charles, M.A.: Intensive insulin treatment in type 2 diabetes. Diabetes Technol. Ther. 7, 818–822 (2005)
5. Clement, S., Braithwaite, S.S., Magee, M.F., Ahmann, A., Smith, E.P., Schafer, R.G., Hirsch, I.B.: Management of diabetes and hyperglycemia in hospitals. Diabetes Care 27, 553–591 (2004)
6. Gautier, J.F., Beressi, J.P., Leblanc, H., Vexiau, P., Passa, P.: Are the implications of the Diabetes Control and Complications Trial (DCCT) feasible in daily clinical practice? Diabetes Metab. 22, 415–419 (1996)
7. Ellmerer, M.: Glucose Control in the Intensive Care Unit: The Rosy Future. In: Diabetes Technology Meeting 2008, Bethesda, Maryland, USA, November 13-15 (2008)
8. Van den Berghe, G., Wouters, P., Weekers, F., Verwaest, C., Bruyninckx, F., Schetz, M., Vlasselaers, D., Ferdinande, P., Lauwers, P., Bouillon, R.: Intensive insulin therapy in critically ill patients. N. Engl. J. Med. 345, 1359–1367 (2001)
9. Van den Berghe, G., Wilmer, A., Hermans, G., Meersseman, W., Wouters, P.J., Milants, I., Van Wijngaerden, E., Bobbaers, H., Bouillon, R.: Intensive insulin therapy in the medical ICU. N. Engl. J. Med. 354, 449–461 (2006)
10. Furnary, A.P., Wu, Y., Bookin, S.O.: Effect of hyperglycemia and continuous intravenous insulin infusions on outcomes of cardiac surgical procedures: the Portland Diabetic Project. Endocr. Pract. 10(suppl. 2), 21–33 (2004)
11. Meijering, S., Corstjens, A.M., Tulleken, J.E., Meertens, J.H., Zijlstra, J.G., Ligtenberg, J.J.: Towards a feasible algorithm for tight glycaemic control in critically ill patients: a systematic review of the literature. Crit. Care 10, R19 (2006)
12. http://www.reaction-project.eu/news.php
13. Plank, J., Blaha, J., Cordingley, J., Wilinska, M.E., Chassin, L.J., Morgan, C., Squire, S., Haluzik, M., Kremen, J., Svacina, S., Toller, W., Plasnik, A., Ellmerer, M., Hovorka, R., Pieber, T.R.: Multicentric, randomized, controlled trial to evaluate blood glucose control by the model predictive control (MPC) algorithm vs. routine glucose management protocols in post cardiac-thoracic surgery patients in the ICU. Diabetes Care 29, 271–276 (2006)

Exploring New Care Models in Diabetes Management and Therapy with a Wireless Mobile eHealth Platform

Jesper Thestrup, Tamas Gergely, and Peter Beck

In-JeT ApS, Jeppe Aakjaers Vej 15, 3460 Birkerød, Denmark, Applied Logic Laboratory.
Hankoczy Jeno Utca 7, 1022 Budapest, Hungary, and Joanneum Research
Forschungsgesellschaft MbH, Elisabethstraße 11a, 8010 Graz, Austria
jth@in-jet.dk, gergely@all.hu, peter.beck@joanneum.at

Abstract. Due to demographic changes, European healthcare systems face two serious challenges: healthcare delivery may become inadequate to perceived needs of the citizens or the cost may spiral out of control. With the decrease in the labour force, there is an urgent need to make more health services mobile allowing citizens with chronic diseases stay longer in the labour markets, reduce the number of lost working days and generally support nomadic working. Mobile technologies have the potential to provide better healthcare while at the same time increasing the working population. However, it calls for dramatic changes of healthcare provisioning and of the care models. The REACTION project develops a mobile, cloud-based platform that provides healthcare services to diabetes patients and caregivers. As part of the project, new chronic care models that support separation of care spaces are proposed.

Keywords: Mobile Healthcare, mHealth, New Care Models, Evolution of Care Spaces, Personalized Healthcare, Wireless Sensors and Devices, Electronic Decision Support.

1   Introduction

In recent years, wireless mobile healthcare (or mHealth) has emerged as an important sub-segment of the field of healthcare practice supported by electronic processes and communication (eHealth). While there is no widely agreed-to definition for these fields, the public health community has coalesced around these working definitions [1]:

- eHealth: Using information and communication technology for health services and information.
- mHealth: Using mobile communications for health services and information.

eHealth and mHealth are inextricably linked. Both are used to improve health outcomes and their technologies work in conjunction. While there are many stand-alone mHealth solutions, it is important to note the opportunity that mHealth presents for strengthening broader eHealth initiatives. For example, a mHealth front-end solution may allow patients to continuously access patient data within a national system, while at the same time being completely mobile. Other mHealth solutions can

K.S. Nikita et al. (Eds.): MobiHealth 2011, LNICST 83, pp. 203–210, 2012.

serve as access point for collecting vital signs and clinical data online and as remote information tools that provide information to patients, healthcare clinics, home providers, and health workers in the field.

In 2012, the European Commission will publish new activity plans aligned with eHealth and mHealth roadmaps for technological research, implementation practice and policy support with the aim to accelerate the establishment, acceptance and wide use of mHealth solutions that will improve disease management globally and support lifestyle changes among citizens.

In this paper we look at an integrated approach to improve long term management of diabetes developed in the REACTION project, which uses wireless technologies for continuous blood glucose monitoring, clinical monitoring and intervention strategies, monitoring and predicting related disease indicators and, ultimately presenting the perspective of automated closed-loop delivery of insulin. This paper will focus on the clinical decision support system implemented with the platform and discuss the need for new or revised care models as a result of the technological progress in mobile communication technologies.

2   About the REACTION Project

The outcome of the REACTION project will be an intelligent, interoperable, cloud-based platform that will provide integrated, professional management and therapy services to diabetes patients in different healthcare regimes including 1) professional decision support for in-hospital environments, 2) safety monitoring for medication dosage and compliance, 3) long term management of patients in Primary Care clinical schemes, 4) care of acute diabetic conditions, and 5) support for self management and life-style changes for diabetes patients. A range of REACTION applications will be developed, mainly targeting insulin-dependent diabetes patients. The applications aim to improve Continuous blood Glucose Monitoring (CGM) and Safe Glycaemic Control for improved insulin therapy management and basal/bolus dose adjustments.

The REACTION platform connects wirelessly to sensors and monitoring devices in the patients' physical surroundings and provides feed back to the patient as well as to informal carers (relatives, neighbours) and to healthcare professionals and emergency teams in the professional healthcare system. The REACTION platform also connects seamlessly to Health Information Systems for health data, health knowledge and decision support, utilizing medical knowledge repositories (e.g. biomedical models). The platform is visualised in Fig. 1.

Wearable, wireless medical sensors are connected in a Body Area Network (BAN) for multi-parametric recording of vital physiological parameters. The BAN interconnects with other wireless sensors in the environment that can record contextual information about the environment and the patients' activities. Devices are interfaced by standardized web services and data may be formatted and pre-processed in the access layer's active nodes/gateways, which operates on e.g. a mobile Smartphone or iPad platform. The gateway can also execute Apps that handle simple episode monitoring and other services, which are needed during periods of non-connectivity.

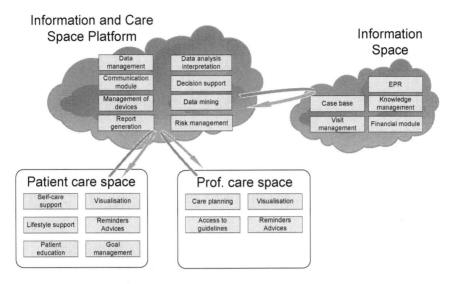

Fig. 1. The REACTION platform integrates information and care spaces

Further, the gateway runs interactive Apps which support self-monitoring and self-management; facilitate access to external Health Information Systems and personalised feedback from health care professionals. For medical devices not capable of operating web services (due to resource constraints or proprietary concerns), the gateway acts as a platform for virtualisation of those devices.

A typical clinical scheme in a mHealth solution will consist of a series of personalised actions, each of which can be described as a service. A typical REACTION clinical care plan will thus be established through orchestration of the relevant cloud based services, executed in a pre-described sequence according to "clinical logistics" modelling traditional clinical workflows. This feature in the REACTION platform introduces higher abstraction mechanisms and thus makes the application developer independent of using a specific programming environment to orchestrate REACTION services.

3   Clinical Practice

The REACTION project implements the service platform in three clinical field trials: Safe Glycaemic Control in the hospital ward; chronic care and lifestyle management in the Primary Care sector and Automatic Glycaemic Control (AGC) with closed-loop feedback. In this paper we will briefly outline the objectives of the in-hospital field trial, which is undertaken by the Endocrinology and Cardiology wards at the Medical University of Graz.

In-hospital hyperglycaemia has been found to be an important marker of poor clinical outcome and mortality among diabetes patients [2] [3]. Other studies have shown that medication errors are common and also associated with poorer outcomes [4]. In a randomized, controlled study conducted in a surgical intensive care unit [5], strict control of blood glucose levels with insulin reduced morbidity and mortality, significantly reducing in-hospital mortality from 11 to 7 percent in the entire study population. The first application domain of the REACTION platform will thus feature a suite of services aiming at Safe Glycaemic Control (SGC) of diabetes patients in the general hospital ward.

The current workflow related to glycaemic management has been analysed and a REACTION application has been developed that offer decision support for professional caregivers on a mobile iPad platform (Fig. 2.).

The application monitors a range of parameters from various sources including glucose level, nutritional intake, administered drugs and the patient's insulin sensitivity. The data are contextualised and algorithms and physiological models will be used to calculate the required insulin doses for SGC. Results will be delivered to physicians and nurses at the point of care.

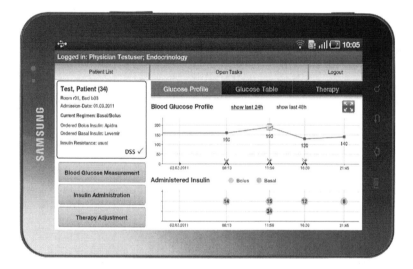

Fig. 2. Decision support application for Safe Glycaemic Control for care givers

The REACTION application will, later in the project, interface to hospital systems and hospital protocols across wards thus supporting integrated care management.

The in-hospital field trial focuses on providing a point-of-care electronic decision support (eDSS) tool for insulin therapy for physicians and nurses not specialised in diabetes. The objective of the first field trial is to compare the efficacy of the eDSS over a published best practice paper-based insulin titration protocol for glycaemic control and evaluate safety and usability of the eDSS. The target group is hospitalised patients with type 2 diabetes for the length of their hospital stay, with a maximum of 21 days.

Technologically, the hospital environment is particularly challenging to mobile communication technologies due to the many sources of electromagnetic disturbances and the potential adverse impact on patient safety caused by possible malfunctioning or black-out.

4   The Need for New Care Models

This first field trial aims to provide insight into the possible enhanced care using the REACTION platform, but do not attempt to establish broad clinical evidence. However, the outlook for enhanced glycaemic control, and thus care management, provoke the entire care model to be reviewed. It is important to remember that mobile health technologies are not objectives, but tools, that should be applied in ways to achieve local, national, and regional health objectives, addressing important public health challenges such as chronic disease management as well as contribute to improving the lives of individuals [6] [7]. Mobile healthcare provides new possibilities for revising central parts of the established care models for chronic diseases.

Planning new care models for the future involving mHealth is highly complex and involves a number of components that play significant roles in the formation of the trends we are looking for, when defining new care models. These factors are several, and include biomedical and clinical R&D, financial incentives, technology development and the socio-economic environment. These factors influence the development and changes in the attitude of the participants of the healthcare system and the ability to carry out changes that are needed to implement a new care model.

Several prediction studies have been published, e.g. by health institutions like the National Health Service (NHS) in the UK, the Australian Centre for Health Research (ACHR) etc. as well as private observers. These studies differ in their approach and in the aspects they consider. Some studies emphasize technological aspects [8] [9]; other focus on healthcare policies [10] [11]. ACHR focus on the aspects of health services [12].

Some common elements can be extracted from these approaches and we have selected those factors which, in our opinion, are significant in the formation of new care models [13] [14] [15]. These factors are (see Fig. 3):

- Systems biomedicine, an important area of the biomedical and clinical R&D
- Care space evolution, integrating many different trends
- The ICT factor, providing information technological support for mHealth
- Personalisation, aiming at the individualisation of the care
- Patient focused organisational re-engineering

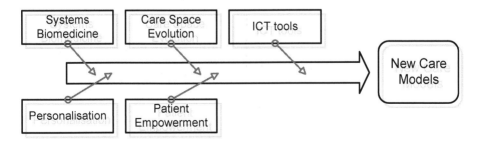

Fig. 3. Significant factors that influence the new care models

Chronic diseases share three important properties: (i) acute and chronic phases alternate during their progress; (ii) their adequate diagnosis and state monitoring require multilevel system biomedical characterization; (iii) their progress may be significantly influenced by the patient's behaviour.

Care for chronic diseases therefore has its own characteristics: (i) it should be continuous, i.e. it should be available between the contact visits and periods of hospitalization and not just during them; (ii) it should be proactive and predictive; (iii) it should consist of professional care provided by medical personnel and self-management provided by the patients themselves; (iv) it should be able to influence the patient's lifestyle; and (v) the care should be dynamic, meaning that all the participants should learn and adapt during the care process. In the medium term this requires re-design and implementation of new care models.

Important aspects of chronic disease management are personalisation, inclusion and patient empowerment. Personalisation and patient empowerment are obvious attributes of mHealth solutions and closely connected to care space evolution, i.e. the changes in physical or virtual spaces where care is provided. In this regard, ICT tools and systems biomedicine act as enabler of mHealth solutions. Care space evolution is thus of particular importance in mHealth care models.

mHealth results in an explicit splitting of the care space into two interrelated spaces, the activity space and the information space. The *care activity space* is the space where the patient is physically located, sometimes coinciding with the professional care space (the hospital or the doctors office) but mostly not (see also Fig. 1.). The *care information space* is a virtual space where patient data and information about the disease are kept and made available through an ICT infrastructure [13] [14] [16].

mHealth solutions makes the patient mobile and thus decouple the actual *activity space* from the traditional physical care space of the professional caregivers. The patient receives care in the home, at work and when travelling, i.e. mostly outside the physical space of the professional healthcare system.

mHealth (and eHealth) solutions also places high emphasis on the sound development of the care *information space*. New healthcare methods generally create a growing volume of data on the one hand, and an ever growing demand for information, on the other hand. The data growth is connected with the development of diagnostic and monitoring methods and tools. The information demand is connected with new decision making methods and with modelling methods required for a better implementation of the necessary actions.

The decoupling of care spaces is thus very important in chronic care, because it allows streamlining the roles of each stakeholder while at the same time providing them with tailored information for each role. mHealth widens and extends the effect of the care space evolution, while at the same time reinforces patient empowerment.

However, a known challenge to be overcome when moving the care activity space out of the doctor's office is the risk of the patient being isolated or alienated. If visits by care givers or family members are substituted with mHealth monitoring tools, patients can become isolated even in the most populated areas, especially senior citizens. It is imperative that a delicate balance is struck between closing the digital divide and closing patients in a virtual prison. Thus, inclusion enhancing methods must be incorporated in the mHealth solutions and the care models at all levels [17].

Overall, the separation of care spaces in chronic disease management provides, if managed correctly, great opportunities for improved care at the point of need as well as organisational streamlining and thus potential for cost savings. However, new clinically accepted care models, which take advantage of the new opportunities provided by the mHealth solutions, must be developed in order to fully explore the benefits while at the same time conserving and promoting inclusion.

5   Conclusion

The intelligent, interoperable platform developed by REACTION will provide integrated, professional, management and therapy services to diabetes patients in different healthcare regimes across Europe. An in-hospital field trial will focus on providing a point-of-care electronic decision support (eDSS) tool for insulin therapy for physicians and nurses not specialised in diabetes.

The outlook for enhanced glycaemic control supported by mobile healthcare applications provokes the entire care model for diabetes patients to be reviewed and further work needs to be done on this aspect.

The trends in care model evolution are influenced by many factors such as biomedical and clinical R&D, financial incentives, technology development and the socio-economic environment. An important aspect of future chronic disease management is that personalization, inclusion and empowerment of the patient has to be an essential part of the care model.

With insufficient impact data about how mobile technologies are influencing health outcomes, it is difficult to identify and replicate best practices. Good quality randomized controlled trials on mobile technologies in the care of chronic diseases with sufficient power and duration to assess long term and cost effects are required.

Of particular interest is the decomposition of the care space into a *care activity space* and a *care information space* and the further decomposition of the care activity space into a mobile care space (the patient's environment) and a traditional care space (the clinic or hospital).

However, the evolution of care spaces and the potential impact on health outcome must be further investigated and evidenced, before future care models will be adopted. The new care models must be clinically accepted, inclusive and correctly managed in order to fully explore the benefits in terms of improved health outcome and reduction in healthcare costs.

Acknowledgment. This work was performed in the framework of FP7 Integrated Project Reaction (Remote Accessibility to Diabetes Management and Therapy in Operational Healthcare Networks) partially funded by the European Commission under Grant Agreement 248590. The authors wish to express their gratitude to the members of the REACTION Consortium.

References

1. Vital Wave Consultation. mHealth for Development. The Opportunity of Mobile Technology for Healthcare in the Developing World. UN Foundation-Vodafone Foundation Partnership, Washington, D.C. and Berkshire, UK (2009)
2. Umpierrez, G.E., Isaacs, S.D., Bazargan, N., You, X., Thaler, L.M., Kitabchi, A.E.: Hyperglycemia: an independent marker of in-hospital mortality in patients with undiagnosed diabetes. J. Clin. Endocrinol. Metab. 87(3), 978–982 (2002)
3. NICE-SUGAR Study Investigators: Intensive versus conventional glucose control in critically ill patients. N. Engl. J. Med. 360(13), 1283–1297 (2009)
4. National Diabetes Inpatient Audit (NaDIA), England (2010)
5. Van den Berghe, G., Wouters, P., Weekers, F., Verwaest, C., Bruyninckx, F., Schetz, M., Vlasselaers, D., Ferdinande, P., Lauwers, P., Bouillon, R.: N. Engl. J. Med. 345, 1359–1367 (2001)
6. Shields, T., Chetley, A., Davis, J.: ICT in the health sector: Summary of the online consultation, infoDev (2005)
7. Mechael, P.N.: Towards the Development of an mHealth Strategy: A Literature Review, WHO (Updated by Sloninsky, D., the Millennium Villages Project 2008) (2007)
8. Chronic Care at the Crossroads Exploring Solutions for Chronic Care Management. Intel Corporation (2007)
9. Healthcare 2015 and care delivery. IBM Global Business Services (2008)
10. The new science of personalized medicine, PricewaterhouseCoopers (2009), http://www.pwc.com
11. HealthCast 2020:Creating a Sustainable Future
12. Georgeff, M.: E-Health and the Transformation of Healthcare. Australian Centre for Health Research Limited (2007)
13. Deutsch, T., Gergely, T.: Cybermedicine, Medicina, Budapest (2003)
14. Gergely, T., Szőts, M.: Quality in Healthcare, Medicina, Budapest (2001)
15. Deutsch, T., Gergely, T., Levay, A.: New Care Model for Chronic Patient Care and its Intelligent Infocommunication System I-II. In: Informatics and Management in Healthcare, vol. 8, Budapest (2009)
16. Gergely, T.: Integrated care space for chronic diseases. In: eHealth Week, Budapest, Hungary (2011)
17. Mordini, E., Wright, D., de Hert, P., Mantovani, E., Wadhwa, K., Thestrup, J., Van Steendam, G.: Ethics, e-Inclusion and Ageing. Studies in Ethics, Law, and Technology 3(1) (2009)

A Mobile Android-Based Application for In-hospital Glucose Management in Compliance with the Medical Device Directive for Software

Stephan Spat[1], Bernhard Höll[2], Peter Beck[1],
Franco Chiarurgi[3], Vasilis Kontogiannis[3], Manolis Spanakis[3],
Dimitris Manousos[3], and Thomas R. Pieber[2]

[1] JOANNEUM RESEARCH Forschungsges.m.b.H.,
Institute for Biomedicine and Health Sciences, Graz, Austria
{stephan.spat,peter.beck}@joanneum.at
[2] Medical University of Graz, Department of Internal Medicine,
Division of Endocrinology and Nuclear Medicine, Graz, Austria
{bernhard.hoell,thomas.pieber}@medunigraz.at
[3] Computational Medicine Laboratory, Institute of Computer Science,
Foundation for Research and Technology - Hellas, Heraklion, Crete, Greece
{chiarugi,vasilisk,spanakis,mandim}@ics.forth.gr

Abstract. In healthcare, the distribution of smartphones and tablet PCs running operation systems like Apple iOS or Android, attracts the interest in many application fields. In this paper we present insights into the development process of a Google Android-based tablet PC system, designed for in-hospital glucose management treatment of acute ill patients with type 2 diabetes. The system provides decision support for insulin dosing and falls within the scope of the Medical Device directive for software which came into effect in March 2010. In order to support usability and fulfill the design requirements, according to IEC 62366 standard, we included physicians and nurses in the design process to implement a modular user interface based on established clinical workflows using mockups and functional prototypes. With this approach we provided a solid foundation for validating our system with the demands of the medical device directive.

Keywords: decision support, diabetes mellitus, mobile device, hospital, medical device directive.

1 Introduction

The development of applications for smartphones and tablet PCs running operation systems like Apple iOS or Google Android attracts the interest in many fields of daily living [1]. In healthcare new applications fields like home telemonitoring or ambient assisted living have been based on mobile devices acting as remote terminals for medical data collection and intuitive user interaction [2], [3], [4]. At hospital wards smartphones and tablet PCs enable clinicians and nurses to treat patients more easily

K.S. Nikita et al. (Eds.): MobiHealth 2011, LNICST 83, pp. 211–216, 2012.
© Institute for Computer Sciences, Social Informatics and Telecommunications Engineering 2012

directly at their hospital beds or to share and present data to clinical personnel for decision making independently on their current location [5].

In this paper we present insights into the development process of a mobile Android-based in-hospital glucose management system for the treatment of acute ill patients with diabetes type 2. The system will provide decision support for insulin dosing and therefore falls within the scope of the revised medical device directive (MDD) [6] for software design and implementation. Thus, we focus our discussion on the design and development process of software as a medical device.

2   Method

This section describes the methodology used for the user-centered design phase of the application. It gives insights into the technical aspects of developing a Google Android-based mobile application and discusses typical questions related to the medical device directive.

2.1   Medical Device Directive for Software

The revised Medical Device Directive for Software came into effect in March 2010. Medical software which falls within the scope of the MDD has to comply with the same rules as medical devices do. The question that must be asked is: If we are going to apply a medical device, operated using a dedicated software component, to a human, does this device complies with the medical device directive? If yes, how can we prove it? Which standards do we have use? Risk analysis, change requests on requirements, software life cycle management and consequently stringent documentation of all activities have to be performed. It is necessary to identify the risks and the actions taken against the risks (for each crucial step). Following standards have been considered as relevant for the development of medical software which falls within the scope of MDD [6]:

- ISO 13485 standard defines the requirements for a quality management system for medical devices.
- IEC 62304 standard has emerged as a global benchmark for management of the software development lifecycle.
- ISO 14971 has traditionally been adopted as the base standard for risk management for medical devices and will also be used for software.
- IEC 62366 and IEC 60601-1-6 provide information about the application of usability engineering for software as medical devices.

The regulations of the medical device directive and their interaction are complex. Thus, an important question for every producer of software in the medical domain is: When is a software component considered as medical device? Often the answer is complicated but experiences show that the following definition is a good starting point. Software can be considered as a medical device for applications which:

- are explicitly mentioned in the medical device directives
- control or influence medical devices
- analyze patient data generated by a medical device and that will be used for diagnosis or monitoring
- are used for diagnosis or treatment of physical or mental diseases

Thus, the purpose of the application, which has to be defined by the producer, has a great influence on consideration and classification of a software application as medical device [6].

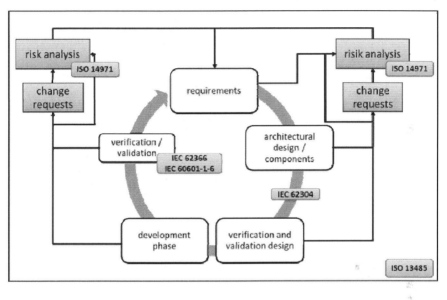

Fig. 1. A simplified graph, assigning relevant standards to the main development process components, of the mobile in-hospital glucose management system is shown in this figure

2.2   Case Study: Mobile Application for In-hospital Glucose Management

Design Phase. Referring to the requirements of the medical device directive (especially IEC 62366 and IEC 14971), which target usability of medical software and consequently the provision of clinical safety for users and patients, we have chosen a user-centered design approach.

A team consisting of technicians, software engineers, physicians and nurses from JOANNEUM RESEARCH and the Division of Endocrinology and Metabolism at the Medical University of Graz defined in weekly meetings the first version of the user interface and the functionality of the glucose management system. We used paper[1] and software mockups as trigger in order to have a basis for discussion and testing. FORTH-ICS provided external reviews for the design phase. We supported the design

---

[1] http://www.artfulbits.com/Android/Stencil.aspx

process by regular risk analysis sessions with all involved stakeholders. Derived risks have been collected and incorporated as change requests into the requirements. In the first iteration, we summarized the elicited requirements in an extensive specification document. A detailed discussion of the design phase can be found in [7].

Development Phase. Due to maintainability and expandability we decided to distinguish between an Android-based user interface and a platform independent backend (Java-based webserver) which contains business logic for the decision support, as well as the data storage and interfaces to the hospital information system. The exchange of data between the backend and frontend components requires mutual authentication and is completely done via encrypted web services to provide data security. The frontend application presents the data, received from the backend, in an appropriate manner to the user and collects new relevant data. The behavior of the frontend application relates in every step to the clinical workflow, which was identified together with end-users in the design phase.

During the development process of the frontend, as well as of the backend, we adapted Atlassian JIRA[2] as a requirement, issue tracker and task management tool for the documentation of each implementation step. JIRA acts as preparation for the needs of the IEC 62304 standard, and for ensuring an overview of open and already completed requirements, development tasks and identified bugs. In JIRA, each of these issues is assigned to an authorized editor who reports after finishing the issue. We connected JIRA to Atlassian Fisheye for source code management and Atlassian Bamboo for continuous integration and release management.

Verification Phase. In addition to the need of a detailed documentation, IEC 62304 and IEC 62366 demands also for verification and validation during the development process. Whilst we use TestNG[3] for the backend to verify the functionality after finishing the implementation of each system unit, we are testing the correct behavior of the frontend application on simulated user interactions. Android offers tools for instrumentation testing, which verifies, that the application prints out the desired output on the screen for every input. We used the free testing tool robotium[4] for the simulation of user interactions. Before we start frontend test cases the database is initialized with test case specific data through a separate web service. We use the Maven-Android plugin[5] as software project management tool for the generation of executable files. Maven for Android organizes application related dependencies and allows the automated execution of the frontend tests.

Thus, we created a flexible and configurable development and verification process, for the Google Android-based development environment, that provides a stable base for developing according to the regulations of the medical device directive.

---

[2] http://www.atlassian.com/software/jira/

[3] http://testng.org/

[4] http://code.google.com/p/robotium

[5] http://code.google.com/p/maven-android-plugin/

3   Results and Conclusion

Already in the early phase of the requirement elicitation process clinicians requested that they would prefer a software system that offers only the required base functionality and an easy to use interface, tailored to the current workflow. Usability of medical device software seemed to be one of the most important aspects, according to the avoidance of critical situations that harm patients or user. In order to avoid poor usability and consequently fulfill the requirements according to IEC 62366, we supported the physicians and nurses to design the user interface based on the established clinical workflow by their own, using paper mockups and functional prototypes. However, each ward in each hospital usually follows its own workflow. Therefore, great attention will be placed in future on the maintainability of the user interface. We are currently implementing a flexible dialog creation mechanism, which dynamically initializes Android dialogs based on a given parameter set. The type of parameter specifies what dialog is required (e.g. free text field, number picker). In a next step a configurable workflow engine component will be implemented for handling application sequences based on workflow patterns.

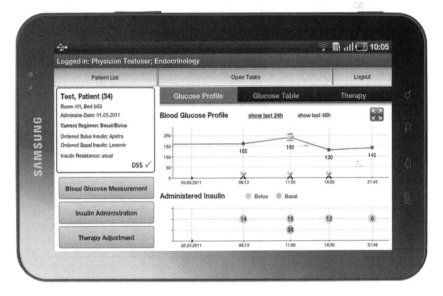

Fig. 2. This screenshot shows the main screen of the mobile glucose management application. The application runs on a Samsung Galaxy Tab with a seven inch touch screen and Android 2.2[6] as operating system. The main screen visualizes the blood glucose profile and the administered medication and provides the main functionalities of the application.

In addition to usability, other challenges appeared during the implementation process. The Android framework provides a wide range of available libraries but it is not explicitly designed to be used for medical applications. We are missing sufficient

---

[6] http://developer.android.com/sdk/android-2.2.html

security components that allow for secure transfer of patient data via web services. A secure SOAP client able to support encrypted communication and mutual authentication of both communication endpoints is currently under development.

The Android development platform provides a basic framework for developing software components according to the MDD directives while overcoming the various limitations on available libraries and, at the same time, offers to designers a powerful testing tool allowing to produce high level validation and verification tests for applications.

The decision support service for insulin dosing is based on a basal/bolus insulin regimen protocol, which has been developed with clinicians in parallel to the development process. Currently, this protocol is validated on paper in a clinical trial at the department of Endocrinology and Cardiology Medical University Hospital in Graz. In a next step, the validated protocol will be implemented into the electronic glucose management system for decision support and another clinical trial will performed. As precondition the system must meet the requirements of the medical device directive to get the approval of the Ethics Commission.

Acknowledgments. This work was partly funded by the E. C. under the 7th Framework Program in the area of Personal Health Systems under Grant Agreement no. 248590.

References

1. Keijzers, J., den Ouden, E., Lu, Y.: Usability benchmark study of commercially available smart phones (Internet). In: Proceedings of the 10th International Conference on Human Computer Interaction with Mobile Devices and Services - MobileHCI 2008, p. 265 (2008)
2. Kollmann, A., Riedl, M., Kastner, P., Schreier, G., Ludvik, B.: Feasibility of a mobile phone-based data service for functional insulin treatment of type 1 diabetes mellitus patients. J. Med. Internet Res. 9, e36 (2007)
3. Boulos, M.N.K., Wheeler, S., Tavares, C., Jones, R.: How smartphones are changing the face of mobile and participatory healthcare: an overview, with example from eCAALYX. Biomed. Eng. Online 10, 10–24 (2011)
4. Weaver, A., Young, A.M., Rowntree, J., Townsend, N., Pearson, S., Smith, J., Gibson, O., Cobern, W., Larsen, M., Tarassenko, L.: Application of mobile phone technology for managing chemotherapy-associated side-effects. Annals of Oncology 18, 1887–1892 (2007)
5. Biemer, M., Hampe, J.F.: A Mobile Medical Monitoring System: Concept, Design and Deployment. In: International Conference on Mobile Business (ICMB 2005), pp. 464–471. IEEE Computer Society, Sydney (2005)
6. Hall, K.: Devloping Medical Device Software to IEC 62304. European Device Technology 1(6) (2010),
   http://www.emdt.co.uk/article/
   developing-medical-device-software-iso-62304
7. Hoell, B., Spat, S., Plank, J., Schaupp, L., Neubauer, K., Beck, P., Chiarugi, F., Kontogiannis, V., Pieber, T.R., Holzinger, A.: Design of a mobile, safety-critical in-hospital Glucose Management System. In: Proceedings of MIE 2011 (2011)

Ontology-Driven Monitoring of Patient's Vital Signs Enabling Personalized Medical Detection and Alert

Anna Hristoskova[1,2], Vangelis Sakkalis[2], Giorgos Zacharioudakis[2], Manolis Tsiknakis[2], and Filip De Turck[1]

[1] Ghent University, Department of Information Technology - IBBT,
B-9050 Ghent, Belgium
{anna.hristoskova,filip.deturck}@intec.UGent.be
[2] FORTH-ICS, Computational Medicine Laboratory,
GR-70013 Heraklion, Crete, Greece
{sakkalis,gzaxar,tsiknaki}@ics.forth.gr

Abstract. A major challenge related to caring for patients with chronic conditions is the early detection of exacerbations of the disease that may be of great significance. The dedicated clinical personnel should be contacted immediately and possibly intervene in time before an acute state is reached, by changing medication, or any other interventions, in order to ensure patient safety. This paper presents an Ambient Intelligence (AmI) framework supporting real-time remote monitoring of patients diagnosed with congestive heart failure. The remote monitoring environment, enhanced with semantic technologies, provides a personalized, accurate and fully automated emergency alerting system that smoothly interacts with the personal physician, regardless his/her physical location in order to ensure in time intervention in case of an emergency. The proposed framework is able to change context at runtime in case new medical services are registered, new rules are defined, or in case of network overload and failure situations.

Keywords: Medical workflows, Ambient Intelligence, Semantic reasoning, Medical Ontologies, Medical Alarm, Quality of Service (QoS).

1 Introduction

During the past decade technology has gradually been moving to the concept of Ambient Intelligence (AmI) in which smart environments help inhabitants in everyday life. AmI supports pervasive diffusion of intelligence in the surrounding environment, through various wireless technologies (Zigbee[1], Bluetooth[2], RFID[3], WiFi[4]) and intelligent sensors. The first applications appearing in the clinical

---

[1] http://www.zigbee.org/
[2] http://www.bluetooth.com/
[3] http://en.wikipedia.org/wiki/Radio-frequency_identification
[4] http://en.wikipedia.org/wiki/IEEE_802.11

K.S. Nikita et al. (Eds.): MobiHealth 2011, LNICST 83, pp. 217–224, 2012.
© Institute for Computer Sciences, Social Informatics and Telecommunications Engineering 2012

domain were mainly focused on addressing the need to better support remote patient monitoring (vital sign monitoring [1], soft copy radiological film review [2]) and provide condition specific diagnostics and treatment [3]. Such e-health applications and wireless medical devices can significantly improve quality of health care and promote evidence-based medicine. Later it became apparent the need to provide services able to interconnect all the fragmented available e-health systems and automation systems, in order to realize integrated platforms facilitating time-critical care in case of an emergency. In this direction our framework provides: i) personalized monitoring of chronic disease patients that is able to detect the patient's health status, ii) intelligent alerting of the dedicated clinician in case of an emergency, iii) dynamic adaptation of the full vital signs' monitoring environment in any available device located in close proximity to the clinician and iv) ontology-based modeling of the patient's and clinician's context and the available devices. To achieve such functionality the following device and technologies were available in our paradigm:

- Wireless or wearable medical devices and sensors acquiring patient's vital signs. In our reference implementation the supported measurements are: Blood Pressure[5] (BP), SpO_2[6], Heart Rate (HR), body weight[7] and 12-lead ECG monitoring[8].
- Indoor Localization System (ILS) [4] consisting of a network of sensors used as anchor points in order to specify the location of a person. Commonly used techniques involve RFID tags, triangulation algorithms based on WiFi signal (strength, angle, distance, attenuation), infrared and visible light communication and ultrasound waves. Depending on the cost, the required precision and the use case-specific parameters, various solutions or combinations can be applied.
- Monitoring application recording the aforementioned bio signals and hosting risk assessment algorithms to enable the alerting process. A full description of this application as applied in a clinical environment is described in [5].
- Ontology-driven application intelligence capable of reasoning on the patient data and available (medical) devices. The applied medical and device ontologies define a formal representation of knowledge by a set of key domain concepts and the relationships between those concepts enabling reuse of the medical knowledge and device interoperability.

In order to illustrate the AmI framework in real use, a specific scenario related to Congestive Heart Failure (CHF) is proposed in the next section.

---

[5] A&D UA-767PBT Blood Pressure Monitor acquiring BP (systolic, diastolic and mean arterial) measurements and HR, transmitted via Bluetooth.
[6] Nonin Avant 4000 Digital Pulse Oximeter providing real time measurements of HR and SpO_2, transmitting via Bluetooth.
[7] A&D UC-321PBT Weight scale measuring the person's weight, transmitting via Bluetooth.
[8] Welch Allyn Cardio Perfect 12 lead ECG Recorder transmitting the recorded ECG via fiber optic cable.

2   Vital Signs Monitoring and Alert Detection Scenario

Our dedicated framework receives real-time patient data and processes them to detect possible deviations from normal values (Figure 1). When a threshold is exceeded (step 1) the dedicated clinician is localized and alerted (step 2, 3) and redirected to a better display (step 4) in order to overview in detail the vital signs (step 5).

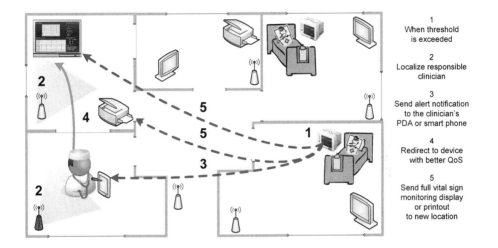

Fig. 1. Localization and notification of the responsible clinician during an emergency

The parameters and default thresholds are illustrated in Table 1. In case of exceeded thresholds, medical personnel are localized and contacted presenting ad-hoc information on the patient's condition on any device within the clinician's reach. In order to further tailor the system to the patient's profile and assist physicians in selecting people who are predisposed by coronary disease, hypertension, or valvular heart disease; we build a CHF related risk profile based on the risk appraisal function proposed in [6] that is based on the Framingham Heart Study [7] (486 heart failure cases during 38 years of follow-up).

The predictors used are based on **Age**, **Coronary heart disease** and **Valve disease status** provided by the patient Electronic Health Record (EHR), as well as on **HR**, on **blood pressure** and on **Body Mass Index** (BMI) provided by the pulse oximeter, the blood pressure monitor and the weight scale respectively. The calculated risk probability may be used to alter the default threshold values (higher risk probability add more constraint on the physiological patterns presented in Table 1).

3   Ontology-Driven Alert Detection

A requirement for an AmI environment monitoring patients with chronic conditions is interoperability between heterogeneous devices and technologies. These

Table 1. Alert detection parameters and corresponding thresholds

Measurement	Monitoring Device	Detection Threshold
low SpO$_2$	Pulse Oximeter	SpO$_2$ < 90%
bradycardia	Pulse Oximeter	HR < 40 bpm
tachycardia	Pulse Oximeter	HR > 150bpm
HR change	Pulse Oximeter	$|\Delta$ HR / 5min$|$ > 19%
HR stability	Pulse Oximeter	max HR variability past 4 readings > 10%
BP change	BP Monitor	systolic or diastolic change > ±11%

devices and the services running on them should be automatically discovered and executed based on semantically-defined features. Additionally an intelligent behavior should emerge through the notion of user (patient or clinician) context allowing for a smart and personalized combination of the available resources.

3.1 Device and Service Ontology

In the presented AmI framework, the available (medical) services are enriched with semantic annotations using the latest version of OWL-S 1.2 [8]. Instead of defining their inputs and outputs using XML Schema types much like WSDL they are expressed by ontological concepts in OWL. For the definition of the service preconditions and effects OWL-S supports the use of SWRL (Semantic Web Rule Language) expressions and built-ins (SWRLB) such as comparisons (equal, less than, greater than, etc), math functions (add, subtract, multiply, divide, etc) [9]. SWRL expressions are used for coding procedural relation in the form of rules. This formal service specification, allows the use of existing description logic reasoners such as Pellet [10] for the execution of data transformations.

The OWL-S service description is extended with the Amigo [11] device ontology. It defines deployment properties between a device and its running services. Amigo provides support for communication protocols such as Universal Plug and Play (UPnP), Service Location Protocol (SLP), Java RMI and Simple Object Access Protocol (SOAP), description of device and user context and QoS information. Using the Amigo ontology one can define a Display service, presented in Figure 2(a) consisting of the standard OWL-S Profile, Process and WSDL Grounding, and additionally specifying a device instance it is deployed on (e.g. LCD TV screen). This explicit deployment specification enables runtime selection of a service depending on the device QoS parameters. The example in Figure 2(b) presents the modeling of a Display service deployed on a LCD TV screen, an instance of a "MediaDevice" concept, having properties such as location, screen parameters, mobility features and device status. This results in the definition of semantically equivalent display services running on different devices. The dynamic selection of a specific service will depend on the semantically defined device parameters.

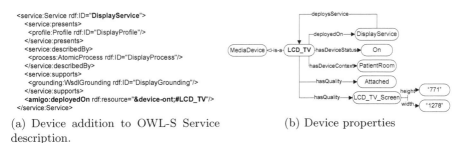

(a) Device addition to OWL-S Service description.

(b) Device properties

Fig. 2. Service deployment on context-rich devices

3.2 Medical Patient State Ontology

The Amigo ontology not only supports the definition of a device context but also of a user context. The user context presented in Figure 3(a) includes among others the user's activity, schedule, and personal details. In order to model the chronic heart conditions of a specific patient it is extended with a medical state concept consisting of a CHF concept (CHF_record) defined as an equivalent concept of the HF ontology [12,13] (Figure 3(b)). This ontology presents a detailed taxonomic overview of the heart failure domain with around 200 classes describing HF related concepts. Examples are "Cardiac_hypertrophy", "Blood_pressure_signs", "Heart_murmurs". The five basic super-classes are:

- **HF_concept:** describes HF terminology, including the risks for CHF, medical synonyms, and types of classification. The classification taxonomy is extended in order to support the previously mentioned Framingham Heart Study risk factors.
- **Patient_characteristic:** contains clinical data in the patient's HF medical record such as demographical characteristics, possible diagnoses, possible signs and symptoms, prognosis and other characteristics. This concept is linked to the Amigo ontology through sub classing of a user's medical state.
- **Testing:** represents knowledge regarding physical examinations and tests performed in medical institutions. Each test relevant to HF has properties that denote the measurements for that test and also which disorders it can detect.
- **Treatment:** consists of medical procedures used in the healing process, including medications, devices, invasive and non-invasive procedures, and recommendations regarding HF.
- **Patient:** reserved for factual knowledge about particular patients.

The new ontology models the patient's context including his medical state and focuses on the CHF scenario. Using these concepts, rules are defined for monitoring the patient condition, suggesting patient classification according to his/her risk profile and specific alerts are sent in case of deteriorated vital signs (e.g. elevated HR).

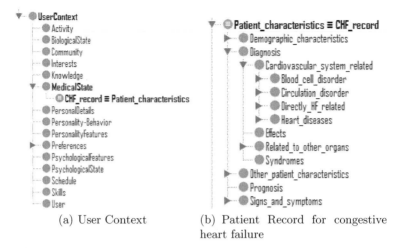

(a) User Context

(b) Patient Record for congestive heart failure

Fig. 3. Ontological definition of a patient's medical state according to a specific model for the heart failure disease

4   The AmI Monitoring Framework

The main building blocks of the AmI environment for monitoring patients with chronic heart conditions are presented in Figure 4. It is designed based on the principles of Service-Oriented Architectures (SOAs) [14], wherein all medical components and devices are implemented as Web services. The Web service technology enables reuse of services for gathering medical data values from patient monitoring devices resulting in a component-based system. Depending on defined or automatically calculated thresholds the acquired patient measurements are sent to the clinician's tablet PC, printer or a nearby TV screen [15]. When needed, medical services are also invoked on request to query for overviews of historical decision outputs and results.

The AmI components subscribe for specific events to the **Execution Environment** managing the processing steps of the monitoring and alerting system. Repository of the available AmI services and their semantic OWL-S

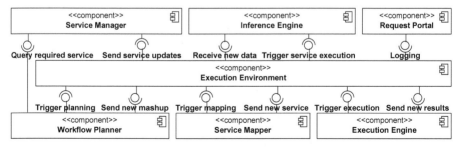

Fig. 4. AmI patient monitoring and emergency detection framework architecture

descriptions is the **Service Manager**. It enables automatic querying for service interfaces or devices offering specific QoS (defined using the extension on the Amigo ontology). In the presented AmI framework a service is viewed as a state-based rule; "IF `Service-Preconditions` THEN `Service-Effects`". These service rules and additional user-defined rules stating the default medical thresholds are uploaded to the **Inference Engine** which uses Pellet reasoning on new patient measurements to evaluate rules. Whenever a service precondition or a user-defined rule is satisfied by the available data, the **Inference Engine** triggers the execution of that service. Services scheduled for execution are passed on to the **Workflow Planner** which decides if additional data is required such as patient data from medical services and responsible clinician contact details. Combining these services a medical workflow is constructed using HTN planning described in [16]. This workflow is translated into an executable composite process by the **Service Mapper** through a specification of the data and control bindings between the services. For each semantic service description an actual executable service instance is selected depending on the device capabilities it is deployed on and the clinician's context (e.g. location). Usually a notification is sent to his personal tablet PC. Through device comparison of QoS properties like screen height and width extra services are triggered to summarize the patient data for a small screen and supply information on device locations with better resolution. Following is the actual execution by the **Execution Engine** handling the invocation workflow of the services. The service results are added to the **Inference Engine** enabling a dynamic system inferring new knowledge at runtime. The **Request Portal** provides an overview of the constructed medical workflow.

5   Conclusion

This paper presents the development of an ambient intelligence framework supporting real-time monitoring of patients diagnosed with congestive heart failure. Services monitoring vital patient signs are transformed into SWRL rules evaluated by a Pellet-based Inference Engine. Planning algorithms are implemented that automatically assemble medical workflows out of existing semantically enriched services. Dynamic adaptation of the constructed medical workflows take into account the clinician's location and succeed in remotely displaying the full vital sign monitoring application in the device of his/her choice, in order to ensure in time intervention in case of an emergency. The proposed framework is able to change context at runtime in case new services are registered, new rules are defined, or failure/overload of the network, through dynamic reconfiguration and personalization of the constructed workflows.

Future work includes the extension of the AmI framework with dynamic distributed deployment making optimal use of the available resources for the execution of the alerting services during an emergency.

Acknowledgment. This work was partially supported by the community initiative program INTERREG III, project "ΥΠΕΡΘΕΝ", financed by the EC through the European Regional Development Fund (ERDF) and by national funds of Greece and Cyprus.

References

1. Gao, T., Greenspan, D., Welsh, M., Juang, R., Alm, A.: Vital signs monitoring and patient tracking over a wireless network. In: 27th Annual International Conference of the Engineering in Medicine and Biology Society, pp. 102–105 (2005)
2. Kostomanolakis, S., Kavlentakis, G., Sakkalis, V., Chronaki, C.E., Tsiknakis, M., Orphanoudakis, S.C.: Seamless Integration of Healthcare Processes related to Image Management and Communication in Primary Healthcare Centers. In: Proceedings of the 18th International Conference EuroPACS 2000, pp. 126–132 (2000)
3. White, L., Terner, C.: E-health, phase two: the imperative to integrate process automation with communication automation for large clinical reference laboratories. Journal of Healthcare Information Management (JHIM) 15(3), 295–305 (2001)
4. Chiou, Y.S., Wang, C.L., Yeh, S.C.: An adaptive location estimator using tracking algorithms for indoor WLANs. Wireless Networks 16(7), 1987–2012 (2010)
5. Kartakis, S., Tourlakis, P., Sakkalis, V., Zacharioudakis, G., Stephanidis, C.: Enhancing the patient experience through Ambient Intelligence applications in health care. In: 5th International Symposium on Ubiquitous Computing and Ambient Intelligence (UCAmI 2011), Riviera Maya, Mexico, December 6-8 (2011)
6. Kannel, W.B., D'Agostino, R.B., Silbershatz, H., Belanger, A.J., Wilson, P.W.F., Levy, D.: Profile for estimating risk of heart failure. Archives of Internal Medicine 159(11), 1197–1204 (1999)
7. Lloyd-Jones, D.M., Larson, M.G., Leip, E.P., Beiser, A., D'Agostino, R.B., Kannel, W.B., Murabito, J.M., Vasan, R.S., Benjamin, E.J., Levy, D.: Lifetime risk for developing congestive heart failure: the Framingham Heart Study. Circulation 106(24), 3068–3072 (2002)
8. OWL-S, Semantic Markup for Web Services, http://www.w3.org/Submission/OWL-S/
9. Horrocks, I., Patel-Schneider, P.F., Boley, H., Tabet, S., Grosof, B., Dean, M.: SWRL: A Semantic Web Rule Language Combining OWL and RuleML (2004), http://www.w3.org/Submission/SWRL/
10. Pellet: OWL 2 Reasoner for Java, http://clarkparsia.com/pellet/
11. Vallée, M., Ramparany, F., Vercouter, L.: Dynamic service composition in ambient intelligence environments: a multi-agent approach. In: Proceeding of the First European Young Researcher Workshop on Service-Oriented Computing (2005)
12. Jovic, A., Gamberger, D., Krstacic, G.: Heart failure ontology. To appear in Bio-Algorithms and Med-Systems (2011)
13. Gamberger, D., Prcela, M., Jović, A., Šmuc, T., Parati, G., Valentini, M., Kawecka-Jaszcz, K., Styczkiewicz, K., Kononowicz, A., Candelieri, A., et al.: Medical knowledge representation within Heartfaid platform. In: Proc. of Biostec Int. Joint Conference on Biomedical Engineering Systems and Technologies, pp. 205–217 (2008)
14. Cândido, G., Barata, J., Colombo, A.W., Jammes, F.: Soa in reconfigurable supply chains: A research roadmap. Engineering Applications of Artificial Intelligence 22(6), 939–949 (2009)
15. Hristoskova, A., Moeyersoon, D., Van Hoecke, S., Verstichel, S., Decruyenaere, J., De Turck, F.: Dynamic composition of medical support services in the ICU: Platform and algorithm design details. Computer Methods and Programs in Biomedicine 100(3), 248–264 (2010)
16. Hristoskova, A., Volckaert, B., De Turck, F.: Framework Managing the Automated Construction and Runtime Adaptation of Service Mashups. In: International Workshop on Semantic Interoperability - IWSI (2011)

A Mobile Reasoning System for Supporting the Monitoring of Chronic Diseases

Aniello Minutolo[1,2], Massimo Esposito[1], and Giuseppe De Pietro[1]

[1] Institute for High Performance Computing and Networking, ICAR-CNR
Via P. Castellino, 111-80131, Napoli, Italy
[2] University of Naples "Parthenope" Department of Technology Naples, Italy
{minutolo.a,esposito.m,depietro.g}@na.icar.cnr.it

Abstract. Advances in health care technologies are radically impacting the management of chronic diseases by providing a new long-term care option that combines supportive systems for monitoring and assessing the patients' health status with activities of daily living. In this respect, this paper presents a mobile reasoning system which can be used to build knowledge-based Decision Support Systems for monitoring and managing ubiquitously and seamlessly chronic patients, specifically designed and developed as a light-weight solution suitable for resource-limited mobile devices. The system is devised to offer knowledge representation and reasoning facilities able to face and efficiently reason on the continuous and real-time flow of data generated by the sensor devices with the final aim of providing answers within a prescribed time and given constraints on the processing power and resources.

Keywords: Decision Support, Mobile Inferential Reasoning, Ontologies.

1 Introduction

Nowadays, advances in health care technologies are radically impacting the management of chronic diseases by providing a new long-term care option that combines Decision Support Systems (DSS) for monitoring patients' health status with activities of daily living so as to promote individual independence and well-being.

In particular, knowledge-based DSSs are more and more widely adopted in such scenarios: they model medical knowledge and experts' know-how for inferential reasoning in order to supply alarms as a response to a worsening of the patient's status, plus suggestions about the actions to do. Such a typology of DSSs, typically associated with desktop systems, are recently undergoing a radical transformation in order to face a set of new challenging scenarios, where information must be supplied, received, and/or used anywhere for supporting individuals or organizations seamlessly and ubiquitously in their decision-making tasks.

In this respect, mobile health DSSs for chronic disease monitoring are increasingly appearing on smart phones or Personal Digital Assistant devices (PDAs), with the aim of facilitating self care and communications with physicians by reasoning on data gathered by sensor devices.

K.S. Nikita et al. (Eds.): MobiHealth 2011, LNICST 83, pp. 225–232, 2012.

This has been facilitated by the spread of pervasive computing – a growth in small sensors, wearable or handheld devices, and wireless networking technologies. However, the focus has been twofold. At the lower layers, the attention has been on getting data from the sensors and on creating architectures and frameworks that will support the integration of their data. At the upper layers, efforts have typically focused on using the sensed information to infer conclusions, e.g. a suggestion or an action to do. This has taken several forms, ranging in complexity from simple numerical threshold type systems to more advanced systems that seek to build rich models of the domain described in semantic web languages and then reason over them.

Mobile reasoning over domain knowledge bases generally involves sensed data that are updated at a relatively low frequency – a person entering a room, a device being turned on, etc., and, in this respect, some mobile reasoning systems have been built to handle sensed data formalized by means of semantic web languages [1-3].

Unfortunately, such mobile reasoning approaches do not scale to real-time and computation intensive applications, where sensed data are updated at a very high frequency, e.g. mobile health DSSs where vital signal data are constantly generated by sensors placed on the body.

In detail, on the one hand, they are not able to handle the huge volume of dynamically changing facts and information, since the lack of operators for updating and/or retracting the existing knowledge forces the adoption of the strategy to rebuild the domain knowledge base in accordance with the incoming new data every time, so generating a recurrent inference processing overhead.

On the other hand, they implement computationally heavy inferential procedures neither characterized by a light-weight and efficient implementation suitable for resource-limited mobile devices, nor optimized to provide inferences within a prescribed time and given constraints on the processing power and resources, so involving that the incoming data rate is much higher than the time taken by the inferential procedure.

According to these considerations, this paper presents a mobile reasoning system which can be used to build knowledge-based DSSs for ubiquitously and seamlessly monitoring and managing chronic patients, specifically designed and developed to i) support the definition, updating and retracting of medical knowledge by means of existing ontology languages, such as OWL (Web Ontology Language)[4] and RDF (Resource Description Framework)[5], and a proposed rule-based formalism including non-monotonic operators; ii) provide an inferential reasoning algorithm based on a lazy evaluation [6] to enable the generation of inferences within a prescribed time and given constraints on the processing power and resources. so reducing the space complexity and improving the time response.

2   Mobile Reasoning System

The proposed mobile reasoning system has been designed to support inferential reasoning procedures in mobile DSSs with the final aim of monitoring the health status of chronic patients. The main components of the system are shown in figure 1.

The *Working Memory* (WM) is the repository in which both medical knowledge and experts' know-how expressed in terms of OWL ontologies are stored.

In more detail, at the system start-up, the *Knowledge Base Manager* (KBM) encodes the terminological knowledge describing the specific domain in well-defined and semantically-rich OWL descriptions, expressed in terms of classes and properties, and then rearranges and stores such descriptions into the WM as a collection of facts, i.e. RDF triples expressed in the form of *<subject,predicate,object>*.

The KBM will dynamically update the WM by inserting assertional knowledge consisting in new facts, represented by individuals (instances of concepts) with the corresponding instances of properties.

More specifically, starting from the patient data coming from sensing devices, a set of individuals of the ontology concepts is instanced. Moreover, the values associated to the input data are also associated to the corresponding properties of the defined individuals. All the individuals are finally added to the WM as new facts. The WM will be also updated with the inferred facts, i.e. the new facts generated at the end of the reasoning process.

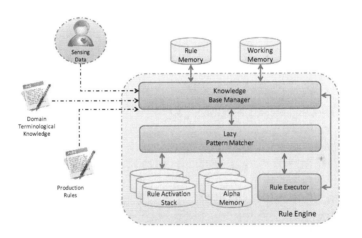

Fig. 1. The main components of the Mobile Reasoning System

The *Rule Memory* (RM) is the repository where production rules (i.e. if-then rules) are stored. The syntax used to formalize the rules has been defined in order to represent, in a natural and understandable manner, procedural knowledge about actions and suggestions to be generated for supporting clinical operators in the management of chronic patients.

Starting from the terminological knowledge expressed in triples, a set of production rules is built, each of them made by a conjunction of condition elements (CE) in its left-hand side (LHS) and a set of actions in its right-hand side (RHS), respectively.

> *name: antecedents -> consequents*
>
> *where*
> > *name   = a string identifying the rule*
> > *antecedents = bodyterm, bodyterm, ...*
> > *consequents = headterm, headterm, ...*
>
> *bodyterm ∈ { (sub,pre,obj), not(sub,pre,obj), functionToCall(arg1,arg2,...)}*
> *headterm ∈ { (sub,pre,obj), functionToCall(arg1,arg2,...)}*

Fig. 2. The rule syntax

The proposed rule-based formalism, showed in figure 2, has been designed in order to extend the expressiveness of ontology languages, allowing the explicit use of non-monotonic operators enabling closed-world reasoning over the WM, such as *retraction*, for updating an assertion according to new input data, and *negation-as-failure*, for determining negative information in the case of not completely represented knowledge.

Three kinds of term can be used in a CE, namely *triple pattern*, *negated triple pattern* and *function call*. A triple pattern is defined as a generalization of a triple object with the additional property that it can contain variables instead of only static values. A negated triple pattern can be used only in the LHS of a rule and it will be interpreted as the test of the absence in the WM of facts matching the triple pattern. Function calls allow to invoke internal procedures able to evaluate logical conditions, to compute arithmetic expressions, and to retract facts.

The *Rule Engine* (RE) is based on a forward chaining scheme, i.e. a data driven method that can be described logically as repeated application of the generalized modus ponens. In other words, available data are supplied as facts and used to evaluate eligible rules and draw all possible new inferred facts. The most computation intensive and resource-consuming part of the inferential reasoning is the matching process, which determines the set of satisfied rules that can be executed given the current set of facts, since it is a hard combinatorial problem.

In order to grant an efficient handling of memory and computational resources and achieve good performance, a *lazy pattern matching algorithm* has been proposed, specifically designed and implemented as a light-weight solution suitable for resource-limited mobile devices.

The idea of the algorithm is to evaluate the so called *rule activations*, i.e. which rules are satisfied and which data satisfy them, with the final aim of computing only one rule activation in each cycle, based on the observation that only one of them is fired anyway.

Differently, other matching algorithms [7, 8] first compute all the applicable rule activations and then execute them according to a selection strategy, and, thus, are characterized by an exponential worst-case complexity in terms of time and space.

The lazy evaluation does not waste time in computing superfluous rule activations which could be never executed and enables to fire the first eligible rule as soon as it has been identified, so as to improve the performance in terms of average response time. Moreover, by avoiding the computation of all possible rule activations, the worst-case space complexity can be reduced to a polynomial bound.

Thus, in the case when applied to remote monitoring scenarios, the proposed lazy algorithm is able of efficiently elaborating the huge volume of vital signal data, for example generated by body sensors, thanks to its reduced space complexity, and, contextually, of facing the real-time and computation intensive requirements by granting better performance in terms of response time.

More in detail, the whole inferential process is described as follows. The *Lazy Pattern Matcher* (LPM) operates on working memory elements (WMEs) expressed as RDF triples characterized by a unique and incremental ID.

In order to determine eligible rule activations, the LPM builds memory structures, named *alpha memories,* for explicitly storing information about the results of *intra-condition* tests. *Intra-condition* tests apply to determine which WME satisfies or violates a specific CE of a rule and store the results into the associated alpha memory. Since the same CE can occur in many rules to be successively evaluated, an alpha memory can be shared between different rules.

A non-negated CE usually contains variable references to be bound in order to assume values according to the matching facts stored into its alpha memory. When the same variable reference occurs multiple times in different CEs of a rule, the consistent binding of that variable must be evaluated by means of *inter-condition* tests that involve multiple CEs. If a variable occurs only once in the LHS, no test is necessary.

Thus, for a rule, the combination of facts stored into non-negated alpha memories which satisfies the inter-condition tests, does not violate any negated CE and verifies the conditions specified in any function call represents a *rule activation.*

To support the research of eligible activations for each rule, i.e. the inspection and combination of facts stored in the alpha memories, trying to find a match for the whole rule, the LPM maintains *a rule activation stack.* When a WME satisfying the intra-condition test of a non-negated CE is found, a new element is pushed onto the stack by the LPM, which contains both the IDs of the last recent facts inserted into every non-negated CE and a reference to that specific CE (*Dominant Reference*).

Each stack element enables to enumerate a subset of potential rule activations by combining the facts stored into the non-negated alpha memories according to the DR and the IDs stored into the current stack element.

After investigating all the available combinations associated to the current stack element, another stack element is popped. When the stack is empty, no combination of facts can be further tried, and, thus, no other rule activation can be found. The rule activation research is computed only if the rule is active and its stack is not empty. A rule is active when its non-negated alpha memories are not empty and its negated alpha memories are empty.

As soon as an eligible rule activation is found, the research of other possible rule activations is paused by storing a pointer to the interruption point into the stack, and, then, the corresponding rule is executed by the *Rule Executor*.

Successively, if the WM is not changed at all, the research is resumed from the last interruption point, i.e. by removing the next element from the top of the stack. Otherwise, if the assertion or retraction of facts has been generated in the WM by the executed rule or by new input data produced by the sensing devices, the LPM recognizes the rules which can be involved by these changes and updates both the alpha memories associated to their CEs and the related rule activation stacks. After that, the LPM determines the new active rules with a non-empty stack, selects one of them and restarts the research of its rule activations. The whole lazy matching procedure is summarized in figure 3.

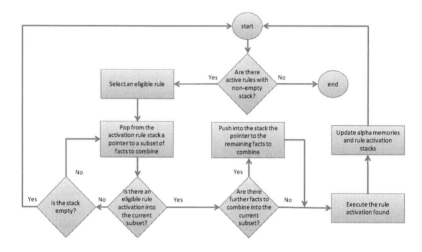

Fig. 3. The main components of the Mobile Reasoning System

It is worth noting that the proposed lazy approach does not save the intermediate results of inter-condition tests, but it recalculates them every time it is required. Other pattern matching algorithms [7] save all partial results of inter-condition tests into so-called beta memories in order to compute them only once and store them for later reuse.

However, the time gained by using such additional memory should be weighed against the inherent cost with respect to the specific application scenario. In particular, applications for chronic patient monitoring, where input data are constantly sent by sensor devices, require a continuous refresh of facts in the WM in terms of assertions and retractions. Thus, beta memories should be frequently updated, so as to make the storage of intermediate inter-condition tests' results useless, and, thus, decrease both space and time performance. As a result, omitting the memory support for inter-condition tests as proposed in the presented algorithm represents a significant factor to improve space and time performance in real-time and intensive applications.

3   Implementation and Results

The overall mobile reasoning system has been implemented for resource-limited mobile devices by using Java 2 Platform, Micro Edition (J2ME) in accordance with the Mobile Information Device Profile 2.0 (MIDP) and Connected Limited Device Configuration 1.1 (CLDP).

As a proof of concept, it has been applied to the case study described in [9] for monitoring cardiovascular diseases. In particular, to test the feasibility of the proposed system, the medical knowledge defined in [9] and formalized in terms of ontologies and rules has been used with the aim of detecting potentially abnormal situations.

Seven rules with an average number of antecedents equal to nine are stored into the *Rule Memory* (for more details about the rules used refer to [9]). The monitored parameters are: the heart rate, the estimated Standard Deviation Normal beat to Normal beat (SDNN), the patient's posture and his/her physical activity.

In accordance with the consideration that the admissible measuring range for the heart rate varies from 30 bpm (i.e. beat per minute) to 240 bpm, the heart rate can change its value up to a maximum of four times per second. This involves that the system has to reason on a variable number of heart rate measures per second, varying from 0 to 4. As a result, the system has been tested supposing the heart rate value is updated at a frequency equal to 1Hz and, also, the other parameters change their values at the same frequency.

Taking into account such considerations, five different measures for each parameter have been inserted into the *Working Memory,* where the different parameters are supposed to have been simultaneously acquired at each measurement. The five measures have been built ad hoc in order to activate rules reporting an abnormal situations only in three out of five cases.

The system has been deployed and tested on two mobile smart phones with comparable hardware features, namely Nokia N8 and HTC Legend, in order to evaluate its behaviour and effectiveness with respect to two different java compliant platforms, i.e. Symbian and Android.

In particular, with respect to both the Nokia N8 and HTC Legend, the overall reasoning time is approximately equal to 56 ms, whereas the response time required to detect and execute the first rule activation, i.e. the first abnormal situation, is more or less equal to 30 ms. So, the response time is sensibly less than the overall reasoning time, which can be reasonably equated to the response time of non-lazy approaches, where a response is generated only at the end of the whole reasoning process.

As a consequence, independently of the mobile device used, the proposed system is shown to be proficiently applicable to the presented case study regarding the cardiac monitoring. Indeed, on the one hand, it meets the real-time performance demand, since its response time is strongly less than the updating frequency of the monitored parameters. On the other hand, its response time is reasonably supposed to outperform the performance which could obtained by using classical approaches, even if this cannot be actually verified since, currently, the code of previous works supporting mobile reasoning are not accessible.

4   Conclusions

In this paper a mobile reasoning system able to build knowledge-based DSSs for monitoring and managing ubiquitously and seamlessly chronic patients has been presented. The system offers knowledge representation and reasoning facilities to face and efficiently reason on the continuous and real-time flow of data generated by the sensor devices.

The core of the reasoning system is a lazy pattern matching algorithm, specifically designed and implemented as a light-weight solution for granting the efficient handling of memory and computational resources and achieve good performance especially in real-time and intensive applications.

The proposed mobile reasoning system has been implemented for resource-limited mobile devices by using Java 2 Platform, Micro Edition and deployed on two mobile devices, namely Nokia N8 and HTC Legend, with comparable hardware features, in order to evaluate the library's behaviour and effectiveness with respect to two different java compliant platforms, i.e. Symbian and Android.

As a proof of concept, it has been applied to the case study described in [9] for monitoring cardiovascular diseases, showing its effectiveness to meet the real-time performance demand of that scenario.

Next step of the research activities will be to minutely compare the mobile reasoning system with respect to other existing one, in terms of performance evaluation and experimental assessment, and to apply it to other mobile health scenarios where patients affected by chronic pathologies have to be monitored.

References

1. Sinner, A., Kleemann, T.: KRHyper - In Your Pocket. In: Nieuwenhuis, R. (ed.) CADE 2005. LNCS (LNAI), vol. 3632, pp. 452–457. Springer, Heidelberg (2005)
2. Ali, S., Kiefer, S.: μOR – A Micro OWL DL Reasoner for Ambient Intelligent Devices. In: Abdennadher, N., Petcu, D. (eds.) GPC 2009. LNCS, vol. 5529, pp. 305–316. Springer, Heidelberg (2009)
3. Kim, T., Park, I., Hyun, S.J., Lee, D.: MiRE4OWL: Mobile Rule Engine for OWL. Accepted for publication at the 2nd IEEE International Workshop on Middleware Engineering, ME 2010 (2010)
4. Patel-Schneider, P., Hayes, P., Horrocks, I., et al.: OWL web ontology language semantics and abstract syntax. W3C Recommendation 10 (2004),
   http://www.w3.org/TR/owl-semantics/
5. Resource Description Framework, http://www.w3.org/rdf/
6. Weert, P.V.: Efficient Lazy Evaluation of Rule-Based Programs. IEEE Transactions on Knowledge and Data Engineering 22(11), 1521–1534 (2010)
7. Forgy, C.L.: Rete: A fast algorithm for the many pattern/many object pattern match problem. Artificial Intelligence 19, 17–37 (1982)
8. Miranker, D.P.: TREAT: A New and Efficient Match Algorithm for AI Production Systems. PhD dissertation, Columbia Univ. (1987)
9. Minutolo, A., Sannino, G., Esposito, M., De Pietro, G.: A rule-based mHealth system for cardiac monitoring. In: The 2010 IEEE EMBS Conference on Biomedical Engineering (IECBES 2010), Kuala Lumpur, Malaysia, November 30-December 2, pp. 144–149 (2010)

Using SOA for a Combined Telecare and Telehealth Platform for Monitoring of Elderly People

Georgios Lamprinakos[1], Stefan Asanin[2], Peter Rosengren[2], Dimitra I. Kaklamani[1], and Iakovos S. Venieris[1]

[1] National Technical University of Athens, School of Electrical and Computer Engineering, Heroon Polytechniou 9, 157 73, Athens, Greece
[2] CNet Svenska AB, Svärdvägen 3b, 182 33 Danderyd, Sweden
glamprinakos@icbnet.ntua.gr,
{stefan.asanin,peter.rosengren}@cnet.se, dkaklam@mail.ntua.gr,
venieris@cs.ntua.gr

Abstract. Remote monitoring is considered one key factor towards the improvement of elderly people's quality of life and the reduction of healthcare costs. This paper discusses the prevailing network and software architectural principles in this sector and presents the way that they are applied into the inCASA platform. We emphasize on how services could be offered in a standard and re-usable way on top of the underlying physical layers. The aim is to apply Service Oriented Architecture (SOA) in a healthcare-based Internet of Things (IoT) environment. This approach supports a combined telecare and telehealth solution for monitoring of frail elderly people.

Keywords: Remote healthcare monitoring, ageing independently, Service Oriented Architecture, Internet of Things, Web of Things, middleware, ubiquitous healthcare environment.

1 Introduction

Ageing population is one of today's major issues. There is a worldwide focus on improving elderly people quality of life while reducing health-care costs. Under these circumstances, Remote Healthcare Monitoring has gained a great interest for many years, with the perspective to offer health services remotely on top of health devices and biometric sensors. These services shall help elderly people to live in an independent way in their own home targeting and provide a significant decrease in their hospitalizations.

From a technical point of view, such environment consists of various devices, sensors, communication links and protocols. New services should be easily deployed to meet ever evolving needs of end users. There is thus a growing interest in applying Service Oriented Architecture (SOA) [1] to ease new services development and deployment in a healthcare-based Internet of Things (IoT[1]) environment.

---

[1] IoT refers to physical objects' virtual representation on the Internet or other network.

K.S. Nikita et al. (Eds.): MobiHealth 2011, LNICST 83, pp. 233–239, 2012.
© Institute for Computer Sciences, Social Informatics and Telecommunications Engineering 2012

There are several similar research and pilot deployment activities, namely MobiHealth [2], CAALYX [3], CommonWell [4] and RENEWING HEALTH [5]. The approach of the inCASA (Integrated Network for Completely Assisted Senior Citizen's Autonomy) project [6] differs from the previous ones, in the sense that it is based on a high level middleware which allows developers to build SOA-based applications on top of device and sensor networks without the need to get involved with low-level network and communication issues.

A key aspect of the inCASA solution is that it combines both a telecare and telehealth perspective, so that movements, activities and habits are monitored as well as vital health signs such as blood pressure, glucose levels, heart rate, weight etc. The service-oriented approach makes it possible to create a uniform way of integrating, accessing and consuming data from devices and sensors in both.

2   The inCASA Platform Architecture

inCASA aims to create citizen-centric technologies and a services network to help and protect frail elderly people, prolonging the time they can live well in their homes. As a key contribution of the platform, it is considered the profiling of user habits and the generation of alarms in case of divergence. This is instigated by the fact that elderly people tend to follow a quite habitual way of living (i.e. daily routines).

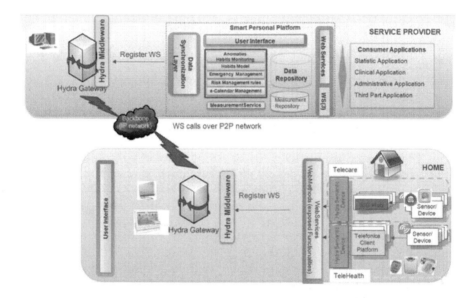

Fig. 1. The overall inCASA platform solution with two sets of middleware installations and physical locations with a Web Service defined structure of communication

The main functional user requirement is the existence of an interface provided to the professional end users where they can observe real-time measurements, extract historical data, get alerts, set and configure parameters per patient. In order to support this, a multi-component reference architecture has been developed (Fig. 1). The inCASA platform is divided into two entities that are communicating with each other through Web Service calls over a Peer-to-Peer (P2P) network provided by Hydra (more explained in the next section):

1. End User's entity where both clinical and environmental data are collected.
2. Service provider's infrastructure entity where data is collected, analyzed, stored and made available to Consumer Applications.

2.1   The inCASA Platform Environment

Firstly, the inCASA platform environment is divided into environment monitoring devices and vital-sign monitoring devices where the first are collected by an Activity Hub [10] while the latter are collected by a Telehealth gateway. The common set of devices/sensors requirements includes accuracy, energy-saving, continuous up-time, configurability and wireless connectivity support. It is also important to prevent physical or logical unauthorized data access by data encapsulation in a standardized communication protocol.

Secondly, the platform environment is comprised by a service-oriented middleware for Internet of Things applications, Hydra, and service consuming subsystems like the Smart Personal Platform and for the inCASA project specially developed Consumer Applications.

2.2   The Hydra Middleware

The Hydra Middleware is the central building block for the Socio-Medical platform in inCASA solution supporting discovery and configuration of inter-connected devices. Hydra is the architectural component of the proposed solution that allows SOA application in a healthcare-based Internet of Things environment [7], [9].

Hydra allows different isolated home networks (for both patients and medical personnel) to be interconnected by implementing an architecture based on P2P technologies [11]. It discovers the different devices (medical sensors and actuators) and selects the most appropriate software components through a proxy that controls the specifics of the communication and data exchange with the device. It also offers the services of the device in a standard and easy-to-consume way, by the use of web services and UPnP (Universal Plug and Play) technologies. This allows for service discovery at the local network level where services are published in the overall Hydra network, that is a P2P network and transparent to the user of the application.

To summarise, it could be said that the Hydra middleware is an intelligent software layer placed between the operating system and applications and it contains a large number of software components (i.e. managers) that handle various processing tasks.

2.3   The Smart Personal Platform

The Smart Personal Platform (SPP) retrieves, stores and analyzes the end user's data received from the inCASA gateway (i.e. Hydra). The SPP has two main roles. Firstly, it collects the monitoring data and creates a habits model for the patient. Secondly, it is focused on the logical processing of incoming monitoring data including extended reasoning mechanisms responsible for comparing of collected data against stored user habits model, to detect deviations.

The SPP is a software application consisting of various modules, namely the Habits monitoring module responsible for the building of "User Habits" profile and the generation of alarms in case of divergence, the Emergency and Risk Management module which analyzes the incoming data to identify the possible actions needed or risks to be handled and the Socio-Medical Calendar module which allows the inCASA operators to arrange appointments and plan for future activities with the inCASA registered patients via the Consumer Applications interface.

2.4   The Consumer Applications

In inCASA, Consumer Applications (CAs) use Web Services exposed by the SPP. These are a set of high level views available to the personnel of the inCASA pilots and are responsible for the rendering of data and alerts for professional GUIs.

Since Consumer Applications are the back-end of the inCASA platform where the operators have access, they are also responsible for the integration of Telecare and Telehealth data into a unified view per patient. Moreover, Consumer Applications expose web services to be called by the SPP upon alarm generation, whose functionality includes on-screen alert and relevant update to the operators or/and to patient's relatives via SMS and e-mail.

3   Middleware as a Key Component towards SOA

Hydra's main role in inCASA is to allow software architecture specifications to be based on its middleware principles in order to ensure communication and interoperability between different modules.

Fig. 2 shows how the different middleware components are embedded in gateways and devices to create the SOA in which the services of each device appear as Web Services and UPnP services. Two Network Managers in two separate locations within same network are using P2P techniques to communicate. A Discovery Manager discovers any new device that appears on the other side by using SOAP tunnelling [11] mechanisms of the Network Manager on first side. It makes a query about the remote device or application. The Network Manager on second side receives the request and resolves it using an internal ID database which results in a local Web Service call to the device's generic Hydra Web Service. This call is then handled by the Application Service Manager that adds a semantic layer and complements the Application Device Manager with a service perspective mapped to the device

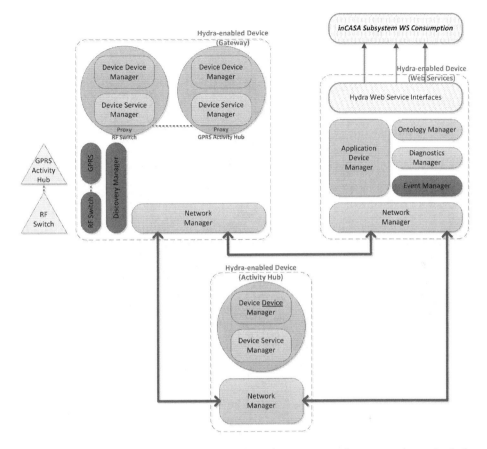

Fig. 2. The Hydra Middleware architecture and its main components for representing a physical device as a web service

functionalities. This underlying communication in Hydra is based on SOA and transparent as resource for model-driven development of inCASA applications [7].

4   inCASA SOA Solution

inCASA's various types of medical devices and sensors are managed by extended device ontology where the semantic representations of devices and their service descriptions are used to generate Web Service interfaces. These allow inCASA programmers to access and use devices using standard web technology.

The inCASA middleware uses SOAP tunnelling to make web services calls between physical devices in two different networks enabling control and access of a device in patient's home from a service provider network. Using the SOA and MDA

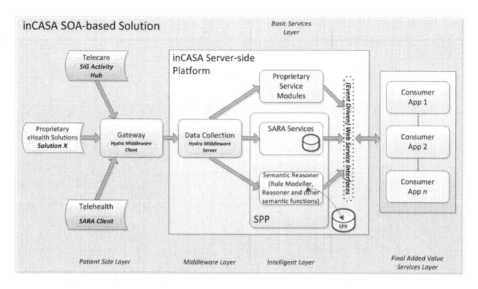

Fig. 3. The inCASA SOA solution with all involved technologies present

approaches in Hydra, inCASA has gained features that allow developers to create any possible ubiquitous services and systems that interconnect devices, people, terminals, buildings, etc. This is realised by providing interoperability at a semantic level by extending semantic web services to the device level. The middleware client that runs on the home gateway uses web services to publish embedded interfaces and services to the inCASA network. In this way inCASA uniformly supports different standards where for instance any Continua/IEEE11073 [8] device can connect to the gateway as well as other types of non-Continua devices.

5   Conclusion

This paper shows how a middleware based approach allows inCASA to implement SOA support to a wide range of applications for ageing independently by remote monitoring. The approach achieves interoperability at service level and not by data exchange mechanisms alone while meeting the requirements for connecting and using a wide range of different devices even though following different types of standards.

Acknowledgment. This work was performed in the framework of CIP-ICT-PSP Project inCASA (Integrated Network for Completely Assisted Senior citizen's Autonomy) partially funded by the European Commission. The authors wish to express their gratitude to the other members of the inCASA Consortium for valuable discussions.

References

1. Microsoft MSDN,
   http://msdn.microsoft.com/en-us/library/aa480021.aspx
2. EU MobiHealth Project, http://www.mobihealth.org
3. CAALYX Project, http://caalyx.eu/
4. CommonWell Project, http://commonwell.eu/index.php
5. RENEWING HEALTH project, http://www.renewinghealth.eu
6. inCASA project (Integrated Network for Completely Assisted Senior Citizen's Autonomy), http://incasa-project.eu
7. Eisenhauer, M., Rosengren, P., Antolin, P.: HYDRA, A Development Platform for Integrating Wireless Devices and Sensors into Ambient Intelligence Systems. In: The Internet of Things, 20th Tyrrhenian Workshop on Digital Communications, Sardinia, Italy (2010)
8. Continua Health Alliance, http://www.continuaalliance.org
9. Hydra Project, http://www.hydramiddleware.eu/
10. Steinbeis, http://stzedn.de/capt2web-sniffer.168.html
11. Lardies, F.M., Antonlin, P., Fernandes, J., Zhang, W., Hansen, K., Kool, P.: Deploying Pervasive Web Services over a P2P Overlay. In: 18th IEEE International Workshops on Enabling Technologies: Infrastructures for Collaborative Enterprises, pp. 240–245. IEEE Computer Society, Groningen (2009)

Adaptive Assistance: Smart Home Nursing

Nikola Serbedzija

Fraunhofer FIRST
Berlin
nikola@first.fraunhofer.de

Abstract. Home nursing gains in significance with the human age being prolonged. More and more people reach the state when they need some assistance in order to live independently at their own. However, home nursing is a resource demanding activity stretching medicare to its limits. In this situation, new technology can help. Reflective assistance is concerned with the construction of flexible 'smart' systems that control the eldercare environment, adapting the ambient to the needs of individuals. New technology transforms a living space to a helpful residence assistant that observes inhabitants and offers aid or calls for it in the case of need. To achieve this goal, systems must be capable of monitoring the behavior of the elderly people and of responding to dynamic changes in their performance or physical and psychological situation. This paper describes an approach to design and develop a home ambient that offers medicare, mobile monitoring, rehabilitation exercises and improved comfort of elderly inhabitants.

Keywords: Medicare, Home nursing, Physiological computing, Patient monitoring, Adaptive systems.

1   Introduction

An important characteristic of smart technology [1] is a seamless and implicit human computer interaction that uses wireless sensor/actuator devices to detect user situation and respond accordingly. In order to offer smart assistance, the system must have some means of assessing the context of interaction without explicit user intervention. This can be done by making both human's behavior and inner state a part of the processing loop, e.g. by deploying the sense-analyze-react principle performing a seamless observation, situation evaluation and active reaction. Having a generic support for such implicit and awareness-rich processing would allow deployment of smart technology in a whole range of medicare areas.

Home health care consists of "a part-time skilled nursing care, physical therapy, occupational therapy, speech-language therapy, home health aide services, medical social services, durable medical equipment (such as wheelchairs, hospital beds, oxygen, and walkers) and medical supplies, and other services" [2]. Some of these services may be performed (in its basic form) without direct human participation. Recently, a new generation of control systems has been developed [3] offering control

K.S. Nikita et al. (Eds.): MobiHealth 2011, LNICST 83, pp. 240–247, 2012.
© Institute for Computer Sciences, Social Informatics and Telecommunications Engineering 2012

strategy enriched by physiological and socio-behavioral analyses. Systems are called reflective as they diagnose users' physical, social and psychological state and react accordingly in a given situation. Such systems may significantly support home nursing, performing tasks where human presence is not necessary thus leaving more time for personal contact during the visits.

2   Theory behind

This approach deploys the concept of a biocybernetic loop [4,5] allowing for a multiple physiological sensing, composite analyses and decision making. The function of the loop is to monitor changes in user state in order to initiate an appropriate adaptive response. It takes results of affective computing and combines it with higher level understanding of social and goal–oriented situations. The approach is multi-modal as it takes into account different kinds of information, processing them in multiple loops at different time scales. There are three major phases of a single loop: sense, analyze and activate. These phases are repeated endlessly, where each consecutive cycle takes into account the effects of the previous one, performing constant self-tuning and self optimization.

The first phase of the bio-cybernetic loop is monitoring of the user in a given situation. The collecting of information can be done by observing: (1) overt actions (e.g. location, looking, pointing), (2) overt expression (e.g. changes in behavior associated with psychological expression), and (3) covert expression (e.g. changes in physiology associated with psychological expression).

The analyses phase of the biocybernetic loop is a process that involves psychology, physiology and behavioral science knowhow. In order to make an effective use of such diverse data, this approach relays on affective and physiological computing results and deploys rule-based reasoning. The loop is designed according to a specific rationale, which serves a number of specific meta-goals, defined by the application needs..

The final phase of reaction is devoted to the adequate system response which is performed through a certain action of the system actuators having further influence on both the user and the controlled situation.

Based on a better understanding of both personal involvement and social and behavioural situation of the user, a reflective system [6] offers adaptive control of different types and different time scales: (1) immediate adaptation supporting safety, (2) short-term adaptation responding to a more complex states that require several steps of self-tuning, (3) long-term adaptation providing individualisation as a process that guarantees that the system has co-evolved with individual user and can target its functioning to specific individual needs.

3   Implementation

Developing reflective software involves tasks like real-time sensor/actuator control, user and scenario profile analyses, affective computing, self-organization and adaptation. To accomplish these requirements, a service- and component-oriented

[6,7] middleware architecture, based on reflective ontology [8], has been designed that promises a dynamic and re-active behavior featuring different biocybernetic loops. According to the reflective ontology, the reflective software is grouped into three layers: (1) Tangible layer - a low-level layer that controls sensor and actuator devices; (2) Reflective layer - with more complex services and components that evaluate user states; (3) Application layer - a high level layer that defines application scenario and system goals; by combining low and high level services and components from other layers, application layer runs and controls the whole system.

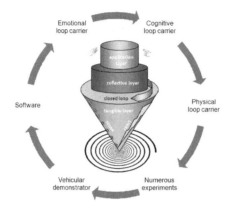

Fig. 1. Reflective architecture with the closed loop

Figure 1 metaphorically illustrates the reflective layered architecture as a spin top, exercising different temporal loops at tangible, reflective and application level. A control loop (initialized with users' profile and scenario settings) starts by sampling the psycho-physiological and other measurements, continues with their analyses and finishes by adaptive system reaction. In a next iteration the system influence (caused by the reaction) can also be sensed and further tuned. The overall design goal is to have a generic modular structure that follows the patterns of immediate, short and long term adaptation and is capable of dynamic configuration and efficient functioning.

4   Application

Reflective technology has been successfully tested in the automotive domain [9] implementing the concept of a "vehicle as a co-driver". Based on a comprehensive driver's psycho-physiological analyses and driving situation evaluation, the reflective vehicle features active support in driving by warning in case of high mental effort or dangerous driving situations, by adapting car entertainment according to driver's mood and by re-shaping the seat according to drivers comfort. However, being genuinely generic and re-usable, reflective framework can be deployed in a range of different application domains.

4.1   Reflective Home Care

A reflective home care for elderly people is currently under development [10]. The goal is to construct a flexible 'smart' ambient to control the home for elderly people offering both medical and rehabilitation services at one side and improving the quality of living at another. The system should transform a home into friendly and supportive home nurse. The functionalities offered include:

- Medical (physiological) monitoring;
- Rehabilitation support
- Seating/laying comfort;
- Monitoring of inhabitant movement;
- Monitoring of home appliances;
- Control of TV and media-rich entertainment;
- Communication with mobile devices.

Enriched with reflective assistance, a smart home plays an active role in everyday life supporting medical diagnoses and check-ups, rehabilitation and or physical exercises, watching health condition as well as emotional and mental state of the inhabitants.

Reflective support for each of the above mentioned functionalities is achieved by embedding numerous sensor and actuator devices into home settings and by deploying reflective control to these devices.

- Medical support consists of heart rate and blood pressure measurements – as regular daily check up, with warnings displayed on TV - and remote mobile monitoring by medical staff.
- Rehabilitation support controls exercise devices (e.g. home cycle or walking track) according to the instruction displayed on television. Physiological monitoring (e.g. heart rate, blood pressure) is done simultaneously guiding the exercise according to the body response.
- Comfort control is done via seating and laying sensors, checking the body pressure at critical points and modifying (if necessary) the shape of the armchair/matrices via air pumps [11].
- Movements control is done via cameras placed at each room and is used only for critical hot line warnings (in case of sudden fall).
- Home appliances are connected to automatic switch-on/off device allowing remote/mobile monitoring and control.
- Entertainment control consists of a mood player [6] designed to control entertainment center according to emotional response.
- Communication control connects to mobile devices allowing for urgent calls and remote monitoring.

With the above described functionality, reflective home performs many of the routine functions of a home nurse. It does daily medical check-up, assists in active exercise and rehabilitation, cares for inhabitant physical, emotional and mental state, reminds and/or overtakes the control over switching on/off the home appliances and supports mobile communication with emergency center, friends and relatives. In summary it overtakes numerous medical, psychological, physical and social functions, complementing the work of a home nurse.

4.2   Reflective Support

The reflective framework [6] with its already existing support for ambulatory psychophysiological measurement and diagnosis offers a ready to use solution for a number of problems that appear in the domain of remote measurements and analyses of user states. This makes the reflective framework a straightforward means for the use within the home care scenarios.

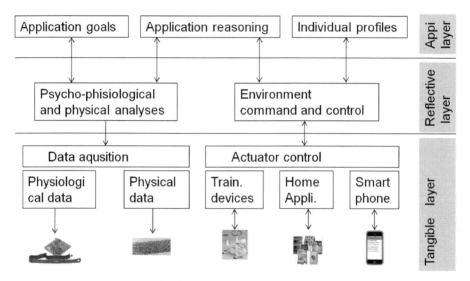

Fig. 2. Home care reflective layers

Figure 2 shows the organization and structure of the home care reflect deployment. The tangible layer provides interface to the required devices. The sensors are: the camera (for location tracking and posture detection) and physiological measurement devices. The devices containing both sensors (report the device state) and actuators are: home exercise tools, kitchen appliances, heating, air-condition, lighting, television (used for visual assistance and entertainment) and the communication module. Tangible interface for the devices from Table 1 already exists and the interface for the devices from Table 2 is currently under development.

Reflective layer fuses the sensor measurements and performs diagnoses of the user condition (e.g. by physiological measurement during exercises). At this level, high-level commands are given to the system control component, which are further decomposed to the concrete actuator devices. The core logic for the event driven behavior and supportive reaction of a biocybernetic loop is given in the reflective taxonomies [8] for the concrete situation. For example, safety actions in case of exceptional values obtained from the sensors are triggered automatically as they are built into the system through inter-relationship among ontology entities and short term biocybernetic loop response (in case that camera through posture recognition detects "an accident" – emergency will be alarmed). The reflective layer also deploys a rule-based reasoning to perform diagnosis and to trigger further system actions.

The application layer maintains a personal database and uses the same rule based engine to make decision how the system should function, according to the predefined system goals. Different factors are used in making the control strategy. For example, if the blood pressure shows "high" value during a routine check-up, an alarm will be raised contacting the medical center. However, if the same value is obtained during the exercise, a warning will be raised instructing to stop with the training and to repeat blood pressure check after 10 minutes. Such a strategy is combined with personalized characteristics as "critical margin" may differ from person to person. In a similar fashion all other system strategies are implemented as rules and are used to control other system components (remote medicare, remote/mobile kitchen control, etc).

Table 1. List of devices with existing reflective controllers

Device	Picture	Description	Model Id.	Data
Force Sensors Cushion		16x16 Force sensors matrix	pressure mat system Pliance X (Novel)	CoP position vs time
Camera		web cameras	Hercules Classic1,3 MgpixVGA	Facial exp.; posture
HR sensor		2 channel ECG sensor cable	Article: NX-EXG2-2A	ECG raw signal
Respiration sensor		RSP sensor (respiration)	Article: NX-RSP-1°	RSP raw signal
Bio-sensors		Multi-channel datalogger (10-chan.)	Nexus10	Bio-sensors raw-data
Skin Conductance		GSR sensor (incl. fingerclips)	Article: NX-GSR-1A	SC raw signal
Skin Temp sensor		Temperature sensor (skin)	Article: NX-TMP-1°	SkinTemp raw signal
Smart phone		Remote monitoring and control		Internal objects

Table 2. List of new devices needed for home medicare

Device	Picture	Description	Model Id.	Data
Training devices		Speed meter,	-	Speed effort
Home appliances		Fridge, cocking devices, air condition	-	Switch On/Off
Entertainment		TV set	-	Multi media
Bubble mattress		Anti Decubitus mattress		Pressure points

The modules for interfacing and high-level diagnoses for the devices from the table 1 could be used without any change. The corresponding software for the devices shown in table 2 is presently under development.

5   Conclusion

The paper presents the use of the reflective approach, a novel technology featuring seamless man-machine interaction in solving problems in domain of health care (home nursing), ambient assisted living and eldercare. A new application domain responds to the challenge of modern societies "growing older" with increasing need to support elderly population in independent living. A number of scenarios are described that are designed to support senior citizens in medicare, comfort living, socializing and home services. The reflective assistant effectively supplements a number of duties that home nurse traditionally performs.

The major reason for developing reflective computing has been to enrich smart systems with genuinely supportive behavior in terms of doing what the users need in a seamless and personalized way. The advantage of this approach is in introduction of an adaptive component-oriented technology that offers a generic solution for a wide application spectrum, from entertainment, via ambient assisted living and medicare up to embedded real-time systems. Reflective computing uses psycho-physiological measurements to determine emotional, cognitive and physical user states that are further used to tune the system reaction. The framework features multiple levels of adaptation and pervasive behavior.

Most of the functions described in this paper have been already developed for another application domain[6,9] and could be re-used directly. Other functionalities (e.g. home exercise control and home appliances control are under development). Looking at the nursing home checklist [2], it may be concluded that reflective support – as a home nurse assistant - fulfills important requirements, a nurse must satisfy: medical competence, pleasant behavior, a good sense for temperature, atmosphere and well being at home, remaining at the same time silent and respectful towards residents, sensitive to emotional, physical and mental state of the people, 24 hours available and ready to communicate with friends, relatives and medical experts. Covering such a spectrum of activities, the system eases the home nurse work and leaves the nurse more time for human contact – the function that technical system can never substitute.

The future work is seen in further extension of the use of reflective technology in medicare domain. This primarily requires refinement and enrichment of the reflective ontology in providing support at all levels of reflective programming: namely interfacing new devices, diagnosing new situations and enlarging its knowledge base. Mobile monitoring is also under development including access to most of information collected by the system via mobile phones, thus binding smart technology with human control and intervention. Improvements in the domain of information protection are also on the research agenda as the system deals with highly sensitive personal information that should not be available for any other but humanitarian use.

References

1. Norman, D.A.: The Design of Future Things. Basic Books, New York (2007)
2. Medicare, http://www.medicare.com
3. Serbedzija, N., Fairclough, S.: Reflective Pervasive Systems. In: ACM Transactions on Autonomous and Adaptive Systems (TAAS) (to appear in 2011/2012)
4. Pope, A.T., Bogart, E.H., Bartolome, D.S.: Biocybernetic system evaluates indices of operator engagement in automated task. Biological Psychology 40, 187–195 (1995)
5. Serbedzija, N., Fairclough, S.: Biocybernetic Loop: from Awareness to Evolution. In: Proc. IEEE Congress on Evolutionary Computation CEC, Trondheim, Norway (2009)
6. Reflect project - Responsive Flexible Collaborating Ambient, http://reflect.first.fraunhofer.de
7. Beyer, G., Hammer, M., Kroiss, C., Schroeder, A.: Component-Based Approach for Realizing Pervasive Adaptive Systems. In: Proc. Workshop on User-Centric Pervasive Adaptation UCPA 2009, Berlin, Germany (2009)
8. Kock, G., Ribarić, M., Šerbedžija, N.: Modelling User-Centric Pervasive Adaptive Systems – The REFLECT Ontology. In: Nguyen, N.T., Szczerbicki, E. (eds.) Intelligent Systems for Knowledge Management. SCI, vol. 252, pp. 307–329. Springer, Heidelberg (2009)
9. Serbedzija, N., Calvosa, A., Ragnoni, A.: Vehicle as a Co-Driver. In: Proc. First Annual International Symposium on Vehicular Computing Systems - ISVCS, Dublin, Ireland, July 22-24 (2008)
10. Serbedzija, N.: Reflective Assistance for Eldercare Environments. In: Proc. of Second Workshop on Software Engineering in Health Care - SEHC, Cape Town (2010)
11. Serbedzija, N., Bertolloti, G.-M.: Adaptive and Personalized Body Networking. International Journal of Autonomous and Adaptive Communications Systems (IJAACS) 6(3) (2011) (to appear)
12. SmartSenior: Independent, secure, healthy and mobile in old age, http://www.Smart-Senor.de/

Ubiquitous Healthcare Profile Management Applying Smart Card Technology

Maria-Anna Fengou, Georgios Mantas, Dimitrios Lymberopoulos, and Nikos Komninos

Wire Communications Laboratory, Electrical and Computer Engineering Department, University of Patras
University Campus, 265 04 Rio Patras, Greece
{afengoum,gman,dlympero}@upatras.gr
nkom@ieee.org

Abstract. Nowadays, the patient-centric healthcare approach is focused on ubiquitous healthcare services. Furthermore, the adoption of cloud computing technology leads to more efficient ubiquitous healthcare systems. Moreover, the personalization of the delivery of ubiquitous healthcare services is enabled with the introduction of user profiles. In this paper, we propose five generic healthcare profile structures corresponding to the main categories of the participating entities included in a typical ubiquitous healthcare system in a cloud computing environment. In addition, we propose a profile management system incorporating smart card technology to increase its efficiency and the quality of the provided services of the ubiquitous healthcare system.

Keywords: cloud computing, user profiles, profile management, smart card.

1   Introduction

Nowadays, the patient-centric healthcare approach is focused on healthcare services related to effective treatment, disease prevention, proactive actions and life quality improvement at the right time, right place and right manner without limitations on time and location. Essentially, the patient-centric approach can be materialized through ubiquitous healthcare. Furthermore, the current trend of outsourcing of IT infrastructure via adoption of cloud computing technology leads to more efficient ubiquitous healthcare systems in which end-user data are stored and processed at many different places distributed geographically [1]. However, many ubiquitous healthcare services built by third-party providers are usually tied to specific scopes and they are not able to be adapted to the unique preferences and interest of the end-users. Because of these limitations, personalized healthcare systems based on user profiles have been recently deployed [2]. Profile management materializes the main functionality of personalized healthcare systems [3]. In addition, profile management integrates the user profiling process consisting of three main steps. In the first step, an information collection process takes place in order raw information related to the user

K.S. Nikita et al. (Eds.): MobiHealth 2011, LNICST 83, pp. 248–255, 2012.
© Institute for Computer Sciences, Social Informatics and Telecommunications Engineering 2012

to be gathered. The second phase includes the user profile construction process based on the gathered user information. Finally, in the third phase, a service makes use of the data in the user profile in order to provide personalized services [4]. During the past few years, a lot of effort has been invested in user profile management systems for ubiquitous healthcare systems. However, these profile management systems incorporate user profiles including limited user information associated with user preferences and interests. Therefore, it is essential the creation of more enriched user profiles for efficient ubiquitous healthcare systems. Furthermore, taking into consideration the challenges that a ubiquitous healthcare system should face in order to provide high quality and reliable services, it is required the creation of profiles for all the participating entities of such a system.

Thus, in this paper, we propose five generic healthcare profile structures that correspond to the main categories of the participating entities of a typical ubiquitous healthcare system in a cloud computing environment. The main objective of these proposed structures is to contribute toward the creation of more enriched profiles for all participating entities of a ubiquitous healthcare system. Additionally, we propose a profile management system incorporating smart card technology in order to increase its efficiency and the quality of the provided services of the ubiquitous healthcare system. Especially, smart cards are proposed to include information referencing to large amount of stored profile data for the creation of enriched user profiles. Profile data should always be available, over all networks, from all supported services enabling service continuity and optimal user experience.

Following the introduction, this paper is organized as follows. In Section 2, the related work of ubiquitous healthcare sector in cloud computing environment and profile management is presented. Furthermore, a brief overview of smart card technology is given. In Section 3, the proposed five generic healthcare profile structures are described. In Section 4, the proposed profile management system is discussed. Finally, Section 5 concludes the paper.

2   Related Work

2.1   Ubiquitous Healthcare in Cloud Computing

Several platforms have been proposed that combine cloud computing and network security concepts implemented in the healthcare domain [1]. MoCAsH [5] is an infrastructure for assistive healthcare. It inherits the advantages of Cloud computing and embraces concepts of mobile sensing, active sensor records, and collaborative planning. The work [6] proposes a system designed and implemented using the cloud computing platform to provide a long term offsite medical image archive solution. Phynx [7] presents the development of an open source software solution that supports management and clinical review of patient data from electronic medical records databases or claims databases for pharmaco-epidemiological drug safety studies. In [8], it is proposed a solution to automate the existing processes for patients' vital data collection that require a great deal of labor work to collect, input and analyze the information. This work is also based on the concepts of utility computing and wireless sensor networks. The information becomes available in the "cloud" from where it can be processed by expert systems and/or distributed to medical staff. Finally, [9] presents the implementation of a mobile system that enables electronic healthcare data storage, update and retrieval using cloud computing. The mobile application provides management of patient health records and medical images.

2.2   Profile Management System

Personalization and user profile management holds the promise of improving the uptake of new technologies and allowing greater access to their benefits [10, 11]. In a typical profile management scenario, there are multiple profile storage locations. Many of these locations will not store the total profile but only components that apply to a device, an application, a network function or service. Different locations may have different persistence and priority levels. Although the user profile data is distributed amongst a range of devices, applications and services, ideally, all profile data should always be available, over all networks, from all supported devices and services, including fixed and mobile services allowing service continuity and optimal user experience [11].

The work [12] proposes a user profile management that allows the creation of an instance of the user profile for each application and for each instance of context. Major enhancement consists in the enrichment of the user profile with dynamic data permitting to characterize the user environment, the serving device and network. Sutterer et al. [13] introduce a user profile management approach for managing and delivering context-dependent user profiles for several applications and decouples the application development from context processing in ubiquitous computing environments.

2.3   Smart Card Technology

Smart cards are plastic pocket-sized cards embedded with either a microcontroller chip or only a memory chip with non – programmable logic. Nowadays, smart cards are used in a variety of applications ranging to identify individuals and get cash from the bank to mobile telecommunications (SIM-cards). Smart cards are favorable for applications that require data confidentiality and data integrity. Data confidentiality guarantees that the data/information contained on the card is accessible only by legitimate users. Data integrity ensures that the data/information stored on the card is always defined and is not changed even if the power to the card is removed during a computation involving data stored on the card.

The operating systems used in the majority of current smart cards are based on the provisions of the ISO/IEC 7816 family of standards. The smart card operating systems are stored in the ROM of the chip. Moreover, the available part of RAM and a small part of EEPROM are used by the operating systems. They implement a standard set of commands (e.g. 20-30) which the card can "understand" [14].

3   Proposed Healthcare Profile Structures

The electronic Medical Health Record (eMHR) is considered as the core component for ubiquitous healthcare systems. Treatments, demographic, examinations and other medical information of each citizen are stored in eMHRs. However, to deliver personalized ubiquitous healthcare services, it is also essential the incorporation of profiles of the participating entities in the ubiquitous healthcare systems.

In a ubiquitous healthcare system, there are the following categories of participating entities:

- Patient: around whom the healthcare services are built depending on his health condition and the services that he is interested on;
- Patient's social network: doctors, healthcare providers, family members, volunteers, etc;
- Healthcare Center: in case the patient"s health condition is critical, he will be transferred to a Healthcare Center;
- Smart Home: a home equipped with advanced technology (e.g. wireless sensor networks) adequate to provide safety and a quality of life to people suffering from chronic diseases (e.g. cardiac) or elders having dementia;
- Office: the area where the patient works;
- Vehicle: the personal vehicle that the patient uses to be transferred.

All these entities are either directly or indirectly related to the patient when his health condition is critical and a ubiquitous healthcare service should be delivered. Based on the features of these entities, we propose a profile structure for each entity. To generate the profile structure of each entity, we used [15] and the method described in [16]. Using this method, the basic steps for the generation of the profile structures are: sampling, analyzing and modeling. Thus, the proposed generic healthcare profile structures are the following:

- User Healthcare Profile: This profile corresponds to the Patient and Patient's Social Network categories;
- Healthcare Center Profile: This profile corresponds to the Healthcare Center category;
- Smart Home Profile: This profile corresponds to the Smart Home category;
- Office Profile: This profile corresponds to the Office category;
- Vehicle Profile: This profile corresponds to the Vehicle category;

The proposed five generic healthcare profile structures are depicted in Fig. 1:

User Healthcare Profiles
Personal Information
Preferences
• Personal Preferences • Professional Preferences • Service Format Preferences • Provided Services Preferences
Terminal Capabilities
Required for Third Party
Current Activity
This field allows the user to indicate what is currently doing
Current Context Info
• Environmental Context • Medical Context • Social Context
User History
This field may contain a URI that indicates where the user's history is stored
Rules
This field contains statements that use as input variable information stored in the profile
Policies
• Security • Dedicated to user

Healthcare Center Profile
Identification Info
Supported Healthcare Service
Available equipment and facilities • Ambulance • Beds
Available professional healthcare staff • Doctors • Nurses • Secretaries • IT supported staff • Administrator Support Staff
Healthcare Center Availability
Statistics (Volume metrics)

Smart Home Profile
Identification Info
Support Healthcare Services
e.g. telemonitoring
Available equipment facilities
Available Home Members/ residents for support
Context Information • Environmental

Office Profile
Identification Info
Support Healthcare Services
e.g. telemonitoring
Available equipment facilities
Available Office Members for support
e.g. secretary
Context Information • Environmental

Vehicle Profile
Vehicle Features
Available equipment and facilities
Support Services
Context Information • Environmental

Fig. 1. Proposed Healthcare Profile Structures

4   Proposed Profile Management System

Our proposed profile management system is applied on an e-Health tele-monitoring system in a cloud computing environment for providing monitoring services to patients suffering from chronic diseases remotely. We consider that the e-Health tele-monitoring system integrates a wide spectrum of participating entities including patients, doctors, nurses, family members, a healthcare center, a smart home, an office and a vehicle. The participating patients, doctors, nurses and family members are considered as the users of the e-Health tele-monitoring system.

In addition, each participating entity has its own profile that follows the structure corresponding to one of our proposed healthcare profile structures according to the main category to which the participating entity belongs. Thus, the proposed profile management system makes use of a number of User Healthcare Profiles, a Healthcare Center Profile, a *Smart Home Profile*, an *Office Profile* and a *Vehicle Profile*. The incorporated data of each profile are stored in distributed databases in the cloud computing environment where the e-Health tele-monitoring system works.

The proposed profile management system of the e-Health tele-monitoring system is presented in Fig. 2:

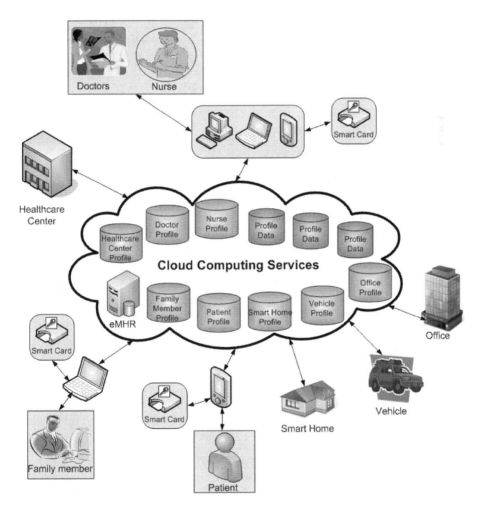

Fig. 2. Proposed Profile Management System

The patients are being monitored by the system in order any critical event to be detected timely. The main objective of the proposed profile management system is the creation of group profiles in order to provide high quality and reliable personalized healthcare services to the patients.

The group profile is created based on the indexes stored in the *User Healthcare Profiles*. Each patient's profile contains an index to a group profile that is triggered whenever an event is detected. The events are related to the patient's health condition and they denote that the patient needs a medical advice or care due to aggravation of his/her health condition. The group is formed through certain management, scheduling and notification events.

Furthermore, in the context of the proposed profile management system, each one of the participating users poses a smart card storing information about the locations of databases where the corresponding profile data are stored. In other words, as it is depicted in Fig. 3, each user's smart card stores URL locations pointing to databases where large amount of profile data are found in order a User Healthcare Profiles to be created.

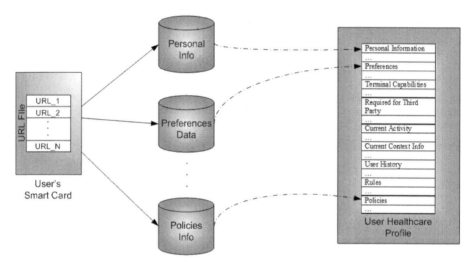

Fig. 3. User Healthcare Profiles in Smart Cards

5   Conclusion and Future Work

In this paper, we have proposed five generic healthcare profile structures for all the main categories of the participating entities of a typical ubiquitous healthcare system in a cloud computing environment. Our vision for these structures is to lead to more enriched profiles for all participating entities. Furthermore, we have proposed a profile management system integrating smart card technology to increase its efficiency and the quality of the provided services of the ubiquitous healthcare system. Finally, as future work, we intend to deploy the proposed profile management system for an e-Health tele-monitoring system in cloud computing environment.

References

1. Löhr, H., Sadeghi, A.R., Winandy, M.: Securing the E-Health Cloud. In: 1st ACM International Health Informatics Symposium, Virginia, pp. 220–229 (2010)
2. Guillén, S., Meneu, M.T., Serafin, R., Arredondo, M.T., Castellano, E., Valdivieso, B.: e-Disease Management. A system for the management of the chronic conditions. In: 32nd Annual International Conference of the IEEE EMBS, Buenos Aires, pp. 1041–1044 (2010)
3. ETSI ES 202 746 V0.0.16: Human Factors (HF); Personalization and User Profile Management; User Profile Preferences and Information (2010)
4. Gauch, S., Speretta, M., Chandramouli, A., Micarelli, A.: User Profiles for Personalized Information Access. In: Brusilovsky, P., Kobsa, A., Nejdl, W. (eds.) The Adaptive Web 2007. LNCS, vol. 4321, pp. 54–89. Springer, Heidelberg (2007)
5. Hoang, D.B., Lingfeng, C.: Mobile Cloud for Assistive Healthcare (MoCAsH). In: Services Computing Conference (APSCC), Hangzhou, pp. 325–332 (2010)
6. Teng, C.C., Mitchell, J., Walker, C., Swan, A., Davila, C., Howard, D., Needham, T.: A medical Image Archive Solution in the Cloud. In: Software Engineering and Service Sciences (ICSESS), Beijing, pp. 431–434 (2010)
7. Egbring, M., Kullak-Ublick, G.A., Russmann, S.: Phynx: an open source software solution supporting data management and web-based patient-level data review for drug safety studies in the general practice research database and other health care database. PDS 19(1), 38–44 (2010)
8. Rolim, C.O., Koch, F.L., Westphall, C.B., Werner, J., Fracalossi, A., Salvador, G.S.: A Cloud Computing Solution for Patient's Data Collection in Health Care Institutions. In: Second International Conference on eHealth, Telemedicine, and Social Medicine, pp. 95–99 (2010)
9. Doukas, C., Pliakas, T., Maglogiannis, I.: Mobile Healthcare Information Management utilizing Cloud Computing and Android OS. In: EMBS, Buenos Aires, pp. 1037–1040 (2010)
10. ETSI TS 102 747 V1.1.1, Human Factors (HF); Personalization and User Profile Management; Architectural Framework (2009)
11. Kovacikova, T., Petersen, F., Pluke, M., Bartolomeo, G.: User Profile Management – Integration with the Universal Communications Identifier Concept. In: 13th WSEAS International Conference on Communications (2009)
12. Chellouche, S.A., Arnaud, J., Négru, D.: Flexible User Profile Management for Context-Aware Ubiquitous Environments. In: 7th IEEE Conference on Consumer Communications and Networking Conference (CCNC), Las Vegas, pp. 980–984 (2010)
13. Sutterer, M., Droegehorn, O., David, K.: User Profile Management on Service Platforms for Ubiquitous Computing Environments. In: 65th Vehicular Technology Conference (VTC), Dublin (2007)
14. Rankl, W., Effing, W.: Smart Card Handbook. Wiley (2003)
15. Draft ETSI ES 202 642 V0.0.28, Human factors (HF), eHealth; personalization of eHealth systems (2010)
16. Casas, R., Blasco Marín, R., Robinet, A., Delgado, A.R., Yarza, A.R., McGinn, J., Picking, R., Grout, V.: User Modelling in Ambient Intelligence for Elderly and Disabled People. In: Miesenberger, K., Klaus, J., Zagler, W.L., Karshmer, A.I. (eds.) ICCHP 2008. LNCS, vol. 5105, pp. 114–122. Springer, Heidelberg (2008)

Identifying Chronic Disease Complications Utilizing State of the Art Data Fusion Methodologies and Signal Processing Algorithms

John Gialelis, Petros Chondros, Dimitrios Karadimas, Sofia Dima, and Dimitrios Serpanos

Industrial Systems Institute/RC Athena,
PSP Bld, Stadiou St., GR26504 Platani Patras, Greece
{gialelis,serpanos}@isi.gr,
{pchondros,karadimas,sdima}@ece.upatras.gr

Abstract. In this paper a methodology for identifying patient's chronic disease complications is proposed. This methodology consists of two steps: a. application of wavelet algorithms on ECG signal in order to extract specific features and b. fusion of the extracted information from the ECG signal with information from other sensors (i.e., body temperature, environment temperature, sweating index, etc.) in order to assess the health state of a monitoring patient. Therefore, the objective of this methodology is to derive semantically enriched information by discovering abnormalities at one hand detect associations and inter-dependencies among the signals at the other hand and finally highlight patterns and provide configuration rulesets for an intelligent local rule engine. The added value of the semantic enrichment process refers to the discovery of specific features and meaningful information with respect to the personalized needs of each patient.

Keywords: patient monitoring, algorithms, data fusion.

1   Introduction

Heart disease is the most important cause of death in many countries. Thus an automated solution of pervasive heart monitoring is required in order to take care of senior chronic heart patients. The electrocardiogram is an important signal for providing information about functional status of the heart. It shows how fast the heart is beating, whether the heart beat is steady or irregular and the strength and timing of electrical signals as they pass through the heart. The processing of the ECG signal in order to extract specific features of the P wave, the QRS complex, the T wave and ST segment, which in conjunction with other physiological parameters such as blood pressure and sweating index, is of great importance in the detection of cardiac anomalies. Furthermore, it is essential to acquire physiological signals from any type of environment - clinical, domestic, rural and urban - accurately and properly classify them so physicians or medical experts are able to correctly evaluate the patient's

K.S. Nikita et al. (Eds.): MobiHealth 2011, LNICST 83, pp. 256–263, 2012.

health status and perform the appropriate actions. Therefore it is essential to follow an approach using various signals and features in order to measure the same underlying clinical phenomenon: the gradually worsening condition of a chronic patient. The aim is to improve the quality and robustness of the indicator, thus improving sensitivity without causing false alarms. The final goal is to come up with a system that measures a number of parameters from easily applicable low-cost sensors, and simultaneously forms a powerful diagnostic tool. The envisaged high level architecture of such a system is depicted in Fig. 1.

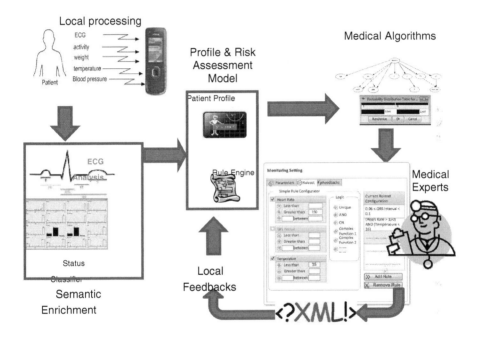

Fig. 1. The envisaged high level architecture of the proposed system

2 Signal Processing Algorithms for ECG Features Extraction

In the clinical domain of patients with chronic heart disease physicians can evaluate patients' health status based on ECG signal analysis. Furthermore, the extraction of ECG features, as they are depicted in Fig. 2, enables the medical experts not only to evaluate heart problems, such as arrhythmia and cardiovascular diseases, but we argue that additional features and meaningful information with respect to the personalized needs of a patient could be discovered.

ECG signals are gathered through appropriate sensors. The measurement of an ECG signal always imposes noise and artifacts within the frequency band of interest.

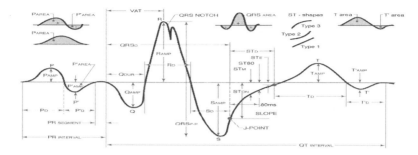

Fig. 2. ECG waveform features

Therefore suitable methods and algorithms have been developed in order to derive the useful information from the noisy ECG signals. Specifically, a preprocessing stage removing noise and/or artifacts from the raw ECG results the starting point of ECG analysis. Wavelet based algorithms can provide efficient localization in both time and frequency [1] thus constitute the most efficient and commonly used approach. The detection of fiducial points [2] which are relevant and strongly correlated to several heart diseases [3] is performed through a parameter extraction stage, as depicted in Fig. 3. This task becomes even more challenging due to the presence of artifacts and time-varying morphology. The key point is to provide a description of the ECG signal in the time-scale domain allowing the representation of the temporal features of a signal at different resolutions.

Fig. 3. ECG signal processing for parameter extraction

Furthermore the basic steps of the ECG signal processing procedure, starting with the QRS complex detection, are depicted in Fig. 4. With respect to the QRS complex, Q and S peaks occur before and after R peak, respectively. R peak is detected on the first scale of Wavelet Analysis. Multi resolution analysis of Wavelet Transform on time-domain ECG is utilized at different Wavelet scales, and the modulus-maxima and zero-crossing approach has been exploited at each resolution level. At the second resolution level, QRS onset and offset (J point) can be identified by using modulus-maxima on the wavelet coefficients. Since wavelet analysis preserves the time domain information of the original signal, the location of these characteristic points (Q, R and S) can be easily obtained from the wavelet domain. Finally, T wave analysis is achieved by considering details at lower frequencies. At the fifth resolution level of wavelet analysis, T wave is detectable using the modulus-maxima based approach on the wavelet coefficients. Appropriate time windows and amplitude thresholds are required in order to get rid of irregular points and misdetections. This is essential so as to provide a robust ECG signal processing methodology.

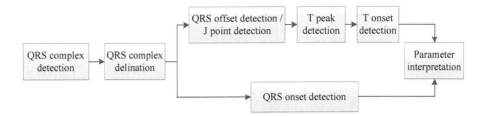

Fig. 4. ECG signal processing procedure

3   Data Fusion Model

An appropriate model describing a fusion process of data obtained from body area sensors, EHR (Electronic Health Records) and/or other historical data has been utilized. Based on these raw data, patient's characteristics and expert's knowledge, a mathematical framework is proposed to extract knowledge. The framework's basic goal is to determine the patient's health status and potentially to be used as a medical decision support tool.

Specifically, medical data, as they may come from different sources, are in different format and types. Typical examples are ECG, medical images, historical data and also habits, such as smoking, exercising etc. Due to the unique nature of these data and in order to extract knowledge, data mining approach is needed as any patterns found should be capable of human interpretation. Moreover relationships or patterns that are extracted may not be commonly accepted or conform to current medical knowledge. Results may indicate an association between the physiological parameters and the class (medical condition of the patient) but there is no implication of cause and effect. The interpretation of these associations is fully up to the experts, so additional visualization and statistical processing is needed to present results in a human interpretable format. Moreover, as the evaluation of the results fully depends on the experts, the proposed methodology can only serve as a decision support tool regarding the prognosis and the diagnosis of the monitoring patient.

There are two main approaches during the design process: supervised learning and unsupervised learning. The basic requirement of classification of patient's health status using supervised learning algorithms is that the dataset must be annotated. The most commonly used algorithms are C4.5 (decision tree approach), Multilayer Perceptron and Naïve Bayes. Unsupervised learning's goal is to determine how a set of data is structured. Clustering and blind source separations are typical unsupervised learning techniques.

Having WEKA [4] available, it was decided to use both supervised and unsupervised machine learning techniques. Besides the typical classification problems, there is an additional need to discover any associations between attributes (physiological parameters). The resulting relations and rules may enlighten hidden interdependencies and provide the experts with meaningful information regarding the patient's health status.

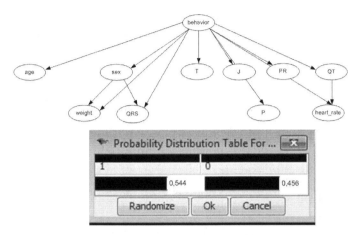

Fig. 5. Screenshot of the Weka tool depicting the developing classifier

In Fig. 5, a screenshot of the Weka tool is depicted. Specifically, the classification process is implemented using NaiveBayes algorithm with 3-parent configuration. The class 'behavior' represents the health status of the patient that can be either normal or abnormal, while the selected attributes that form the annotated dataset are general personal data (age, sex, weight), physiological monitoring parameters (body temperature), as well as extracted ECG features (QRS duration, J point, P wave, Q-T duration, Heart Rate, P-P duration). The training of the model ended up to the structure depicted and indicates some kind of relationship between P wave and P-P duration as well as QT duration and heart rate regarding the ECG extracted features. During testing and run of different algorithms and approaches, it was made clear that results were strongly depended on the dataset we were using, meaning that the algorithms' performance is highly correlated to the size and the quality (missed values) of the used dataset. Therefore, the above findings may trigger the expert for further investigation of the proposed associations outside the scope of the machine learning approach.

4   Methodology for Patient's Personalized Monitoring

The aforementioned data fusion techniques aim to support the doctors and the medical experts in general and also to potentially contribute in discovering additional personalized medical features that will provide information about patients' health status. Under this scope, personalized monitoring of patients could be conclusively accomplished as a workflow mechanism, and lead the doctors to a decision driven by patient's unique parameters as depicted in Fig. 6.

As previously described, the applied data fusion models regularly investigate sensor data along with ECG extracted features for irregularities, associations and inter-dependencies so as to detect abnormal patterns. Once an irregularity or an

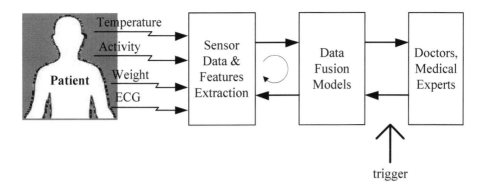

Fig. 6. Workflow mechanism exploiting data fusion assets for conducting doctors personalized monitoring

abnormal pattern is discovered, a trigger signal is provided to the medical expert. After this, the medical expert is able to re-assess the set of rules for the specific patient through a simple set of user interfaces (UIs) as depicted in Fig. 7.

The first step for the re-assessment involves the adding or the subtracting of physiological parameters to be monitored depending on the findings of the data fusion model, designated as "Machine Learning (ML) Findings".

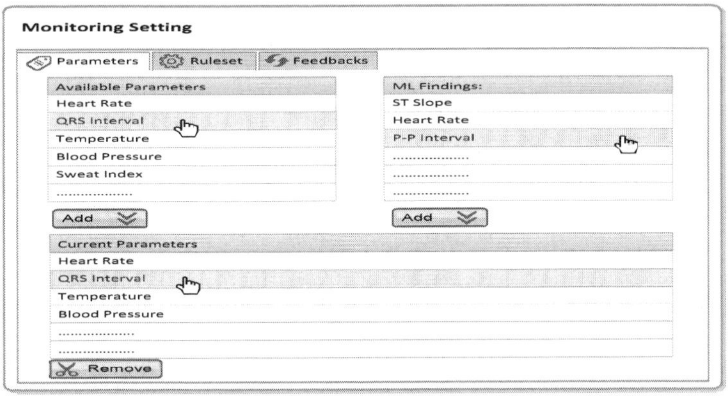

Fig. 7. Adding physiological parameters to be monitored

At a second step, the definition of a set of rules, according to the doctor's expertise and experience, complements patient's treatment towards personalization. The personalization concept lies on the definition of different physiological parameters ranges that correspond to the normal state for each patient. The key feature of personalization is associated with the ability to create rules with the selected physiological parameters as presented in Fig. 8. So far the proposed methodology will provide the doctors and medical experts a ruleset configuration covering almost any

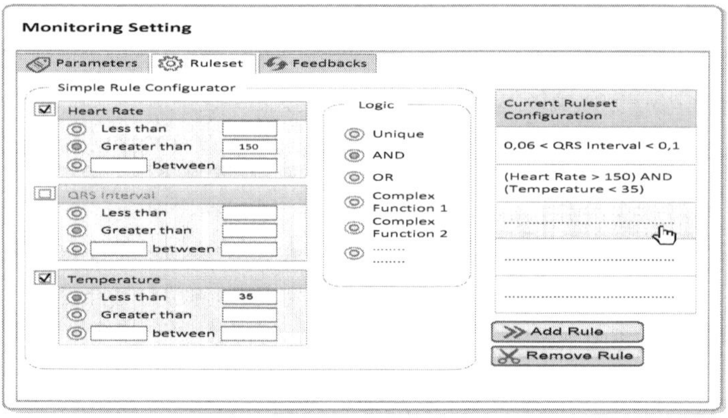

Fig. 8. Creation of simple rulesets

potential irregularity for a given patient, since this ruleset will be constructed based not only on doctors' aspect but also on identified abnormalities issued by the "Machine Learning Findings" as described previously.

Eventually, the constructed ruleset configuration, derived in an interoperable XML-based format such as PMML v4.0 Standard [5], is forwarded to an appropriate, operated on patient's portable device, application that monitors and records his health status. Moreover, doctors and medical experts are also able to predefine a set of feedback actions that will be executed from that application in case of a hit in any of the rules within the defined ruleset, as depicted in Fig. 9.

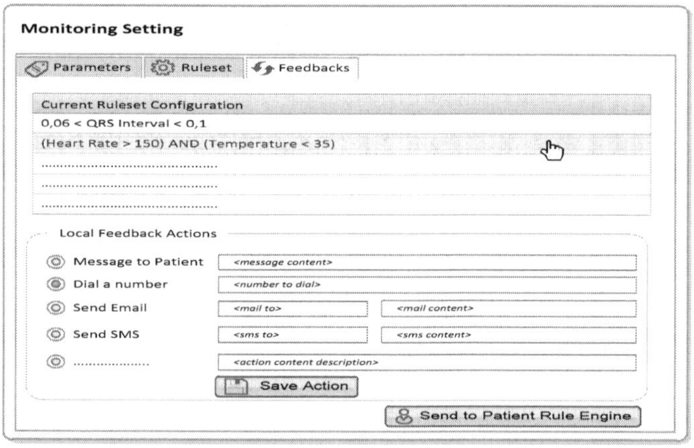

Fig. 9. Defining feedback actions and exporting configuration rulesets

5   Conclusion

In this paper high level system architecture for identifying patient's chronic disease complications is depicted. Furthermore, a methodology was proposed. This methodology which consists of two tasks: application of algorithms on primary sensor readings (e.g. ECG) in order to derive semantically enriched information and a data fusion model that uses the above extracted information as well as data from other sensors (body temperature, environment temperature, sweating index, etc.) in order to assess the health state of a chronic heart disease person. Our goal is to investigate data irregularities, associations and inter-dependencies as well as to detect abnormal patterns and provide interpretation of continuous data which can support an intelligent rule engine machine efficiently.

Acknowledgment. The presented work is taken place in the framework of the ARTEMIS JU Project CHIRON (ARTEMIS - 2009-1 100228 - http://www.chiron-project.eu).

References

1. Li, C., Zheng, C., Tai, C.: Detection of ECG characteristic points using wavelet transforms. IEEE Transactions on Biomedical Engineering 42(1), 21–28 (1995)
2. Martinez, J.P., Almeida, R., Olmos, S., Rocha, A.P., Laguna, P.: A wavelet-based ECG delineator: evaluation on standard databases. IEEE Transactions on Biomedical Engineering 51(4), 570–581 (2004)
3. Choi, S.: Detection of valvular heart disorders using wavelet packet decomposition and support vector machine. Expert Systems with Applications 35(4), 1679–1687 (2008)
4. WEKA, http://www.cs.waikato.ac.nz/ml/weka/
5. PMML v4.0 Standard,
   http://www.dmg.org/v4-0-1/GeneralStructure.html

A Support Vector Machine Approach for Categorization of Patients Suffering from Chronic Diseases

Christos Bellos[1], Athanasios Papadopoulos[1], Roberto Rosso[2], and Dimitrios I. Fotiadis[1]

[1] Foundation for Research and Technology - Hellas, Biomedical Research, Ioannina Greece
[2] TESAN Telematic & Biomedical Services S.p.A., Vicenza Italy
{cbellos,thpapado,Fotiadis}@cs.uoi.gr
rosso@tesan.it

Abstract. The CHRONIOUS system is an open-architecture integrated platform aiming at the management of chronic disease patients. The system consists of a body sensor network collecting patient's vital signals, a Personal Digital Assistance (PDA) for the real-time data analysis based on a Decision Support System (DSS) and a central system for the deeper analysis of patient's status and data storing. The DSS combines several data sources to decide upon the severity of patient's current health status. The first pilot study has been designed and carried out using patients suffering from Chronic Obstructive Pulmonary Disease (COPD). The DSS facilitates a one-against-all multi-class Support Vector Machine (SVM) classification system. The performance of the categorization scheme provides high classification results for most of the patient's health status levels. The involvement of a larger number of patients might increase further the performance of the system.

Keywords: Personalized treatment, Wearable Monitoring, Management of Chronic Diseases, Real-time classification.

1 Introduction

The CHRONIOUS [1] system provides a general patient's management scheme which can be configured and controlled remotely by medical experts or clinicians. The CHRONIOUS system has been designed in order to be flexible so that through the integration of sensors and services covers both patients and caregiver's needs. Its functionalities include the collection of data by sensors which are integrated into a jacket, the control of vital signals, dietary habits and plans, as well as the management and analysis of drug intake, environmental parameters and activity information. Abnormal health conditions are identified using intelligent algorithms running on a portable Personal Digital Assistance (PDA) device and reported to the responsible healthcare professionals supporting decision making and analysis of data.

The flow of data in the proposed system, which is integrated into a mobile application embedded in a PDA device, includes three phases; data preprocessing, analysis and decision making. Data preprocessing is focused on the improvement of

K.S. Nikita et al. (Eds.): MobiHealth 2011, LNICST 83, pp. 264–267, 2012.

signal quality removing several types of noise and artifacts. In the data analysis various types of signals are captured in order to make in real-time a preliminary assessment of the patient's health status. The Data Fusion component fuses all acquired data and forms a multi-dimensional vector to feed the Decision Support System (DSS), triggering the decision making phase.

2   Materials and Methods

The aim of the developed DSS component is twofold; initially constructs the training model and trains the algorithms and afterwards classifies the health episodes according to four different levels of severity on an application embedded in the PDA. A supervised Support Vector Machine (SVM) classifier has been implemented. Apart from the increased performance, SVMs have the ability to handle high dimensional data efficiently, using only a subset of training records, called support vectors, in order to represent the decision boundary.

The specific categorization difficulty that the SVM faces is characterized as multi-class due to the four levels of severity that clinical experts have annotated the health episodes and respectively the available data. The SVMs were originally designed for binary classification [2]. Several methods [3] have been proposed to effectively extend it to multiclass classification where a classifier may be constructed by combining several binary classifiers while others propose to consider all classes at once. In the CHRONIOUS system, the one-against-all SVM classification approach has been implemented.

In the performed analysis three different kernels [4] are used; radial basis function (RBF), Polynomial and Sigmoid and their respective parameters (C, degree and gamma) have been optimized using the differential evolution [5] (DE) method. In order to identify the most important clinical-pathological information for individualized health status identification, two feature selection [6] algorithms, Correlation-based Feature Subset Selection and the Gain Ratio attribute evaluation algorithm were applied, to rank the importance of the features and clinical parameters in the studied disease. The Correlation-based Feature Subset Selection algorithm has been selected from the correlation-based methodology and the GainRatioAttributeEval algorithm has been selected from the ranking methodology. The first algorithm evaluates the worth of a subset of attributes by considering the individual predictive ability of each feature along with the degree of redundancy between them. The latter algorithm evaluates the worth of an attribute by measuring the gain ratio with respect to the class.

3   Results

In the performed analysis we have recorded data from COPD patients using the wearable jacket [8]. The Feature Extraction process has extracted 18 features from 16 patients in a pilot hospital. The applied correlation-based algorithm reduces the number of features to eight, while the ranking algorithm reduces the number of

Table 1. Selected features after the reduction of the initial feature's pool

Feature	Correlation-based algorithm	Ranking algorithm
Mean QR Distance	x	
Mean Heart Rate	x	x
Respiration Rate	x	x
Inhalation Duration	x	
Exhalation Duration	x	x
Oxygen saturation	x	x
Body Temperature	x	
Environmental Temperature	x	x

features to five (Table 1). The features are acquired through a standardized protocol which contains: 6 minutes walking, 45 minutes supine position and 45 minutes standing position [7].

In Table 2 the results of the classification are presented after applying different kernels of the implemented SVM using a dataset which has been constructed using real patients' data.

Table 2. Correctly classified instances and comparison between different kernels

Class - Level of classified severity	Applied kernel	Without Feature selection (%)	Correlation-based algorithm (%)	Ranking algorithm (%)
Level 1	RBF	97	97	**99**
	Polynomial	86	83	84
	Sigmoid	87	88	87
Level 2	RBF	86	87	**91**
	Polynomial	78	78	67
	Sigmoid	65	63	56
Level 3	RBF	90	91	**94**
	Polynomial	86	86	89
	Sigmoid	83	77	81
Level 4	RBF	89	90	**93**
	Polynomial	79	74	73
	Sigmoid	88	86	88

The dataset is randomly split into training and testing dataset using the 10-fold cross-validation method. The levels in the first column of Table 2 represent the different annotated levels of severity as well as the different implemented SVM (since we followed the one-against-all approach, we developed four different binary SVMs). The second column displays the three different applied kernels for each SVM while the last three columns display the percentage of the correctly classified instances for different datasets (entire dataset, optimized dataset after applying the Correlation-based algorithm, optimized dataset after applying the GainRatioAttributeEval algorithm). In this preliminary analysis we notice that the RBF kernel provides more accurate results for the tree different datasets as well as for all four levels which represent the four different developed SVMs.

4   Conclusions

The CHRONIOUS COPD system provides an estimation of the severity of the patient's condition or detects possible critical health episodes. The aim of the developed SVM is to limit the decision error and to increase the system accuracy. In our preliminary analysis we have obtained positive results of the performance and the accuracy of the implemented system.

Comparing the results acquired from the three different datasets that have been formed and facilitated, we conclude that the performance of the RBF kernel followed by the application of Ranking feature selection algorithm provides the highest categorization result. As for the next steps, the employment of larger datasets as well as the implementation of additional classification methodologies will improve the performance of the CHRONIOUS system.

Acknowledgements. This work has been partly funded by the European Commission through IST Project FP7-ICT-2007–1– 216461, www.chronious.eu.

References

1. Bellos, C., Papadopoulos, A., Fotiadis, D.I., Rosso, R.: An Intelligent System for Classification of Patients Suffering from Chronic Diseases. In: 32nd Annual International Conference of the IEEE Engineering in Medicine and Biology Society, Buenos Aires, Argentina (2010)
2. Hsu, C., Lin, C.: A Comparison of Methods for Multiclass Support Vector Machines. IEEE Transactions on Neural Networks 13(2) (2002)
3. Kumar, M., Gopal, M.: Reduced one-against-all method for multiclass SVM classification. Expert Systems with Applications (2011), doi:10.1016/j.eswa.2011.04.237
4. Ayat, N.E., Cheriet, M., Suen, C.Y.: Automatic model selection for the optimization of SVM kernels. Pattern Recognition 38, 1733–1745 (2005)
5. Efrén, M.M., Jesús, V.R.: A Comparative study of Differential Evolution Variants for Global Optimization. In: GECCO 2006, pp. 485–492 (2006)
6. Taniar, D.: Data Mining and Knowledge Discovery Technologies. part of the IGI Global series named Advances in Data Warehousing and Mining (ADWM) vol. 2 (2007)
7. Bellos, C., Papadopoulos, A., Rosso, R., Fotiadis, D.I.: Extraction and Analysis of features acquired by wearable sensors network. In: 10th International Conference on Information Technology and Applications in Biomedicine, Corfu, Greece (2010)
8. Papadopoulos, A., Fotiadis, D.I., Lawo, M., Ciancitto, F., Podolak, C., Dellaca, R.L., Munaro, G., Rosso, R.: CHRONIOUS: A Wearable System for the Management of Chronic Disease. In: 9th International Conference on Information Technology & Applications in Biomedicine, Larnaca, Cyprus (2009)

Combined Health Monitoring and Emergency Management through Android Based Mobile Device for Elderly People

Miklos Kozlovszky[1], János Sicz-Mesziár[2], János Ferenczi[2], Judit Márton[2], Gergely Windisch[2], Viktor Kozlovszky[2], Péter Kotcauer[1], Anikó Boruzs[2], Pál Bogdanov[2], Zsolt Meixner[2], Krisztián Karóczkai[1], and Sándor Ács[1]

[1] MTA SZTAKI, LPDS,
Kende str. 13-17, H-1111, Budapest, Hungary
[2] Obuda University , John von Neumann Faculty of Informatics, Biotech group
Bécsi str. 96/b., H-1034, Budapest, Hungary
{m.kozlovszky,kotcauer,karoczka,acs}@sztaki.hu,
{sicz-mesziar.janos, windisch.gergely}@nik.uni-obuda.hu,
{marton.judit, kozlovszky.viktor,boruzs.aniko, bogdanov.pal,
meixner.zsolt}@biotech.uni-obuda.hu, john.ferenczi@citromail.hu

Abstract. We have developed a combined Android based mobile data acquisition (DAQ) and emergency management solution, which can collect information remotely from patient and send the information towards to the medical data and dispatcher centre for further processing. The mobile device is capable to collect information from various sensors via Bluetooth and USB connection, and further more able to capture and forward manually initiated alarm signals in case of an emergency situation. Beside the alarm signal the system collects and sends information about the patient's location, and it also enables two ways audio communication between the central dispatcher and the patient automatically. The developed software solution is suitable for different skilled users. Its user interface is highly configurable to support elderly persons (high contrast, huge characters, simple UI, etc.), and also provides advanced mode for the "power" users. The developed system becomes part of our testing program, which is carried out in our Hungarian Living Lab infrastructure. The combination of a mobile DAQ device and mobile emergency alarm device within a single software solution enables care givers to provide better and more effective services in elderly patient monitoring.

Keywords: elderly people monitoring, Living Lab, combined mobile data acquisition and emergency alarm software.

1   Introduction

With mobile devices used as data acquisition (DAQ) systems we are able to collect vital information about the elderly and demented patients remotely. Through the co-operation of commercial companies, universities and other non-profit organizations the direct goal of the AALAMSRK [2] project is to develop an integrated,

K.S. Nikita et al. (Eds.): MobiHealth 2011, LNICST 83, pp. 268–274, 2012.
© Institute for Computer Sciences, Social Informatics and Telecommunications Engineering 2012

standardized dementia and health monitoring system (ALPHA system) supported by innovative, modern measurement and info-communication technologies. By the integration of medical expertise and developing assisted living patterns (ALPs), the realized system offers personalized monitoring solution for monitoring and prevention of elderly people, particularly who suffer from neurological diseases such as stroke, dementia or depression.

Considering the real social and market demands and the needs of the health care service provider segment [1], the general project aim is to improve the quality and cost effectiveness of health care services by developing service models, methods, tools, products and services. A consortium led by GE Healthcare - a unit of General Electric Company-, also includes two Hungarian healthcare companies Mednet 2000 Ltd. and Meditech Ltd., and three universities: the University of Pannonia, the University of Szeged and the Obuda University. The consortium is doing research and development of remote telemonitoring system that monitor both activity levels and vital signs such as blood pressure and heart rate, alerting caregivers about potential health issues or emergency situations.

2   Living Lab Infrastructures

Main novelties of the AALAMSRK project are that it brings into the patient's home the medical knowledge and assistance and also it supports new potential opportunities to capture insight medical knowledge with its effective non-stop health monitoring methods. The monitoring is done by the standardized, well-defined environments (so called Living Labs). The Living Labs are supporting all the R&D tasks of the medical, engineering and business (marketing) work packages and also provides evaluation and test environment for new hypotheses and results. Beside a normal full functional HomeHub system which is running on a commodity PC, we have developed an Android based HomeHub using a mobile device (See Fig.1.). This

Fig. 1. In our Living Labs PC based and mobile HomeHubs are collecting sensor data

mobile HomeHub is targeting only limited functionalities of the full solution (due to the smaller screen size and fewer hardware interfaces), but it can extend the usability with additional special features, such as mobility, location awareness and small size.

2.1   Living Labs

The established Living Lab environments are located in three different regions within Hungary (capital city, middle size city and rural area), thus the type of patient environments (living space size, accessibility, communication infrastructure, etc.) are totally inhomogeneous (see Fig. 2.).

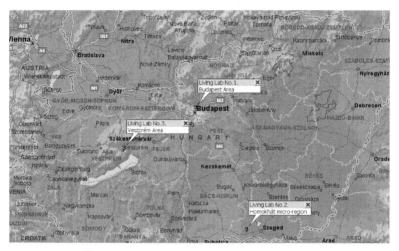

Fig. 2. Used Living Lab infrastructures (No.1.:metropolitan, N.o.3: city, N.o.2:rural)

During the test periods we have learned a lot from the different environment features, and revealed many aspects of various issues concerning sustainability, usability, etc.

3   Mobile and Station like DAQ Systems – a Comparison

In the homes of each monitored patient different type of sensors have been deployed and -with our android mobile and PC based HomeHub software solution- we are collecting information from these sensors about:

- patient movement within the house, with the usage of wall mounted sensors
- patient medication, with sensors of the medicament dispenser
- patient's eating habit, with sensors placed on the refrigerator
- patient activity with so called Actigraph, which is a watch like sensor on the patient's wrist

- patient's blood sugar level
- patient's blood pressure
- patient's weight

Both android mobile based and PC based HomeHub solutions provide:

- Health status visualization (limited) at the HomeHub
- Multi-language support at the HomeHub
- Silent sensor DAQ mode (automatic sensor data collection) via Bluetooth

General features of the PC based software solution are:

- Manual DAQ mode (optionally GUI initiated sensor data collection) via USB or Bluetooth or Zigbee
- Automatic data compression and encryption during data transmission towards the data center
- Automatic sensor data pre-evaluation at the HomeHub
- Health status visualization (with statistics and data mining facility) at the data center
- Emergency alarm

General features of the android mobile based software solution are:

- Manual DAQ mode (optionally GUI initiated sensor data collection) via Bluetooth
- Automatic data compression and encryption during data transmission towards the data center (limited)
- Automatic sensor data pre-evaluation at the HomeHub
- Full featured mobile emergency alarm

4   Mobile Device User Interface

The software is suitable for different skilled users. Its user interface is highly configurable to support elderly persons (high contrast, huge characters, simple UI), and provides advanced mode for the "power" users. We have identified and defined during our Living Lab experiments multiple user skills and hw/sw utilisation levels:

- Elderly persons without any IT knowledge - the front-end can be configured to hide completely the underlying mobile device (all menus and icons, see Fig 3.), in such case the device can act as an intelligent mobile emergency signal device, which can be called from the central dispatcher. Such oversimplified mobile HomeHub, collects and forwards sensor information in stealth mode in the background and contains on its screen only a big red panic button. The emergency alarm can be initiated by a pre-defined utilisation pattern (long button push for 3-5 seconds, repetitive button push for 3-5 times).

Fig. 3. Screens of the mobile Android based HomeHub solution

- For normal and expert end users the in-build additional sw/hw functionalities of a normal Android mobile phone are available (menu sets, SMS , dialing, applications, etc.).

5   Location and Sudden Event Monitoring

In a sudden panic situation the patient can manually (or in future based on some of the sensor data even automatically) activate an alarm with the mobile device. When an alarm signal initiated the central dispatcher is able to receive information about location from the received GPS and GSM/GPRS cell information immediately after the alarm signal is arrived. The automatically established two way voice communication can help to understand the context of the sudden event and refine or drive the problem solving procedure. The dispatcher center has access to all the information about the patient (his/her health/medication status, location, etc.), which is vital in emergency situations.

Some features of the developed software solution are:

- Emergency call soft key
- Automatic location data forwarding (GPS coordinates + GSM/GPRS cell information) via SMS and/or data channel
- Emergency phonebook (with multiple entries)
- Automatic call pick up after emergency call
- Voicemail prevention during emergency call
- Automatic event logging
- Compatibility with large number (about 97%) of available Android based mobile devices (Android version 2.1, and above)

5.1   Emergency Signal – in the Panic Situation

The initiated emergency alarm information is automatically forwarded to the data center/dispatcher. The mobile based HomeHub provides location information of the event. The dispatcher service agent receives this information parallel within a predefined SMS and within a web service call. Both the SMS and the web service call contain: timestamp, the user ID, the phone's IMEI number, the GPS coordinates and the available cell information. The back end of the web service is appending the event data with additional information (google maps link) and sends towards to the dispatcher's Living Lab event tracking subsystem [3] (See Fig. 4.).

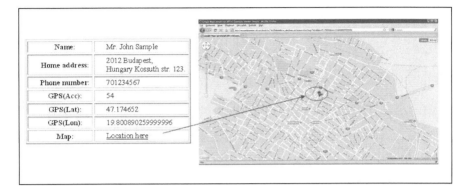

Fig. 4. Received information by the dispatcher about the panic event (via email)

6   Evaluation of the System

Both the mobile and the PC based HomeHub have been tested on various levels from different perspectives by different purposes. Functionality tests, security and performance test have been carried out to provide quality control of the developed software solutions. During our system evaluation we have redesigned the whole HomeHub-data center communication, because the mobile android based HomeHubs have difficulties to use the same JAVA communication interfaces due to the limited available API. In the mobile HomeHubs the central databases are accessed directly through a lightweight web service like interface, and the

Usability tests in our Living Lab environments have been used extensively to receive feedbacks from patients. As a result, in the mobile based HomeHub solution we had to monitor not only the status of the software, but also some mobile hardware specific parameters remotely (such as: battery level), and we had also to redesigned the whole user interface of the handheld device to support elderly persons with low IT skill sets. According to the distributed and collected/evaluated surveys our android based mobile HomeHub solution is capable to provide seamless remote monitoring of elderly persons not only at home, but also abroad. It provides important feedbacks about health status to the patient, and opens up a seamless, location aware, reliable mobile communication channel in emergency situations. As future work we are trying to include fall detection to capture emergency situations and activate an alarm with the mobile device automatically.

Acknowledgments. This work makes use of results produced by the AALAMSRK project (Research & development of assisted living patterns and their integration into decision support system to promote the quality of life at home), Hungarian National Technology Programme, A1, Life sciences, the "Development of integrated virtual microscopy technologies and reagents for diagnosing, therapeutical prediction and preventive screening of colon cancer "Hungarian National Technology Programme, A1, Life sciences, (3dhist08) project and the ÓE-RH 1104/2-2011 project. Authors would like to thank for their financial support hereby.

References

1. Wimo, A., Jönsson, L., Gustavsson, A.: Cost of illness and burden of dementia in Europe - Prognosis to 2030 (October 27, 2009), http://www.alzheimer-europe.org/ Our-Research/European-Collaboration-on-Dementia/ Cost-of-dementia/Prognosis-to-2030
2. Proseniis- Innovation for healthier senior Age, http://www.proseniis.hu/ (acc. August 9, 2011)
3. Kozlovszky, M., Meixner, Z., Windisch, G., Márton, J., Ács, S., Bogdanov, P., Boruzs, A., Kotcauer, P., Ferenczi, J., Kozlovszky, V.: Network and service management and diagnostics solution of a remote patient monitoring system. In: Lindi 2011 Conference, Budapest, Hungary (2011)

An ISO-Based Quality Model for Evaluating Mobile Medical Speech Translators

Nikos Tsourakis[1] and Paula Estrella[2]

[1] ISSCO/TIM/ETI, University of Geneva, Switzerland
[2] FaMAF, Universidad Nacional de Córdoba, Argentina
Nikolaos.Tsourakis@unige.ch, pestrella@famaf.unc.edu.ar

Abstract. Medical translation systems present an intriguing research area as language barriers can become life-threatening when health issues come into place. There is however a lack of common evaluation techniques, making the fair comparison of such systems a difficult task. In this work we try to remedy this deficiency by proposing a quality model based on the ISO/IEC 9126 standard that could serve as a comparison basis among homologous systems. We focus on the mobile world believing that it suits patients' needs better, as they experience diverse scenarios along the pathway to healthcare. Our work involves the definition of the quality characteristics of the model along with the quantification of their importance based on two target groups of users (12 doctors and 12 potential patients) that demonstrate different needs and goals towards the system.

Keywords: Medical Translation, Quality Model, ISO 9126, Mobile Translators.

1 Introduction

Language barriers often cause inconvenience but when medical issues are involved can become life-threatening. Quantitative studies, e.g. [1], have shown that lack of a common doctor-patient language correlates with an increased probability of negative outcomes. Unfortunately, trained medical translators are both scarce and expensive. Even if a universal speech-to-speech translator still seems an insurmountable problem, the substantial gap between the need for and availability of language services in health care could be bridged through effective medical speech translation systems, such as [2]; a system like this would be far more useful to users if it was available on a hand-held device. Indeed, different systems already are efforts towards the deployment of mobile speech-to-speech translation applications [3], [4], [5].

During the lifecycle of these systems authors provided evaluation results leveraging various computer and human centered metrics. Despite some early efforts towards a common evaluation framework [6] we argue that there is a lack of such methodology that would provide a fair comparison framework for different mobile medical translation systems. Additionally, the lack of appropriate quality assessment techniques can deteriorate user satisfaction. As quality is hard to assess and assure,

K.S. Nikita et al. (Eds.): MobiHealth 2011, LNICST 83, pp. 275–283, 2012.

several models try to address software quality issues by employing a set of quality attributes, characteristics and metrics [7], [8], [9], [10]. In this work we discuss how to evaluate mobile medical translators with a quality model based on ISO/IEC 9126 [11]. Unlike other models, it enjoys the benefits from being an international standard and as it is generic, it can be applied to any kind of software product.

Our work had two stages. Initially we had to create the quality model per se, defining the quality characteristics that constitute the model, either by selecting them among those proposed in ISO/IEC-9126 or by introducing new ones. In the second phase we asked two target groups of users that demonstrate different needs and goals towards the system (12 doctors and 12 potential patients), to quantify their preferences concerning which attributes (i.e. quality characteristics of the model representing desired features of a system) are more important.

The paper is organized as follows: in Section 2 we discuss how a system like this should be realized in a hospital environment. In Section 3 we decompose the model according to our case study. Section 4 presents our methodology for ranking the quality characteristics and Section 5 summarizes the results along with a short discussion. The final section concludes.

2   The Pathway to Healthcare

The path to healthcare as described in [12] may involve different stages besides the typical diagnosis scenario between the doctor and the patient. As illustrated in Fig. 1, we can imagine a patient interacting with other staff in the hospital, for example with a secretary at the welcome reception desk, with a nurse during an examination procedure or hospitalization, etc. All these diverse scenarios indicate just the gamut of possible situations.

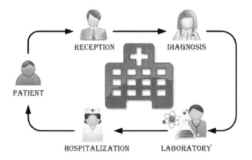

Fig. 1. A typical pathway to healthcare

A fundamental question introduced by Somers [12] is: "Who is the primary user of such a system, the physician or the patient?" On the one hand, there is the doctor who usually has high level education and interacts with the system on a daily basis and conversely, the patient who may use the system solely once in his life. There is no single answer to this question: while many efforts have put the doctor in charge of the dialogue e.g. [2], [13], others have followed a parity oriented approach, where two separate graphical user interfaces are offered for each one of the two parties [5]. Thus, we should address the following:

- Quality of translation. The genre of the task requires safe critical high quality translation.
- Heterogeneous interaction. Users may interact with different personnel or in diverse environments.
- Mobility. Interaction happens with a mobile device, which per se involves special consideration.
- User physical constrains. Patient's physical disabilities can pose hurdles to the efficient usage of the system.
- Wireless interconnection. The network can cause delays or even connection failures.
- Application availability. It could be preinstalled on a hospital's device or users could install it on their own device.

3   Decomposition of Our ISO Model

A generic quality model is proposed by ISO/IEC 9126 [11] for the evaluation of any software product, thus it can also be used in the evaluation of mobile medical speech translators. We described each external attribute with a friendly, application-specific statement to help doctors or potential users first understand attributes and then weight them; this is based on the hypothesis that these users are not necessarily familiar with the ISO terminology. In the second phase the quality characteristics were compared in pairs and we extracted the corresponding weights by adopting a methodology similar to [14], which uses a mutual comparison method [15] and multi-criteria Analytical Hierarchy Process (AHP) [16]. The customized definitions are summarized below; unless otherwise stated all the quality characteristics were adapted form the ones proposed in ISO/IEC 9126.

1.   Functionality

Suitability. The system can be seen either as a replacement when no interpreters are available or as a palliative before resorting to an interpreter. It should therefore support as many languages, domains and diverse usage environments as possible.

Accuracy. In this specific application domain the translation between languages needs to be produced in the most reliable and robust way, achieving high quality.

Interoperability. The system should facilitate interoperability of the different nodes in the pathway, e.g. information about prescribed medication or treatment must be available to the corresponding personnel or system.

Security. The system should guarantee that the information gathered during the interaction is stored and accessed in a restricted manner. Treatment of sensitive medical data should be carefully considered.

Traceability (added attribute). User's activity along the pathway should be traced by the system and may be used to identify possible problems (e.g. delays), perform correct pricing of praxes, etc.

Exploitability (added attribute). Users may be forced to wait their turn for an examination or wait between examinations. This idle time can be used for familiarizing with the application, and thus fostering user's trust.

Controllability (added attribute). If the doctor gives instructions he should also be control of the dialog flow but during diagnosis the weight of control should be equilibrated between the two parties. This also conforms to current clinical theory of patient-centered medicine [17].

2. Reliability

Maturity. The special nature of the application demands zero faults therefore the system should aim to minimize the frequency of failures.

Fault tolerance. In case of any faults the system should resort to a backup plan, e.g. trained personnel could take over control and interact with the patient.

Recoverability. The software should be able to recover after a failure either by incorporating a logging mechanism or by storing user's data locally or remotely.

1. Usability

Understandability. Users should understand what the system is supposed to do. Short introductions should be provided along with context dependent prompts. Cultural limitation or physical disabilities should also be addressed.

Learnability. As in any spoken dialogue application, it should give users immediate feedback on the system's intended coverage, particularly when recognition fails.

Operability. As the end user may use a system like this only once in his life the interaction should be based on simplicity.

Attractiveness. Due to the limited lifecycle of the application (used only in a hospital environment), the issue of attractiveness becomes of lesser importance. However, the success of the system may depend on relevant factors.

Uniformability (added attribute). Following the path of healthcare each user should experience a uniform interaction. This will minimize the effort of learning how the system works in different situations and will cause less confusion.

Trustability (added attribute). The system should offer results that are predicable and don't engender any surprise to end-users, so that patients establish trustful relations towards the system.

Customizability (added attribute). As users may vary in a range of literate to complete illiterate, the system should take this into account as well as other special needs (weak sight, hearing problems, etc).

Privacy (added attribute). The system should contemplate issues of privacy, e.g. patients can be reluctant to talk in front of other, even in front of relatives, be embarrassed when using the system unsuccessfully, etc.

2. Efficiency

Time behavior. Time management is very important as the diagnosis should be made as quick as possible. It should also be dependant on the usage scenario, as it may be more urgent to complete a task at the reception than at the laboratory.

Resource utilization. The system should target to efficient utilization of resources (e.g. battery life, wireless connectivity, data access) and also fair sharing among users.

3. Maintainability

As our work is pertinent to external evaluation we won't delve into these attributes, which reflect mainly a technical (i.e. internal) viewpoint, such that of developers.

4. Portability

Adaptability. It should be adaptable to a number of platforms. Proprietary solutions may narrow the possible options, so the design should take into account forthcoming technologies and open source alternatives.

Installability. If end users decide to install the system on their own device this should be as transparent as possible considering that the application's life time could be limited to just the time the patient stays in the hospital.

Co-existence. The system should successfully co-exist with other independent systems working in a common environment and sharing common resources. Issues of conflicts may include the bandwidth usage, interference problems, other wearable medical devices, etc.

Replaceability. As most of the times upgrading the existing software should not be performed by end-users, issues of replaceability are not a subject of their concern.

5. Compliance (for all characteristics)

In our scenario it could happen that the interaction in environments that impose zero noise level may be prohibited, thus being subject to specific hospital regulations. Furthermore, issues related to interference in specific areas should be considered.

4   Relative Importance of Attributes

The relative importance of each quality characteristic in the model is dependent on the user and, as expected, user's perception about product quality varies across user types. For example, end users typically value usability more than developers do. In order to quantify users' preferences we polled two groups that have different needs and goals towards the system. The first group included 12 professional doctors having a different specialization background (excluding specializations that don't involve direct contact with patients) and a second group of 12 non-doctors with different higher academic background. All participants were between 20-40 years old and gender was approximately balanced across conditions.

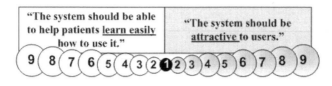

Fig. 2. Learnability vs. attractiveness

For the construction of the survey we considered the fact that participants have limited or no experience with speech-to-speech translation systems, they have no familiarity with the ISO hierarchy and terminology and that they have a busy schedule, so we limited the time devoted to the survey to around 15-20 minutes. Participants were asked to express their opinion by choosing a number in a scale of 1-9 favoring the feature they liked most. An example is shown in Fig 2.

The mutual comparisons were limited within attributes in the same category. For n characteristics at a given category $n(n-1)/2$ mutual comparisons are needed (e.g. 21 comparisons for *functionality*). Our analysis was based on the Analytical Hierarchy Process [16], which shapes a problem in a hierarchical structure. Accordingly, we shape our ISO model in three levels, where the goal (first level) is to evaluate a mobile speech translator; the quality characteristics of the model constitute the second level (e.g. *functionality*) and the sub-characteristics the third one (e.g. *accuracy*).

5   Results and Discussion

The results of applying the AHP are presented in Table I. For the first group (doctors) the weights are depicted in the left side and for the second group (patients) in the right one with the grey background. From the high level attributes of the model *functionality* seems to be the most important for physicians (38.53%) followed by *reliability* (18.04%) and *usability* (17.47%). For patients *efficiency* shows the highest weight (30.32%) and surprisingly *usability* the lowest (9.53%). One explanation could be that patients value features related to performance (e.g. response time) more than the ones related to ease of use. Paradoxically, *compliance* (with hospital regulations) is considered more important by patients than by doctors.

At the second level, results corroborated our intuition that *accuracy* is of utmost importance for both target groups (26.29% and 39.17%); it is followed by *security* (22.27% and 17.01%) and last by *exploitability* (5.40% and 4.26% respectively). Also, physicians seem to care about privacy issues more than patients do (22.63% vs. 14.9%), whereas the latter prioritize *customizability* (related with users' special needs). *Attractiveness*, receives the lowest rank among all evaluators.

Another interesting finding is that physicians consider *co-existence* very important, as a newly introduced system shouldn't affect the systems already deployed. Patients on the other hand prioritize the *replaceability* of the system, despite the fact that they are normally not involved in this process. Finally, both groups agree on the sub-attributes of *efficiency* (clearly favoring *time behavior,* the ability of the system to respond quickly), and they also adopt the same stance for *reliability* prioritizing the elimination of failures (*maturity*). Moreover patients seem to consider the *recoverability* more important than the *fault tolerance.*

Table 1. Weighted Quality Model

Quality factor	Weight %		Quality sub-factor	Weight %	
	Doctors	Patients		Doctors	Patients
Functionality	38.53	10.88	Suitability	18.18	12.28
			Accuracy	26.29	39.17
			Interoperability	10.38	13.37
			Security	22.27	17.01
			Traceability	9.26	8.61
			Exploitability	5.40	4.26
			Controllability	8.22	5.30
Reliability	18.04	13.47	Maturity	55.80	48.70
			Fault tolerance	38.50	6.20
			Recoverability	5.70	45.10
Usability	17.47	9.53	Understandability	4.31	7.63
			Learnability	10.24	9.61
			Operability	15.83	17.42
			Attractiveness	2.37	3.30
			Uniformability	9.43	3.97
			Trustability	11.87	19.91
			Customizability	23.32	23.26
			Privacy	22.63	14.90
Efficiency	5.64	30.32	Time behavior	78.38	69.70
			Resource utilizat.	21.62	30.30
Portability	6.40	15.74	Adaptability	22.14	18.95
			Installability	14.31	15.54
			Co-existence	40.08	26.65
			Replaceability	23.47	38.86
Compliance	13.92	20.06			

As in every human assessment, coherence and consistency are an important matter. This issue is taken into care of calculating a *Consistency Ratio* (CR) provided by AHP (the lower the better), which quantifies how much the evaluators' judgments fulfill the *transitive* property (i.e. if a>b and b>c then a>c). Initially, we started with more than 12 subjects in each of the two target groups; the survey was sent to 16 doctors and 20 potential patients. However, when calculating the CR on the entire dataset for each group we found its CR to be too high, risking useless results. There is also a correlation of individual consistency with the overall consistency of the model. As we eliminated participants with the highest CR, the overall CR (the one obtained after averaging their answers) dropped to an acceptable 10% (approximately). Hence, the study was limited to the 12 most consistent physicians and to the 12 most consistent patients.

It is worth noting that not every subject chosen for the study had CR < 10%. We believe that this is strongly dependent on the number of items under comparison. *Usability* for example, demands 28 pair-wise comparisons, hindering consistent subjective judgments. Another potential source of inconsistency is the formulation of each statement: for high level characteristics the statements embedded multiple concepts and it was there where we encountered most of the inconsistencies. We also observed that physicians exposed lower levels of inconsistency.

Finally, upon completion of the survey each participant was asked to express his/her opinion on different topics, to propose enhancements or to point out deficiencies. In Fig. 3 we present the answers to four of the questions, being *intention of usage, intention of buying, preference over human interpreters, efficient use of system by patients.* Even if both groups seem eager to use a system like this; patients seem reluctant to buy it (only 31% are positives). Less than 31% in both target groups express a clear preference (by answering "Yes") for the system over a human interpreter and lastly, less than half of the participants believe that the system can be used efficiently by the patients.

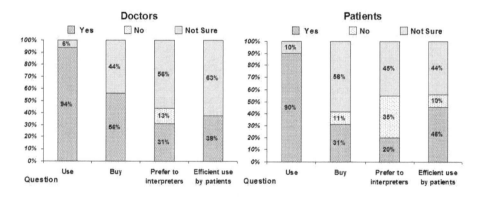

Fig. 3. Subjective opinions of both target groups

6   Conclusions

We defined an ISO quality model suited for mobile medical speech translation systems, which can help evaluators, compare similar systems on a common evaluation ground and could also help developers focus on those aspects of quality that users deem important.

We tried to address some issues related to the design and the implementation of surveys for acquiring the relative importance of attributes and sub-attributes (i.e. weights). We also provided some guidelines that might be useful to others intending to use a similar protocol: formulation of statements, number of comparisons, understanding the tenor of the problem, number of participants and scale for the comparisons, are some of the factors that should be carefully considered.

Finally, the next step of this work involves further decomposing each sub-characteristic if necessary and the accumulation of relevant metrics. Additionally, including weights for the internal attributes is important for the completeness of the model.

References

1. Flores, G.: The impact of medical interpreter services on the quality of health care: A systematic review. Medical Care Research and Review (2005)
2. Bouillon, P., Flores, G., Starlander, M., et al.: A Bidirectional Grammar-Based Medical Speech Translator. In: SPEECHGRAM Workshop, Prague Czech Republic (2007)
3. Zhang, Y., Vogel, S.: PanDoRA: a large-scale two-way statistical machine translation system for hand-held devices. MT Summit, Copenhagen, Denmark (2007)
4. Gao, Y., Zhou, B., Zhu, W., Zhang, W.: Handheld Speech to Speech Translation System. Automatic Speech Recognition on Mobile Devices and over Commun. Networks (2008)
5. Tsourakis, N., Bouillon, P., Rayner, M.: Design Issues for a Bidirectional Mobile Medical Speech Translator. In: SiMPE Workshop, Bonn, Germany (2009)
6. Rayner, M., et al.: A Small-Vocabulary Shared Task for Medical Speech Translation. In: Workshop on Speech Processing for Safety Critical Translation, Manchester, UK (2008)
7. Chidamber, S.R., Kemerer, C.F.: A Metrics Suite for Object Oriented Design. IEEE Transactions on Software Engineering (1994)
8. Wakil, M.E., Bastawissi, A.E., Boshra, M., Fahmy, A.: Object Oriented Design Quality Models – A Survey and Comparison. In: Int. Conference on Informatics and Systems (2004)
9. Hyatt, L.E., Rosenberg, L.H.: Software Metrics Program for Risk Assessment. Elsevier Acta Astronautica 40, 223–233 (1997)
10. Bansiya, J., Davis, C.G.: A Hierarchical Model for Object-Oriented Design Quality Assessment. IEEE Transactions on Software Engineering 28, 4–19 (2002)
11. ISO/IEC. 2001. ISO/IEC 9126-1:2001 (E) - Software Engineering - Product Quality - Part1: Quality Model (2001)
12. Somers, H.: Theoretical and methodological issues regarding the use of Language Technologies for patients with limited English proficiency. TMI (2007)
13. Narayanan, S., et al.: The Transonics spoken dialogue translator: An aid for English-Persian doctor-patient interviews. Dialogue Systems for Health Communication (2004)
14. Behkamal, B., Kahani, M., Akbari, M.: Customizing ISO 9126 quality model for evaluation of B2B applications. Journal of Information and Software Technology (2008)
15. Firyaki, F., Ahlatcioglu, M.: Fuzzy stock selection using a new fuzzy ranking and weighting algorithm. Applied Mathematics and Computation Journal (2005)
16. Saaty, T.L.: The Analytical Hierarchy Process, 2nd edn. McGraw-Hill, New-York (1994)
17. Stewart, M., Brown, J.B., Weston, W.W., McWhinney, I.R., McWilliam, C.L., Freeman, T.R.: Patient-centered medicine: Transforming the clinical method (2003)

An Interdisciplinary Approach to Emergency Responder Mobile Technology Design

Patricia Collins[1] and Sean Lanthier[2]

[1] Carnegie Mellon University Silicon Valley, Moffett Field, CA 94035
[2] IC[3], LLC, Los Altos, CA 94022
`patricia.collins@sv.cmu.edu`, `seanlanthier@me.com`

Abstract. Using an interdisciplinary approach to requirements engineering for an emergency responder support system ensures that system capabilities as well as such considerations as usability and utility are based on the clear needs of emergency responders and other stakeholders. The described methodology integrates brainstorming, competitive assessment, user interviews, scenario development, use case development, mobile device user interface mockups, and emergency responder validation of the resulting requirements. The approach was employed to consider diverse stakeholders in emergency response situations.

Keywords: first responder, emergency responder, paramedic, interdisciplinary design, requirements engineering, emergency response, disaster response.

1   Introduction

Getting requirements right for emergency responders mobile technology needs can save lives and reduce injuries. Therefore, it is worth the effort to take an interdisciplinary approach to gathering, validating, and analyzing candidate requirements for products and services that aid emergency responders.

Carnegie Mellon University Silicon Valley (CMUSV) has undertaken such an approach to requirements engineering for emergency responder mobile technology design, following an Agile software engineering methodology. The aspects of the methodology which meet the Agile guidelines [3] include ongoing customer collaboration, rapid prototyping, accommodation of change, and a focus on individuals and their interactions in a high-performing team.

1.1   Software Engineering

Laplante notes that software engineering is a "systematic approach to the analysis, design, assessment, implementation, test, maintenance and reengineering of software, that is, the application of engineering to software."[9] It differs from computer science as an academic discipline in that software engineering is focused on methodologies, processes, and techniques for each aspect of the lifecycle, while computer science is

K.S. Nikita et al. (Eds.): MobiHealth 2011, LNICST 83, pp. 284–291, 2012.
© Institute for Computer Sciences, Social Informatics and Telecommunications Engineering 2012

the study of theoretical concepts in computer-based computation and information management. The study of emergency responder mobile technology needs took place within the context of an Agile software engineering methodology. This paper describes only the requirements engineering aspect of that lifecycle.

1.2   Requirements Engineering

Software requirements engineering includes the acquisition, validation, analysis, selection, and managed evolution of software requirements for targeted software-based systems. In Agile approaches to software engineering, requirements engineering is an ongoing process, with refinements and modifications occurring just in time for implementation of the corresponding part of the system. However, most Agile approaches do not explicitly address the initial exploration of requirements, which is part of the product visioning process. [12] The early development of prioritized candidate requirements is, nevertheless, essential to understanding the user's needs. CMUSV focused its requirements engineering efforts on the identification of requirements for a product line that would support emergency responders with mobile technology. Towards this end, the team engaged in an interdisciplinary approach to requirements engineering. This approach enabled small requirements engineering teams to gain a well-rounded understanding of the domain and the potential product space, before launching into Agile development.

2   The Interdisciplinary Approach

An interdisciplinary approach is best matched with the goals of deep understanding of design requirements. User-centered design often focuses on interviews with and observations of the target users. [4, 8] This narrowly scoped approach does provide a clear understanding of the user's current work flow, task sequences, and work environment. However, the opportunity to discover creative and innovative solutions is often limited to addressing the identified "breakdowns" in the current way of doing things. To support more innovative solutions, one can combine an assortment of requirements engineering approaches. The following subsections describe the interdisciplinary nature of the various approaches used to discover and validate software requirements.

2.1   Competitive and Analogous System Analysis

Requirements engineers (REs) can begin their investigation by looking at competitive systems. Using an analytic approach, the RE reviews the brochures, technical specifications, and other information about each competitive system. In the case of mobile technology for emergency medical responders, the current market is mostly limited to isolated applications. Therefore, the analysis involved studying partial solutions and identifying candidate features for a more comprehensive solution. [10] In the early stages of the competitive analysis, the REs added all relevant features to the candidate requirements list. Then, as the team's notion of the scope of the product and product line evolved, they marked these candidate requirements with a proposed priority.

Analogous systems are systems that serve a somewhat different purpose than that of the proposed product, but they are instructive to consider for relevant features. [Eclipse 2010] Analyzing analogous systems involves a creative and even playful mindset. For example, mobile technology for emergency responders almost certainly involves communication and collaboration support. Analogous systems might include meeting management tools, such as Avaya's Flare [2] or Adobe's Acrobat Connect Pro [1]. These systems do not solve the emergency responders' problems, but they are used for other kinds of communication and collaboration. Each RE team looked at these systems (and others) as sources of inspiration, rather than as competitive solutions, identifying features that might translate into corresponding features in the imagined emergency medical services product line. In fact, while the analysis of competitive applications yielded only requirements for point solutions, the study of analogous systems identified the need for a unifying platform that could enable such point solutions to be integrated.

2.2   User Interviews and Observations

Using ethnographic techniques as defined in the contextual design methodology [4], we conducted interviews with fourteen emergency responders. In some cases, it was possible to observe emergency responders at the fire station, where we could gather additional information about the emergency responders' culture and environment. In all cases, the interviews relied on prepared sets of open-ended questions, improvised as necessary to get at the heart of issues that emergency responders currently face in their jobs. About half of each interview focused on the participant's current job-related tasks. The other half of each interview focused on the participant's thoughts about how technology might improve his ability to carry out his job safely and efficiently.

2.3   Brainstorming

Brainstorming uses yet another kind of intellectual skill. The REs conducted their brainstorming activities after completing the competitive systems analysis, analogous systems analysis, and user interviews and observations. By this time, the REs had enough understanding of the domain that they could use intuition and inspiration to come up with additional features for the emergency responder mobile technology.

2.4   Scenario Development

In order to validate the requirements with stakeholders, the REs next developed scenarios of use. [13] This was carried out on a creative writing exercise that relies on the RE to tell a story, in which personas (based on character sketches) accomplish a goal, using the envisioned emergency responder mobile technology. The scenario describes the essential features or capabilities of the system in terms of how the user might interact with the system. In the process of developing the narrative, it's common for the RE to discover additional requirements. This generally occurs when the writer reaches a point in the narrative where the emergency responder needs to communicate or collaborate in a way that the existing requirements do not address. Nevertheless, the primary value of the scenarios is that they can be used with stakeholders to ensure that the requirements match the emergency responders' needs.

2.5   Use Case Development

Once exemplary scenarios have been vetted by stakeholders, requirements engineers may choose to analyze the software implications of each scenario in terms of use cases. [5] In this project, the REs identified critical and complex use cases to be "fully dressed." This practice involves selectively diving into the fine points of how the user and system interaction should take place.   For more straightforward use cases, Cockburn recommends more "casual" or "brief" coverage of the use case— identifying the primary flow of interaction between the user and the system, without the need to document the more detailed paths of execution that the user and system might undertake. This is in keeping with the Agile philosophy in which the details of the requirements are not created early in the software engineering lifecycle, but rather are elaborated just in time for design and development. In this project, the requirements engineers were all experienced software engineers. Therefore, they were able to review each other's use cases to analyze the implications for software design. And while use cases are most commonly referenced by software engineers, they are readily understandable by target users.

2.6   User Interface Mockups

To complement the scenarios, the REs developed sketches of key user interface (UI) considerations. Today, there are many open source tools available to support rapid development of user interface mockups for mobile technology (e.g., Balsamiq). In the emergency responder application domain, however, UI mockups may involve more than capturing a graphical user interface. In some circumstances, like densely smoky fires, a visual interface may be unusable. Therefore, the REs needed to mock up audio dialogs that might be used when visual interaction is unfeasible. These mockups must be validated with target users. The REs discovered the need for significant refinement of their early mockups after running through a scenario with illustrative GUI mockups, reviewed by actual emergency responders. Practical issues, such as the use of thick gloves during some emergency response activities (e.g., extracting an injured passenger from a car), meant that the capacitive touchscreen might not be a usable part of the interface. Other user feedback identified the need to very simple screens with maximally differentiated "buttons" (e.g., for vacate, utilities on/off, and man-down alerts).

3   Discussion

Each requirements engineering technique contributed uniquely to the acquisition or validation of requirements for emergency responder mobile technology needs. Competitive systems analysis led to the identification of features and capabilities such as situational awareness support and UI navigation support. In fact, the envisioned product line design was significantly influenced by the competitive analysis. With the wealth of emergency medical services (EMS) applications available, the REs recommended that the platform support integration of best-in-class EMS apps, rather than building proprietary versions of commercially available functionality. Medscape, for example, has produced a wealth of applications for EMS personnel, including a drug reference, drug interaction checker, disease and condition reference, and procedures and protocols reference. [11]

Analogous systems analysis clarified thinking about communication and collaboration technology. As an example, some online meeting collaboration tools support maintenance of a complete log of information exchanged during a meeting. The REs recognized the potential value of maintaining a rich log of incident information that could later be referenced for after-incident critiques and for training. This novel capability for EMS personnel translated into clear system requirements.

User interviews revealed details of the protocols the technology would need to support. For example, when paramedics have search and rescue responsibilities during a fire or hazardous materials (hazmat) incident, they must carry out a manual check-in protocol, the Personnel Accountability Report (PAR). Currently, this protocol involves the use of push-to-talk radios, with the incident commander (IC) initiating the PAR request to each team captain. The team captain then checks with each person on the team. Due to the bandwidth limitations of the radios, the report-back involves slow, sequential, verbal acknowledgement of status. The interviews, however, revealed that first responders and ICs would be well served by an automated PAR protocol, one that automatically generated the PAR requests at fixed intervals, tracked each first responder's acknowledgement, and reported the information to team captains and IC. The target users also provided guidance on the prioritization of capabilities like real-time alerts, multi-way real-time voice communications, and multimedia information sharing. Interviews with paramedics and emergency response medical doctors uncovered the potential benefits of real-time image sharing, where the paramedic could send a photo of a patient at the incident scene to the medical doctor for timely consultation and collaborative assessment of the situation.

Brainstorming, done in the context of what had already been learned about emergency responders' needs, enriched the set of candidate requirements with such feature ideas as searchable incident logs and the identification of nonfunctional (quality) requirements, such as software availability and reliability. The potential danger of RE brainstorming is that the engineers are not experts in EMS or overall emergency response. Therefore, each brainstormed requirement needs to be validated with target users. In this project, the entire set of nonfunctional requirements were reviewed and refined with the help of a veteran firefighter/paramedic. It will be essential to develop prototypes that meet these nonfunctional requirements, because first responders and ICs are often kinesthetic learners—people who understand best when they have the opportunity to interact with a tangible implementation. (In fact, in further work that built on this project, a small team developed a rapid prototype of the alert and acknowledgement capabilities and validated what would be an acceptable real-time performance with the same veteran firefighter/paramedic.)

When it came time to validate the candidate requirements, the REs focused first on the development of personas. By creating characters that spanned the diversity of technology aptitudes and attitudes of real emergency responders, it was possible to test out the feasibility of adoption of the gathered requirements. For a technology-naïve, 55-year-old paramedic persona named John, the question in determining the usability of a feature became, "Would this feature be easy for John to use?" This proved to serve as a reality-check on some of the more grandiose and technologically complex candidate requirements. In numerous conversations with first responders, the REs heard repeatedly that it is essential to keep the UI as simple as possible. Furthermore, the first responders cannot be distracted by their mobile phones from carrying out their primary responsibilities.

REs developed scenarios for medical emergencies, as well as other types of emergency incidents. By writing a complete story of a realistic emergency incident, each RE was able to recognize when a requirement was really unnecessary for that type of situation. As well, the unfolding story identified missing requirements. For example, one RE team created a rich incident scenario that involved an explosion and fire at a pharmaceutical plant. As the scenario unfolded, it became clear that off-duty EMS responders might need to be called to respond to such a major incident. This resulted in a novel set of requirements for how EMS personnel get timely notification and give timely responses when they are off duty.

The development of use cases was probably the most technical of the requirements engineering tasks, because the use case developers had to think in terms of the software system requirements in greater depth. While this approach to modeling requirements has a certain appeal to some detail-oriented software developers, Agile methodologies encourage a lighter-weight documentation, such as the user story. [Cohn] In hindsight, significant effort could have been reduced by adopting user stories instead of use case documentation, without a reduction in understanding of the problem space. A follow-on prototyping effort for some of the highest priority requirements revealed that it was easy enough to prototype, evaluate, and revise the user-system interaction design without reference to a use case document.

While it is very helpful to go through a scenario with target users in order to confirm the vision of the emergency responder mobile technology, we found that UI mockups were also helpful in getting target users to provide specific feedback on the user-facing requirements for the system. Like early prototypes, the UI mockups give the target user something specific to react to. For those who say, "I'll know what I like when I see it," UI mockups are especially important. Mobile screen mockups of maps, for example, demonstrated the need for simple zoom capabilities. The small footprint of the mobile screen made it clear just how little information could be provided at once and still be visible and interactive.

The REs were fortunate to have a 31-year-veteran firefighter/paramedic as the "onsite customer" representative. This provided a sanity check on the evolving product vision, requirements, and prototype.

The primary challenge to implementing this interdisciplinary approach is finding requirements engineers with the aptitude and inclination to tackle the wide variety of tasks that are involved in generating a comprehensive, validated, prioritized set of system requirements. Nevertheless, this project, which involved a class of Carnegie Mellon University Software Engineering graduate students, demonstrated that these skills are readily learned and applied in a very short period of time (seven weeks from requirements engineering inception to completion of all deliverables).

4   Conclusions

A multi-pronged approach to requirements engineering results in a richer set of functional and nonfunctional requirements. With such an approach, it is much easier to identify the requirements of an assortment of stakeholders, to ensure a more innovative and powerful solution, and to validate that the requirements will meet the stakeholders' needs. The use of an *interdisciplinary* approach accommodates the variety of ways in which stakeholders may best understand requirements, whether that means reading narrative scenarios, viewing mockups, walking through use cases, or reviewing lists of requirements statements.

The methodology described in this paper has been demonstrated to support a variety of stakeholders in the requirements engineering efforts for mobile technology support for emergency medical services. The deliverables from this project have been successfully applied to the development of a first responder application, software system architecture, and technology roadmap.

Acknowledgements. Professor Reed Letsinger, co-designer of the Carnegie Mellon University Silicon Valley Requirements Engineering course; the CMUSV Software Engineering students of the Spring 2011 Requirements Engineering course; and the fourteen first responders, fire chiefs, and other emergency response stakeholders who participated in interviews and requirements validation activities.

References

1. Adobe Connect 8, http://www.adobe.com/products/adobeconnect.html
2. Avaya Flare Experience Guided Tour, http://www.avaya.com/usa/campaign/avaya-flare-experience-guided-tour/
3. Beck, K., et al.: Manifesto for Agile Software Development. Agile Alliance (2001), http://agilemanifesto.org/
4. Beyer, H., Holtzblatt, K.: Contextual Design: Defining Customer-Centered Systems. Morgan Kaufmann, San Francisco (1997)
5. Cockburn, A.: Writing Effective Use Cases. Addison-Wesley Professional, Boston (2000)

6. Cohn, M.: User Stories Applied: For Agile Software Development. Addison-Wesley Professional, Boston (2004)
7. Guideline: Requirements Gathering Techniques. Eclipse (November 20, 2010),
   `http://epf.eclipse.org/wikis/openup/`
   `core.tech.common.extend_supp/guidances/guidelines/`
   `req_gathering_techniques_8CB8E44C.html`
8. Holtzblatt, K., Wendell, J.B., Wood, S.: Rapid Contextual Design: A How-to Guide to Key Techniques for User-Centered Design. Morgan Kaufmann, San Francisco (2004)
9. Laplante, P.: What Every Engineer Should Know about Software Engineering. CRC Press, Boca Raton (2007)
10. Leffingwell, D., Widrig, D.: Agile Software Requirements: Lean Requirements Practices for Team, Programs, and Enterprises. Pearson, Boston (2011)
11. Medscape App for Android. WebMD,
    `http://www.medscape.com/public/android`
12. Pilcher, R.: The Product Vision. The Scrum Alliance (January 9, 2009),
    `http://www.scrumalliance.org/articles/115-the-product-vision`
13. Sutcliffe, A.: Scenario-based Requirements Engineering. In: Proceedings of the 11th IEEE International Conference on Requirements Engineering, pp. 320–329. IEEE, Washington

TOPS - System for Planning and Providing the Health and Social Services at the Home Environment of Clients

TomášVáňa, David Žák, and JiříLebduška

University of Pardubice, Faculty of Electrical Engineering and Informatics,
Studentská95, 532 10 Pardubice, Czech Republic
tomas.vana@student.upce.cz,david.zak@upce.cz,
jiri.lebduska@student.upce.cz

Abstract. In the health care field there is a growing demand for services provided in a clients' home environment. The high quality of services and responsibility for health and lives of patients require a professional approach and an appropriate set up of business processes within organizations providing these services. The aim of this paper is to describe the system for support of planning, recording, reporting and invoicing of the health and social services at home environment of clients.

Keywords: information system, mobile communication, planning, invoicing, health services, social services, home.

1 Introduction

Modern information and mobile technologies allow the creation of comprehensive solutions that can improve coordination and communication between employees and reduce paper work in many fields of human activity. Quite logically, these solutions are deployed first in engineering, where they are used by technically educated workers. Application of these solutions in other spheres then puts high demands on intuitive, good ergonomics and high stability of all elements of the solution.

Organizations offering home nursing and care services have qualified staff for providing specialized services. Based on contracts with health insurance companies or clients these organizations ensure the execution of these services by authorized personnel at clients' home environment.

The ambition of presented TOPS[1]project is to develop and offer a complex information and telecommunication solution to these organizations that cover all common and specific needs for providing services mentioned above, for example:

---

[1] TOPS - label for System for Planning and Providing the Health and Social Services at the Home Environment of Clients.

K.S. Nikita et al. (Eds.): MobiHealth 2011, LNICST 83, pp. 292–299, 2012.

- client evidence,
- care orders in the form of vouchers or contracts,
- staff evidence including their education and attendance,
- scheduling clients' visits,
- data acquisition about provided services directly at clients' home,
- care reporting,
- invoicing of provided services to health insurance companies or to clients.

2   Analysis

The existing solutions e.g. Pečovatelská služba[2], Alora[3] are mainly offered as installation package. Most of organizations providing health care are nonprofit. This is the reason they do not have resources for providing particular IT solution. Presented project is offered as supported service. It allows reducing cost for purchasing and maintenance.

Based on the analysis of needs and processes in these organizations the following basic requirements for the developed system were established:

- cover the entire set of activities in home care and nursing services from client intake, care planning, care reporting, recording all information, conducting long-term and short-term care plans, rating and invoicing,
- filed staff will report provided care directly in the client's home,
- synchronizations of data between stationary systems and mobile devices every few minutes,
- work on a mobile device has to be allowed even when the transmission network is not available (offline),
- use of a solution cannot be tied either to the existence of IT professionals in customer organizations or to buying the new equipment,
- data security,
- implementation of many different user roles,
- implementation of different rating rules and requirements for billing,
- creation of statistical reports,
- updates in line with legislative requirements,
- the ability to modify solution according to customers' requirements.

To successfully address the above-mentioned requirements the necessity of implementing these key intelligent features has been identified:

- proper allocation of care to the client visits based on contracts or vouchers,
- implementation of tools for working with multiple visits in scheduler,
- synchronization of data between the stationary system and mobile devices with minimal volumes of transferred data,

---

[2] Pečovatelská služba, Petr Zajíc software.
[3] Alora Home Health Software, Alora Healthcare Systems.

- automatic validation of reported data about provided care,
- implementation of a universal billing.

3   System Architecture

The TOPS architecture (Fig. 1) shows global concept of the whole solution. The architecture has three main parts: mobile devices for field-workers, PC for staff working in service centers and server infrastructure.

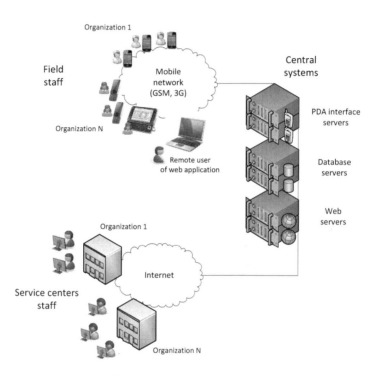

Fig. 1. TOPS Architecture

The base of the server infrastructure is a central database server. This server forms the lowest layer of the entire system.

Database servers use Oracle platform which is appropriately equipped to ensure safe and reliable operation of the system and necessary computing power. During the TOPS development the emphasis on solving data security was accentuated because sensitive personal details about clients, their diagnosis, etc. are stored in central database and transmitted to smart mobile devices (PDA).

The main user interface for service centers staffis a web application, which includes:agenda, care planning, preparation and printing of contractual documents, reportsand invoicing. This application is equipped with many dynamic components using AJAX technology and communicates directly with the central database via Web servers.

Field staff use mobile devices such as PDA. PDAs are operated with the Android operating system and the TOPS mobile client. These mobile devices communicate with central systems via a public mobile network (e.g. 3G, HSDPA, EDGE,GPRS) in a separate APN[4]. It allows separating a communication inside TOPS project from the other data communication in mobile network [1]. Communications with client application on mobile devices cater PDA interface servers.

4   PDA Mobile Devices

A mobile device is given to a field staff that provides health and social care at a client's home. An integral part of the client application development was an analysis of the field staff needs, not only in terms of accurate records of provided care, but also in communication with coordinating person during operating changes in the daily plan. For the coordinating person application provides actual information about the staff location and actions in progress. PDA also use camera as barcode reader. Barcode reader is useful to identify client [2] or to verify that worker really made the activity.

Thanks to the touch screen and easy control it allows the quick and intuitive auditing for the health status of clients, actions in progress or drugs that were administered. These records could be later showed, edited and billed to clients or insurance company using TOPS web application. Field worker can also use information about his planned and realized visits, addresses and access details to the client and contact information of his family members.

4.1   Synchronization

Unlike web application, client application for mobile device does not communicate directly with the central database but it uses local database. Local database content is synchronized with a central database server via PDA interface servers. The main purpose of this solution is to secure EHR[5] in situations when online connection is not available to central systems via mobile network (e.g. in sparsely populated areas).

Synchronization mechanism was developed as a part of a project. Synchronization uses XML standard for communication and minimizes the volume of data synchronized between mobile device and central database. The synchronization procedure is activated on mobile device every 5 minutes, therefore the central systems and mobile devices work with almost online information and can react very quickly to possible changes of daily schedule.

---

[4] APN - Access Point Name identifies an IP Packet Data Network, that a mobile data user wants to communicate with.
[5] EHR – Electronic Health Record.

4.2   PDA Task Processing

Task transferred to the PDA (Fig. 2) contains also EHR needed for its execution. After a manual acceptance of the received task, device starts to record a field worker motion trajectory during transport stage. Time period of different stages (transportation, particular actions, waiting for client, study of medical records, etc.) is automatically measured.

Fig. 2. PDA application screens. From the left side: daily task list, one task view - list of operations, daily record of the visiting client.

By using simple dialogs the worker may confirm the actual number of executed actions, medicine consumption and eventually values of temperature, blood pressure, glucose, etc. The client application on PDA is also able to browse daily records, comments and measured values retrospectively. Qualified staff can therefore assess an acute deterioration or improvement of client´s health. At the end of a visit the employee can also write daily record.

4.3   Data Security

Every employee log on to the application with his unique user account name and password. During synchronizations only data filtrated on the basis of user permissions necessary to carry out planned activities are transferred to mobile devices. [3]

Data cannot be misused by unauthorized people even in the case when mobile device is stolen or lost. Downloaded data are encrypted. The number of invalid attempts to log on is limited. Data is erased from mobile device in the situation when this limit is exceeded.

5   Web Application

The web application is designed primarily for coordinating staff and managers. Advantage of the web solution is the ability to use the application practically from anywhere and from any platform. There is no need to install application on computer. Internet connection and installed web browser are the only needed requirements.

Application is composed from modules: Scheduler, Cards, Codebooks, Maps, Reports, Invoicing and Administration. User access to concrete modules, organizations and its parts (institutions and workplaces) and service types is limited by user roles and licenses. Application design is modular and there is possibility to add the necessary modules for future functions or standards dynamically, e.g. [4]. In the following chapters there are brief characteristics of main modules.

5.1   Module Scheduler

Module Scheduler (Fig. 3) is the planning calendar for managing employee schedules.

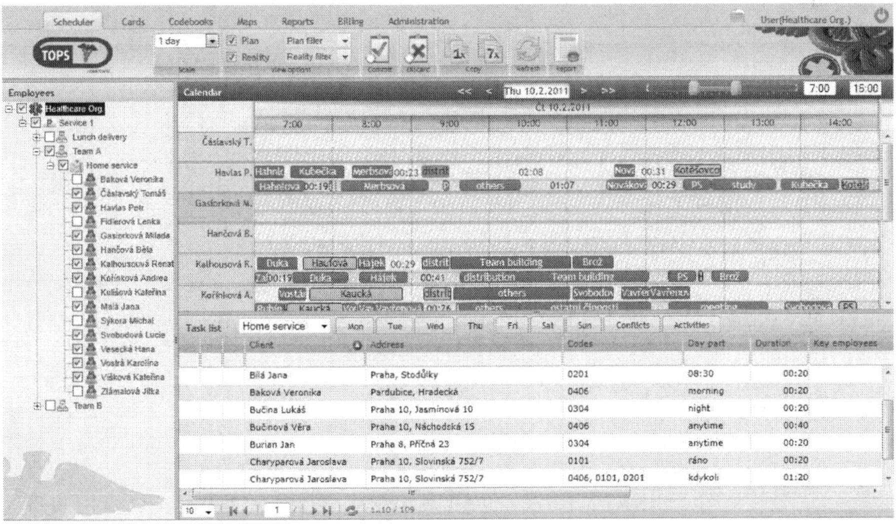

Fig. 3. Web application – module Scheduler. Hierarchical structure of organization and employees on the left side allows to select employees and to display their planning calendars in the central part of the window. Task list placed in the bottom displays tasks prepared for scheduling.

Color of task informs about task state. The coordinating employee responsible for planning activities can therefore simply recognize if the task has already been transferred to the PDA, whether activity starts or whether it was successfully or unsuccessfully completed. In the necessary cases this employee can make operational rescheduling by moving task with mouse to the different person calendar or to different time.

The module Scheduler can display employees' calendars in three regimes: plan only, reality only and both together. Simultaneous display offers a visual comparison of the plan and reality. This module includes intelligent functions like plan transfer to different employee, quick creating of the new plan based on history and care orders, etc.

The system automatically generates the tasks based on data from client's contracts or vouchers for each day to the task list. The task list is essentially a table where one row corresponds to one task or activity. This arrangement allows viewing relevant information for planning, such as the address of the client, required time for the visit or the expected duration of task. The tasks are automatically removed from the task list after their transfer with mouse to the planning calendar.

5.2   Module Reports

The module Reports is the extension of application which is designed primarily for managers. This module allows creating different kinds of reports and statistics. All reports are displayed in the tables (also called grids) within the web application. The reports can be exported into files in formats CSV, XLS, XLSX.

The key features of reports are dynamics and flexibility. Reports are represented by database queries. Any report definition can be uploaded into the application dynamically at any time without changes of database structures.

In the module Reports there is the strict verification of user rights which limits not only access to concrete report, but also restrict a set of processed data (rows and columns of reports). Similar security verifications are implemented in all modules of TOPS application.

The module includes the ability to create press kits. The application allows defining a very wide range of output formats - from simple formats such as CSV or TXT through XLS to complex outputs which include text, tables, images and graphs - e.g. formats PDF, DOC, RTF, XLSX, XLSM.

The most typical reports are monthly statement of work, an overview of monthly billing, an overview of care provided to clients, different annual statistical reports, etc.

5.3   Module Invoicing

The system can perform the rating of all services provided to clients. The unique mechanism for care and nursing services rating was developed. User can define several different types of rating rules which have an impact on the overall outcome of an invoicing process.

Calculations can be based on the number of performed operations, kilometers, time spent, amount of consumed material, weight of washed and pressed clothes, flat fee or various combinations of these parameters. The module was built with respect to user friendliness and simplicity.

An optional addition to the invoicing process is the ability to create reports for health insurance companies. These report scan be uploaded to the systems of health insurance companies.

5.4   Module Maps

The TOPS application allows working with maps, in which following flags (listed below) can be shown for selected employees and time period:

- actual location based on GPS data from employee PDA,
- locations of planned/performed visits,
- trace of employees' movement,
- user-defined points of interest (health facilities, pharmacies, supermarkets, cook shops etc.).

Module Maps uses Google Maps API. GPS data about employees' locations are sent to the central database during the synchronization process. Module Maps is a useful tool for utilization of employees' working time because it helps to respond effectively to any request for changes in plans.

6   Conclusion

The article describes the system TOPS. TOPS is a complex telematics solution within the field of e-health specifically designed for providing the nursing and social services at the clients' home environment. Powerful and robust solution was created by usage of advanced mobile and information technology which helps with evidence, planning, reporting and invoicing of provided services. These features allow employees to save time on administrative work and spend more time on professional and productive activities. This solution brings significant economic effect to the organizations, because the productivity of field workers after deployment TOPS increased by 10% to 15%.

The complex solution described above is offered as a service. It also allows the quick implementation of requirements following from relatively frequent legislative changes in this area. TOPS running as a service has become operationally and financially affordable for large, medium and small organizations that provide health and social care.

References

1. Mouly, M., Pautet, M.B.: The GSM system for mobile communications, p. 701. CELL & SYS, Palaiseau (1992) ISBN 0-945592-15-9
2. Ohansson, P.E., Petersson, G.I., Nilsson, G.C.: Personal digital assistant with a barcode reader— A medical decision support system for nurses in home care. International Journal of Medical Informatics 79(4) (2010), http://www.sciencedirect.com/science/article/pii/S1386505610000171 (cit. August 23, 2011)
3. Stajano, F., Anderson, R.: The Resurrecting Duckling: Security Issues for Ad-Hoc Wireless Networks. In: Malcolm, J.A., Christianson, B., Crispo, B., Roe, M. (eds.) Security Protocols 1999. LNCS, vol. 1796, pp. 172–182. Springer, Heidelberg (2000)
4. Botsivaly, M., Spyropoulos, B., Koutsourakis, K., Mertika, K.: A Homecare Application based on the ASTM E2369-05 Standard Specification for Continuity of Care Record. In: AMIA Annu. Symp. Proc. 2006, Athens (2006)

A Data Synchronization Framework for Personal Health Systems

Davide Capozzi and Giordano Lanzola

Department of Computer and Systems Science
University of Pavia
Via Ferrata 1, 27100 Pavia, Italy
{davide.capozzi,giordano.lanzola}@unipv.it

Abstract. This paper illustrates the design of a multi-platform synchronization framework which is particularly useful for speeding up the implementation of Personal Health Systems on mobile devices. Those devices turn out to be of great help since in order to transfer any data available at the patient site to the clinic and vice-versa a solid networking infrastructure and data exchange protocol is needed. The framework we developed extends an open source platform available on the market by empowering it with new features that better decouple domain specific data from the underlying transport logic. In the last part of the paper two prototypes exploiting the framework are described.

Keywords: Healthcare telemetry and telemedicine, Measurement and monitoring technologies, Mobile devices for patient monitoring, Transmission of patient data.

1   Introduction

The increasing aging of the population combined with many unhealthy lifestyles being adopted nowadays and resulting in an augmented prevalence of obesity are acting as a modern plague in most of the western countries. In fact they are responsible for an increased incidence of chronic disorders such as coronary artery disease, congestive heart failure or diabetes which account for the majority of the medical expenses [1]. It is now clear that such a current trend cannot be sustained any longer [2] and new and more effective ways of coping with chronic diseases should be pursued in order to reduce long-run medical expenses and prevent the onset of those complications which frequently result into specific treatments and hospitalizations pressing on the health care budgets.

With respect to this concern, there is a growing interest about Personal Health Systems (PHSs) in the technological communities which has been stirred up recently. This term refers to devices made available by the joint achievements in microelectronics and nanosciences and exploiting the Information and Communication Technologies (ICT) to provide applications supporting the personalization and individualization of the treatment process [3].

K.S. Nikita et al. (Eds.): MobiHealth 2011, LNICST 83, pp. 300–304, 2012.

2   Materials and Methods

Mobile phones, Personal Digital Assistants (PDAs), Smart-phones, and tablets nowadays have such a great variety of technical features that allows to choose each time the product fitting any given application at best, but it becomes a serious drawback inasmuch the plain connectivity is of concern.

According to the models advocating the decoupling and separation of concerns among the different components building up a system, we envisioned instead a layered architecture where data exchange is supported by an underlying layer shared among all platform and supporting their interoperation platforms [4].

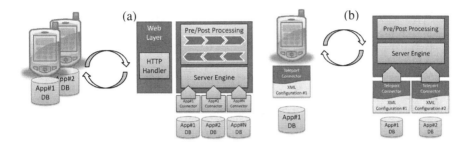

Fig. 1. (a) The synchronization framework. (b) The Teleport Connector within the framework.

What we need is a synchronization framework that can easily be accessed from different devices running different Operating Systems (OSs), store data on different formats and provide some programming facilities to allow us customizing and extending its basic functionalities. Figure 1(a) shows the idea of a web-based synchronization framework that could be exploited as a transparent two-ways-data-exchange layer by a PHS application. On the left there are two smartphones running two different PHSs having each one its own Data Base (DB); each time an application needs to be synchronized, it starts an HTTP connection towards the remote server that exposes an HTTP Handler in its Web Layer. The majority of the synchronization platforms available on the market adopt at this level an open communication protocol named SyncML [5]. SyncML [6] is an open industry initiative supported by hundreds of companies including Ericsson, IBM, Lotus, Matsushita, Motorola, Nokia, Openwave, and Starfish. It seeks to provide an open standard for data synchronization across different platforms and devices.

To minimize communication time, the standard assumes that each device maintains information about modification flags for each of its records with respect to every other device on the network. For this reason, in the architecture a Server Engine is needed that takes the burden of managing stored records in terms of handling record IDs, detecting and trying to resolve conflicts among records, as well as keeping trace of record modifications. The extensibility of the platform is represented by connectors plugged on the bottom of the server: each one for a different PHS application. Through the connector, the Server Engine interacts each time with a different application DB, since its structure and the domain knowledge are enclosed in that component. Last but not least, a pre/post processing function block is needed in the architecture, in order to resolve any data format conflict between clients and server.

For the implementation of our synchronization framework addressing PHSs we chose the open source Funambol platform [7], since it better captures our needs and reflects the architectural features described above.

3   Results

The main goal of our synchronization framework is to decouple the transmission of data between client and server from the management and the storage of data on both sides. That important feature enables us to reuse the solution for any PHS application that, in this way, exploits the synchronization framework for exchanging its data transparently with a remote server anytime this is necessary, without the requirement to be connected continuously to the internet.

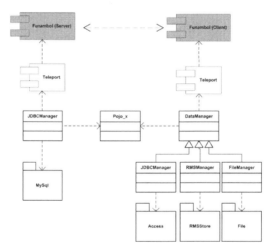

Fig. 2. The UML architecture of the Teleport Connector

We implemented that feature exploiting the extendibility of the Funambol platform, developing a generic connector named Teleport Connector (TC) that interfaces the whole architecture as displayed in Figure 1(b). TC can be exploited both on the client-side interfacing PHS applications and on the server-side connecting to PHS remote DBs. This component responds to the tasks of encapsulating the application data into a generic format, shared between client and server, sending them through the HTTP connection, described in the paragraph above, and accessing the application specific DB exploiting an XML Configuration file provided with the application itself.

For the data encapsulation we designed generic POJOs (Plain Old Java Object) that enclose data in a collection of key-value couples disregarding their types and thus gaining generality. POJOs are the only data type that TC can manage. In this way the TC could transmit and receive POJOs without being aware of what they contain.

In order to exploit the SyncML protocol for exchanging data, TC has been equipped with the capability to encode a POJO into a plain-text document complying with an XML based formalism, since XML can be natively managed by those mobile platforms running a Java Virtual Machine, such as J2ME or Android. For any other mobile platform we have provided a custom linear text encoding, independent of the XML formalism.

Furthermore, the TC is able to access the application DB without containing any wired knowledge about the domain specific data structure; this can be performed through the parsing of an application specific XML configuration file that encapsulates the definition of the data structure. Moreover, TC not only is totally independent from application data but also from the physical/logical support where information is stored. In order to save its generality, TC can't be dependent on those technologies, so that a data store abstraction layer has been introduced. Figure 2 shows in more details the last feature described.

On the topmost part of the figure the Funambol platform components (the server on the left side and the client on the right) are displayed. From those the TC architecture develops for both server and client, sharing the most part of the code located in the Teleport package. On the client side, the data store abstraction layer is represented by the component DataManager used directly from the Teleport package to access data and, in the meantime, extended by many custom DataManagers. Each particular DataManager, such as JDBCManager, RMSManager or FileManager, takes the burden of interacting with a specific data store technology, such as respectively an MS Access DB, an RMS Record Store or even simply a file. On the server side, since the data store technology has been established to be a MySQL database, we streamlined the architecture, so that just a JDBCManager is required to represent the data store abstraction layer.

4   Conclusions

This paper described the design of a multi-platform synchronization layer which is particularly useful for speeding up the implementation of PHSs. The synchronization layer described has been utilized for the implementation of two separate applications running on different devices. Since the overall focus of the paper is just on the synchronization layer and its constraints and does not allow for an extensive description of the applications, we just mention briefly each of those applications including a reference to a publication where more detailed descriptions are available [8].

The first prototype is meant for managing uremic patients who are experiencing renal failure as a major complication of diabetes and are thus undergoing Peritoneal Dialysis (PD). Thus it is mandatory for those patients to strictly control blood pressure and weight, and regularly keep informed their treating staff about any variations. The application for acquiring data is implemented on a Smart-phone running Symbian-Os and supporting the J2ME development platform. It has been designed to acquire data directly from a blood pressure monitor and a scale exploiting a Bluetooth wireless connection and uses the device display just for a minimal interaction with its users which are likely to be elderly people.

The second prototype is meant instead to support in a medium sized randomized controlled trial for patients undergoing an Artificial Pancreas (AP) therapy. The requirements for the clinical trials see the patients undergoing an AP therapy at their domiciles while the treating staff at the clinic should be able to follow in almost real-time the evolution of the patient's clinical state. AP units are available so far only as research products and run on Personal Computers. Thus in that case decoupling the synchronization layer from the application one has been most useful for adding networking capabilities to the AP units without having to modify their code. With the only knowledge that AP units saved their data to a local database implemented with Microsoft Access, it was quite easy to establish a link with the server to ship those data to the clinic. On the way back in this case is sent information concerning directives for the AP unit and informational messages for the patient.

For both prototypes the synchronization server is running on a PC and the underlying database is implemented using MySql Server.

References

1. Levit, K., Smith, C., Cowan, C., Sensenig, A., Catlin, A.: Trends - Health spending rebound continues in 2002. Health Affairs 23(1), 147–159 (2002)
2. Fogel, R.W.: Forecasting the cost of US Health Care in 2040. Journal of Policy Modeling 31, 482–488 (2009)
3. Maglaveras, N., Bonato, P., Tamura, T.: Special Section on Personal Health Systems. IEEE Transactions on Information Tehcnology in Biomedicine 14(2), 360–363 (2010)
4. Lindholm, T., Kangasharju, J., Tarkoma, S.: Syxaw: Data Synchronization Middleware for the Mobile Web. Mobile Networks & Applications 14(5), 661–676 (2009)
5. Agarwal, S., Starobinski, D., Trachtenberg, A.: On the scalability of data synchronization protocols for PDAs and mobile devices. IEEE Network 16(4), 22–28 (2002)
6. SyncML Specifications, http://www.syncml.org/downloads.html
7. Fornari, F.: Funambol Mobile Open Source (Paperback), ch. 10. Packt Publishing (2009)
8. Capozzi, D., Lanzola, G.: Utilizing Information Technologies for Lifelong Monitoring in Diabetes Patients. J. Diabetes Sci. Technol. 5(1), 55–62 (2011)

Service-Oriented Middleware Architecture for Mobile Personal Health Monitoring

Matts Ahlsén[1], Stefan Asanin[1], Peeter Kool[1], Peter Rosengren[1], and Jesper Thestrup[2]

[1] CNet Svenska AB, Svärdvägen 3b, 182 33 Danderyd, Sweden
{matts.ahlsen,stefan.asanin,peeter.kool,peter.rosengren}@cnet.se
[2] In-JeT ApS, Jeppe Aakjaers Vej 15, 3460 Birkeroed, Denmark
jth@in-jet.dk

Abstract. Developers of applications for health and wellness monitoring are facing a diversity of protocols, standards and communication mechanisms for collecting data from heterogeneous sensors, devices and services, as well as when exporting data to various health and wellness services and systems. The REACTION platform addresses this using a middleware approach which leverages the development tasks to a service-oriented level allowing developers to use open standard technologies like web services. The REACTION SOA (Service-Oriented Architecture) approach offers a scalable and inter-operable platform for use in different healthcare settings. The REACTION applications are based on numerous individual services that can be developed and deployed to perform clinical monitoring and feedback tasks, execute distributed decision support and security tasks, support work flow management, and perform event handling and crisis management.

Keywords: Remote Healthcare Monitoring, Diabetes, SOA, Internet of Things, Middleware, Semantics.

1 Introduction

The REACTION project [1] aims to research and develop an intelligent service platform that can provide professional, remote monitoring and therapy management to diabetes patients in different healthcare regimes across Europe. The platform is designed to help developers create support for both carers and patients in the management of diabetes but has the possibility of managing other chronic diseases. Today developers of applications for health and wellness monitoring are facing a diversity of protocols, standards and communication mechanisms both when collecting data from heterogeneous sensors, devices and services, as well as when exporting data to various health and wellness services and systems. For example, IEEE11073 standards are being used by the Continua Alliance [14], while there are numerous of legacy devices using proprietary interfaces both over Bluetooth as well as through USB or serial port. In the wellness sector there are many ANT+ [5] sensors available, and recently several WIFI-based health devices [8] with associated Internet services have been introduced. In addition to this several cloud-based health services offers services for uploading and analysing the collected health data [7], [13].

K.S. Nikita et al. (Eds.): MobiHealth 2011, LNICST 83, pp. 305–312, 2012.

The REACTION platform addresses this heterogeneity by applying an "Internet of Things" perspective on medical device connectivity. It uses a middleware approach which leverages the developer´s tasks to a service-oriented level allowing developers to use open standard technologies like web services. The middleware approach also makes the applications independent of the underlying device and service protocol level and ensures interoperability as well as re-usability, since new devices can be deployed and/or old ones replaced without the applications have to be re-built.

The REACTION middleware builds on results from the integrated project Hydra which researched Open Source middleware for Internet of Things. The Hydra middleware incorporates support for ontology-driven discovery of devices, P2P (Peer-to-Peer) communications, use of semantic technologies for code generation [2],[4],[11]. The REACTION project extends and adapts the middleware to device connectivity in the health and wellness sectors. This allows developers to rapidly create health and wellness applications as collections of services which can be orchestrated to perform desired workflows supported through the platform spheres (Fig. 1), which improves development efficiency while delivering trusted and reliable patient-oriented services.

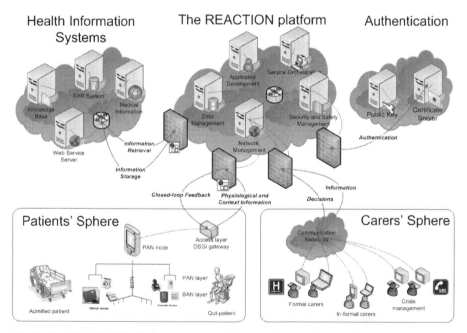

Fig. 1. The REACTION platform concept with two sphere (patient and carer) depicting the dynamics of a SOA-based framework for remote patient monitoring

Wearable medical sensors are connected in a BAN (Body Area Network) for multi-parametric recording of vital physiological parameters. The BAN interconnects with other sensors in the environment that can record contextual information about other vital parameters and the patients' activities in a PAN (Personal Area Network). A local REACTION gateway also handles episode monitoring, alarms and services needed during periods of non-connectivity. The gateway further manages personalised patient feedback from health professionals adapted to user terminals and self-monitoring and autonomous regulation of the connected devices in the BAN. The gateway can be remotely configured and managed through the service-oriented architecture.

Health Information Systems (HIS) will be integrated in REACTION applications in the form of services, these services can subsequently be orchestrated into workflows through the Service Orchestration subset.

2   Usage Scenarios

This chapter describes some typical monitoring applications that could be built using the REACTION platform. The purpose is to illustrate the use of different technologies in a chronic disease management perspective. We assume that the patient is at home and REACTION is integrated in the home environment. The first two scenarios take part as diabetes' co-morbidities.

Hypertension: The application developed monitors the patient with ECG (Electrocardiogram), using a wireless electronic patch, ePatch [15], which is part of the REACTION project. This ePatch contains sensors and a ZigBee chip. The patch transmits data to the Reaction gateway hosting the REACTION client software. The data is rendered in a user interface to allow clinicians to analyse it. The ePatch allows the user to move freely around all day. The patient has also been advised to exercise, so he makes long walks every day. Using an ANT+ enabled GPS-watch he can then upload the time and length of each walk into an account in a personal cloud service, like HealthVault [7]. Together with the watch he sometimes uses a heart rate monitor attached to his breast to measure the pulse. Monitoring hypertension may help the care of diabetes patients.

Obesity: A Bluetooth-enabled weight scale based on the IEEE 11073 [6] standard is installed in the bathroom. Every morning the patient is reminded by the REACTION application running in his smart phone to perform the weight measurement. The patient steps on the weight scale and the measured weight is then automatically transferred to a stationary or portable PC in the home. The patient has an account with a cloud based health service, like WiThings [8], which he uses to keep track of his weight to see if there is any progress in losing weight with the new exercise and diet program he is using. The REACTION software connects to the cloud health service, using credentials the patient has provided at the time of set up. Many diabetic patients also have dietetic issues and so monitoring weight may assist in personal satisfaction and disease management.

Diabetes: The patient has been ordered to measure his blood pressure and glucose level every evening. Regardless if a device is Continua-certified or not the REACTION platform is able to discover and interface it. Unlike weight measurements the blood pressure and monitored glucose levels are exported to a primary care centre after being aggregated by REACTION data fusion engine together with data related to food intake, e.g. carbohydrates. The blood pressure monitor he has is an older model so it is not Continua-certified. He measures his blood pressure sitting in the living room sofa, while watching TV. The blood pressure values are received by the REACTION client in the same way as the weight in previous, but the values are exported to a primary care centre, where a REACTION server receives the values and stores it in a local EPR. In addition to this, the patient has, for his personal interest, configured REACTION to export the blood pressure values to a health service, HealthVault.

The most critical vital sign for the patient is the glucose, which requires him to use a glucometer. This is a cheaper model that is not wireless but attaches to the USB-port of the computer. He uses a blood sugar test strip which is inserted into the USB-based glucose meter. The glucose values are collected by REACTION and transmitted to the primary care centre. All communication with the remote server is done using IHE-PCD formats [16]. The glucose values are not transmitted to cloud services since the patient has defined a security policy that these values cannot be exported anywhere else except to the primary care centre. But before sending the glucose values, REACTION also performs data fusion and attaches information related to the patient's food intake this day, such as the level of carbohydrates, etc. This has been manually entered by the patient during the day. REACTION also adds today's weight before everything is sent to the primary care database. If the patient forgets to take his blood pressure and/or glucose level, REACTION will interface with other devices at home, for instance flash the lights in the living room three times every hour. This is configurable by setting up rules which are executed by a rule engine. In fact, it is possible for REACTION, due to its P2P service-oriented architecture, to connect to other homes, for instance the home of a relative and alert them of that something went wrong at the patient´s home.

3   Service-Oriented Middleware

The REACTION platform is the central production environment for the deployment of REACTION applications consisting of five subsets providing different functionalities:

- *Data Management* implements data manipulation, data fusion [3], event handling and data transport.
- *Service Orchestration* orchestrates available services in a pre-described sequence for execution.
- *Network Management* is responsible for the communication between devices, persons and external repositories.
- *Security Management* manages security models over user devices.
- *Application Development* is an open SDK (Software Development Kit) for model-driven development of applications. Context awareness is achieved through semantic annotations from patients' devices, environment and from historical data in EPRs (Electronic Patient Records).

The REACTION platform architecture (Fig. 2) is component based. Components relevant for the SOA approach are described in the following subchapters.

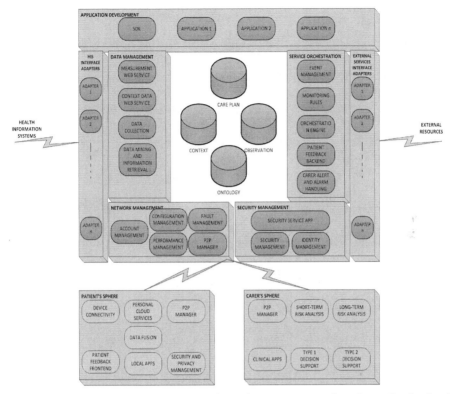

Fig. 2. The REACTION platform with generic service components where the application level provides an orchestration mechanism for design of end-user applications

3.1   Medical Device Connectivity Kit

The service-oriented middleware is made available to developers through an Open Source Medical Device Connectivity Kit (DCK) that enables developers to rapidly and seamlessly integrate medical Continua (i.e. IEEE 11073), medical non-Continua and wellness devices (e.g. exercise machines) into any development environment by the use of XML and web services. Using the DCK a developer can support traditional push-based remote patient monitoring applications, where a device pushes a measurement through to an observations database on the clinic side, local client applications that renders data to the user either in the home or as an app in a smart phone and clinical applications to analyse collected data.

Fig. 3 shows the internal structure of the DCK. In the bottom a Protocol Module ensures that the local environment is searched for both new and existing devices and where a Health Device Profile (HDP) exists by a device the module is able to communicate with it over a low level radio.

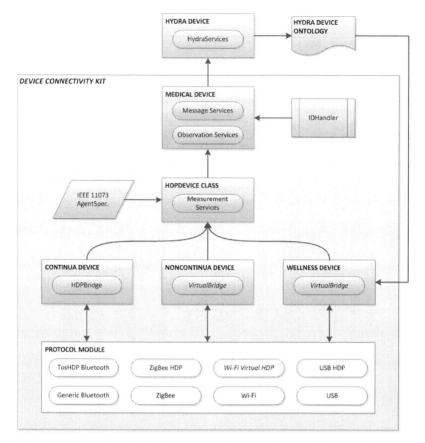

Fig. 3. The HDPDevice class holds Measurement Services that can either use the IEEE 11073 Agent Specialisations or relate the services to those described in the Device Ontology

Where HDP is missing the Protocol Module will ensure QoS by virtually replacing the HDP by imitating its functionality. This enables the communication between the module and the bridge in the different device types. Each device type announces what type of sensor device is trying to communicate whereas the HDPDevice compares this information to the IEEE 11073 specialisations available in a library. If such device exists as part of the standard the HDPDevice will establish a wide range of characteristics relevant for this sensor device type and enable a push approach where the sensor data is forwarded as part of Measurement Services.

On the top, the super class Medical Device sets parameters such as patient ID and other therapy related data through the IDHandler. This helps in constructing a complete HL7 (Health Level Seven) message based on the ORU-R01 format [12]. Finally, the Hydra Device class allows making all underlying functionalities and services available through Web Services.

3.2   Data Fusion and Service Orchestration

The REACTION applications are based on numerous individual services that can be developed and deployed to perform clinical monitoring and feedback tasks, execute distributed decision support and security tasks, support work flow management, and perform event handling and crisis management.

SOA is a collection of services that communicate with each other. The aim of a SOA approach is to have interoperable and loosely coupled services distributed in the network. In this context, a service is a function that is well-defined, self-contained, and does not depend on the context or state of other services.

In the REACTION platform, each device is enabled to offer Web Services that can be consumed by other devices, services or applications through an overlying mobile P2P network. Every service offered by a physical device is identified and invoked using SOAP (Simple Object Access Protocol) messages that are transmitted over the P2P network to create a robust connection, and WSDL (Web Service Definition Language) to define the interface of the services, no matter the implementation language used.

Ensembles of REACTION services are orchestrated by a specific high-level workflow based on BPEL (Business Protocol Execution Language). The workflow will be specified in the application and interpreted by the Orchestration Manager. The Orchestration Manager will make sure the different services available are executed in the described sequence. This component introduces higher abstraction mechanisms and makes the application developer independent of using a specific programming environment to orchestrate REACTION applications. It will also eliminate the interdependencies of services, solve conflicts of services and provide the most flexibility environment needed to realise service-oriented applications.

4   Conclusion and Future Work

In this paper we have described how a service-oriented middleware approach can be employed to facilitate development of health and wellness applications allowing semantic interoperability of heterogeneous devices, services and applications. Specifically, REACTION provides an integrated development and run-time environment for the improved long-term management of diabetes and other chronic diseases. We have explained how tasks such as connecting and collecting values from a medical devices, data fusion from many sensors, service orchestration, export of medical data can be realised using a service-oriented approach. An Open Source Medical Device Connectivity Kit will make the middleware functionality available for developers, and allow creation of applications based on open standards.

Acknowledgment. This work was performed in the framework of FP7 Integrated Project Reaction (Remote Accessibility to Diabetes Management and Therapy in Operational Healthcare Networks) partially funded by the European Commission. The authors wish to express their gratitude to the members of the REACTION consortium for valuable discussions.

References

1. REACTION Project, http://reaction-project.eu/
2. Eisenhauer, M., Rosengren, P., Antolin, P.: A Development Platform for Integrating Wireless Devices and Sensors into Ambient Intelligence Systems. In: Proceedings of the 6th Annual IEEE Communications Society Conference on Sensor, Mesh and Ad Hoc Communications and Networks (SECON), pp. 1–3. IEEE Press, Rome (2009)
3. Lee, H., Park, K., Choi, J., Lee, B., Elmasri, R.: Issues in Data Fusion for Healthcare Monitoring. In: PETRA 2008. ACM, Athens (2008)
4. Hydra Project, http://www.hydramiddleware.eu/
5. ANT+, http://www.thisisant.com/pages/technology/what-is-ant-plus
6. Carroll, R., Cnossen, R., Schnell, M., Simons, D.: Continua: An Interoperable Personal Healthcare Ecosystem. IEEE Pervasive Computing, Mobile and Ubiquitous Systems 6, 90–94 (2007)
7. Microsoft HealthVault, http://www.healthvault.com
8. Withings, http://www.withings.com
9. Bluetooth SIG, http://www.bluetooth.com
10. ZigBee Alliance, http://www.zigbee.org
11. Lardies, F.M., Antolin, P., Fernandes, J., Zhang, W., Hansen, K., Kool, P.: Deploying Pervasive Web Services over a P2P Overlay. In: 18th IEEE International Workshops on Enabling Technologies: Infrastructures for Collaborative Enterprises, pp. 240–245. IEEE Computer Society, Groningen (2009)
12. Health Level Seven, http://www.hl7.org
13. HealthGraph, http://developer.runkeeper.com/healthgraph/overview
14. Continua Alliance, http://www.continuaalliance.org
15. ePatch,
    http://www.madebydelta.com/delta/
    Business_units/ME/Body_sensors/ePatch.page
16. IHE, http://www.ihe.net

Inactivity Monitoring for People with Alzheimer's Disease Using Smartphone Technology

Nicola Armstrong[*], Chris Nugent, George Moore,
Dewar Finlay, and William Burns

Computer Science Research Institute, University of Ulster,
Jordanstown, Northern Ireland
armstrong-n1@email.ulster.ac.uk,
{cd.nugent,g.moore,d.finlay,wp.burns}@ulster.ac.uk

Abstract. Worldwide the number of old and older people is increasing alongside the increase in average life expectancy. Due to this increase the number of age related impairments within the older society, in addition to the prevalence of chronic disease are also heightened. One of the most widespread chronic diseases is dementia, specifically Alzheimer's disease (AD). AD is a brain related condition which impairs a person's memory, thought and judgment. The aim of the current research has been to identify and alleviate a set of problems related to AD using smartphone technology. In order to determine if the level of support for those suffering from AD can be improved, our current work investigates the use of activity/inactivity monitoring using various smartphone services. Inactivity levels are being monitored in order to detect if a smartphone handset has been misplaced unintentionally, and to avoid any impact this may have on smartphone services. Specifically, GSM signal strength, Wi-Fi signal strength and accelerometer data are considered. Three smartphone applications have been developed and tested on a cohort of 8 healthy adult users as part of a pre-study investigation. Results from the pre-study indicate that the optimal approach to detect inactivity on a smartphone handset was via GSM signal strength coupled with accelerometer data.

Keywords: Alzheimer's disease, Smartphone Technology, Assistive Technologies.

1 Introduction

In 2008 the number of old people (aged 60+) worldwide was estimated at 506 million [1]. By the year 2050 this figure is expected to increase significantly and reach 2 billion [2]. Alongside the increase in the number of old people the average life expectancy has also increased, from 55 years old in the early 1990's to around 80 years old at the end of 2007 [3]. As the size of the older population continues to rise, the risk of developing age related impairments and chronic disease is also growing.

---

[*] Corresponding author.

K.S. Nikita et al. (Eds.): MobiHealth 2011, LNICST 83, pp. 313–321, 2012.
© Institute for Computer Sciences, Social Informatics and Telecommunications Engineering 2012

Chronic diseases can be described as diseases of long duration and generally slow in progress [4]. Figures published by the World Health Organisation estimate that 75% of the older population have at least one chronic disease [5]. One of the most common chronic diseases within the older population is dementia [6], a term that describes a number of brain related diseases. The most prevalent form of dementia is Alzheimer's disease (AD) [7]. Symptoms of AD include changes in cognitive decline, decreased judgment and a decreased ability in thought. In 2010 the number of persons with AD (PwAD) worldwide stood at 6.5 million, with on average £25,472 being spent on care costs per person each year [8] [9]. Based on these figures it seems evident that efforts should be made in order to help further assist PwAD on a daily basis. One possible solution to this problem is the use of Information and Communication Technology (ICT) in the form of smartphones.

2   Smartphone Technology

The use of mobile phone technology has steadily become a part of everyday life within today's society. In 2009 the number of mobile phone subscriptions worldwide was estimated at 4.6 billion [10]. Modern mobile phones are now referred to as smartphones and offer a wide variety of functionality over and above traditional voice communications. Smartphones now typically include full internet access including email, short messaging service (SMS), multimedia messaging service (MMS), a build-in camera and video recorder, Wi-Fi, GPS, video conferencing and a wide range of downloadable software applications. Another major benefit of smartphone technology is the ability to send and receive a wide variety of information in real-time. Due to this level of functionality, the use of smartphone technology is now being used within a variety of domains including disease management; health monitoring and health interventions, to name but a few [11]. Clinicians regularly use smartphone to view patient test results, drug information, as a personal organiser and to keep in contact with work colleagues and patients on a daily basis [12].

The overarching aim of this research has been to alleviate a set of problems associated with AD using services deployed through smartphone technology [13]. More specifically, we investigate as part of a pre-study, solutions that may provide higher levels of support through context aware based applications. The aim of the current pre-study is to examine ways of monitoring a person's/smartphones activity in order to detect inactivity within the home environment. This is being carried out in order to eliminate the smartphone handset being misplaced and to avoid the impact this would have on the associated smartphone services. When inactivity is identified, an intervention via a reminder message alert will be triggered. This will be performed through the use of non-invasive smartphone technology via built-in services.

3   Smartphone Applications for PwAD

As part of our ongoing research based on the unmet needs of PwAD, as identified by Lauriks *et al.* [14] as the need to remember, the need for social contact, the need for safety and the need for support regarding activities of daily living (ADL), we have previously developed a suite of smartphone applications to help assist with these unmet needs [13]. These prior applications included an ADL reminder application, a picture dialing/SMS application, a geo-fencing application and a one hour reminder application. In order to test the design and functionality of the smartphone applications they were evaluated by a control group which consisted of 15 healthy adult participants. Results from this previous research had demonstrated that participants felt comfortable interacting with a smartphone [13]. Nevertheless, one problem that was identified from this phase of the work was, due to memory impairments of PwAD, there was a high probability that a participant may forget to lift the handset or misplace it, hence compromising the utility of the applications.

In order to combat this risk, an activity/inactivity monitor application has been incorporated within the set of applications. If inactivity is detected by the handset, it may be assumed that the smartphone handset is not being carried by a participant/PwAD and therefore they may need reminded to lift the device. Ideally when inactivity is detected the handset should automatically trigger an alert in order to allow participants/PwAD to avail of the smartphone services. We then decided to look at various ways of monitoring activity from the handset itself. The rationale for this was to offer the ability for the smartphone applications to be used as a standalone device, which can either be used both inside and outside of the home environment. This is also the least intrusive way of collecting data. Example systems that unitize the use of smartphone technology in order to monitor patient activity include the Context Aware Remote Monitoring Assistant (CARMA) [15], a system which uses various built in sensors including accelerometer data on a smartphone handset in order to monitor a person's daily activity levels. Likewise Zhang et al. [16] have developed a system to help monitor a person's activates. The system can detect whether a person is walking, standing, sitting or lying via accelerometer data collected on a smartphone handset. The aim of the current work has been to detect inactivity as opposed to activity on a smartphone handset through alternative built in services on a smartphone handset alongside accelerometer data.

4   Methods

In order to monitor activity/inactivity directly from the smartphone we decided to consider the built-in services already available within the handset. The smartphone of choice was a HTC touch HD device, running the Windows Mobile 6.1 Operating System. This handset was selected due to its rich features and large 3.8 inch touch screen interface. The smartphone applications were created using Visual Studio 2008 and developed using C#. The following three services were considered:

4.1   GSM Signal Strength

An application was created that records the phone's current signal strength directly from the handset every 5 seconds. For the purposes of the study the recordings were automatically saved to local storage on the phone handset. The motivation for this was based on the hypothesis that if the signal strength on the handset did not change over a predetermined period of time, this may indicate that the handset had been misplaced or was left in a stationary position. This could subsequently form the basis to pre-empt the triggering of an alert.

4.2   Wi-Fi Signal Strength

Wi-Fi signal strength on the handset was also considered to ascertain if the handset's inactivity could be detected. Using the same methodology as the GSM application, we created a Wi-Fi application. Wi-Fi signal strength was sampled and recorded every 5 seconds and stored on the phone.

4.3   Accelerometer Data

Thirdly and finally the accelerometer data on the smartphone was considered. The accelerometer data was sampled every 0.5 seconds as opposed to the 5 second rate with the previous measures, in order to detect any sudden movement changes.

4.4   Combination of GSM/Wi-Fi/Accelerometer data

All three smartphone applications were tested simultaneously. This was carried out in order to determine if the applications above could be used to complement each other.

In order to test the applications individually, we positioned the smartphone handset in 4 different specific marked locations. We decided to position the smartphone handset at 4 different corners within one large room (Figure 1), 2 corners by the window and 2 corners nearer the door (the room remained out of bounds during testing in order to avoid any potential disruption). Each application, after being placed within the room ran for 10 minutes.

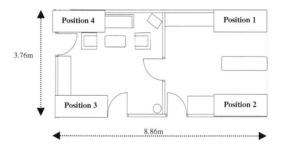

Fig. 1. Floor plan of room used to carry out testing

5   Results

To allow comparison of results and to reduce bias the testing phase was repeated 4 times using two different HTC handsets, a HTC Touch HD and a HTC HD 2. Two different mobile phone networks were also used throughout testing, one with a frequency of 900MHz and another with a frequency of 2100MHz.

5.1   GSM Signal Strength

Results from the testing of the signal strength application demonstrated correlation with environmental surroundings (refer to Table 1). There did appear to be a pattern of little variance when the device was stationary (refer to Figure 2a). Nevertheless, this is not always the case, as in some instances there was a fluxion in recorded GSM signal strength when the phone was stationary. Examples of specific instances include when the phone was positioned at a window in an empty room with no disruptions (refer to Figure 2b).

Table 1. Average signal strength from each position for each handset and network (N)

	HTC Touch HD (N1)	HTC Touch HD (N2)	HTC HD 2 (N1)	HTC HD 2 (N2)
Position 1	85.03	99.79	64.17	100
Position 2	74.18	100	69.12	100
Position 3	79.76	75.43	73.14	100
Position 3	84.86	76.22	98.80	99.66

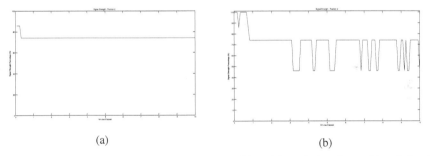

(a)   (b)

Fig. 2. Signal strength on smartphone handset positioned by window in position 2 for 10 minutes on network 1 using (a) HTC Touch HD (b) HTC HD 2

5.2   Wi-Fi

After testing the GSM signal strength application, the Wi-Fi application was then evaluated. Results from this application demonstrated that although the handset was placed in a stationary position, inside an empty room with no interruptions, the recorded Wi-Fi signal strength did fluctuate, regardless of the handset's activity. This was the case throughout all Wi-Fi evaluations, and evident throughout all data collected during the testing phase. For this stage of evaluation we tested the Wi-Fi application using 2 different HTC handsets and excluded the two different mobile phone networks, average results can be viewed below (refer to Table 2). An example of an instance is also presented in Figure 3a.

Table 2. Average Wi-Fi strength from each position, for each handset

	HTC Touch HD	HTC HD 2
Position 1	-70.62	- 68.56
Position 2	-72.05	- 71.34
Position 3	-72.82	- 69.45
Position 3	-72.57	- 71.87

5.3   Accelerometer Data

Lastly, we tested the accelerometer application created for the smartphone using the same methodology as before. Results from this phase of testing showed very accurate results regarding the handset's activity. An example of an instance is shown below (refer to Figure 3b).

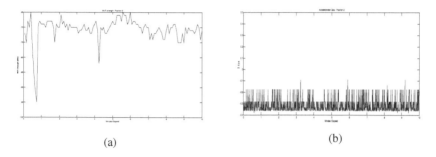

(a)   (b)

Fig. 3. (a) Wi-Fi strength (b) Accelerometer data on smartphone handset while positioned by window in position 2 for 10 minutes on smartphone

5.4   Three Services Combined (GSM/Wi-Fi/Accelerometer Data)

After the testing phase detailed above was complete, all three applications were considered together within one application. In order to do this GSM, Wi-Fi, and accelerometer data were collected concurrently. To test this application we carried out a further 8 evaluations on a control group of adult participants. Each evaluation lasted 30 minutes. Throughout the evaluation phase participants were asked to hold the smartphone on their person (place of their choice) and sit/walk around/place the phone stationary in front of them on a desk, for 10 minutes each time. Results from both the sitting phase and walking phase were as predicted varied. In order to compare results an example data set for the Wi-Fi, GSM and accelerometer data from a 10 minute sitting test for all three applications is presented in Figures 4a, 4b and 5a. A further example data set for the Wi-Fi, signal strength and accelerometer data while a person is walking around for 10 minutes is presented in Figures 5b, 6a and 6b.

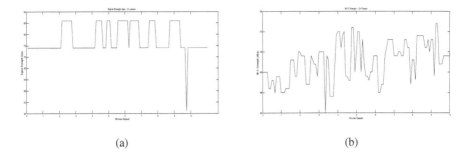

(a)   (b)

Fig. 4. (a) Signal strength (b) Wi-Fi from a sitting (on person) test over a 10 minute period

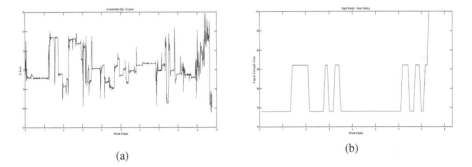

(a)   (b)

Fig. 5. (a) Accelerometer data from a sitting (on person) test over a 10 minute period (b) Signal strength data from a person walking around over a 10 minute period

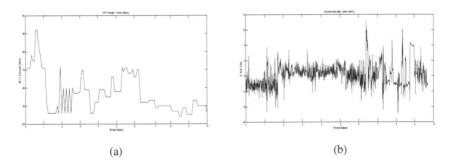

(a)   (b)

Fig. 6. (a) Wi-Fi (b) Accelerometer data from a person walking over a 10 minute period

Nevertheless, results from the phone being positioned on its own proved very promising, even with disruptions in and around the environment. Example results from a 10 minute stationary test for all three applications carried out at the same time are presented in Figures 7a, 7b and 7c.

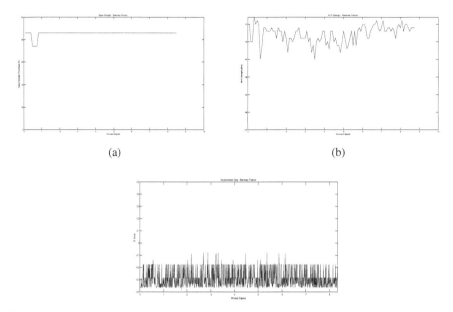

Fig. 7. (a) Signal strength (b) Wi-Fi (c) Accelerometer readings from stationary smartphone over a 10 minute period

Overall, results from this testing phase show that as before, the accelerometer data application is very accurate in detecting activity/inactivity on a smartphone handset. The signal strength tests have also shown promising results. It is therefore though that by coupling both applications, the GSM signal strength and the accelerometer application, it may be possible to detect inactivity within the home environment with an even higher degree of accuracy. The Wi-Fi application has been discounted as a reliable way of detecting inactivity, due to the high level of change in recordings.

6   Conclusion and Further Work

As the old and older society continues to grow, the increase in chronic disease will also ascend. Although there is no cure for chronic disease with the help of assistive technologies conditions such as memory impairment may be effectively managed. Not only do assistive devices provide support for PwAD, they may also relieve caregiver stress and burden. Further work focusing on the signal strength and accelerometer applications will need to be carried out in order to determine if a standalone smartphone can actually detect inactivity within with home environment, in a non-intrusive manner. We will also be incorporating the alert function within the application; this will be triggered when inactivity has been detected. We believe that through the use of ICT in the form of smartphone technology, assistive technologies such as smartphones can be incorporated easily into everyday life [17].

References

1. Kinsella, K., He, W.: An Aging world: 2008, International Population reports, US Census Bureau (June 2009)
2. United Nations Department of Economic and Social Affairs, Report Population Aging (2006)
3. Communication from the Commission to the European Parliament, the Council, the European Economic and Social Committee and the Committee of the Regions – Ageing well in the Information Society – A i2010 Initiative – Action Plan on Information and Communication Technologies and Ageing {SEC(2007)811} /* COM/2007/0332 final
4. World Health Organisation, Health and Chronic disease, http://www.who.int/topics/chronic_diseases/en/
5. Busse, R., et al.: Tackling Chronic Disease in Europe, Strategies, interventions and challenges, Observatory Studies Series No. 20 (2010)
6. Alzheimer's Association, 2009 Alzheimer's disease Facts and Figures. Alzheimer's & Dementia 5(3) (February 2009)
7. Communication from the Commission to the council and the European Parliament, in a European initiative on Alzheimer's disease and other dementias, Brussels, COM (2009) 380 final, {SEC(2009) 1040}{SEC(2009) 1041} (July 22, 2009)
8. Alzheimer's Association, 2010 Alzheimer's disease Facts and Figures. Alzheimer's & Dementia 6 (February 2010)
9. Improving Dementia service in Northern Ireland: A regional strategy (September 2010)
10. World Health Organisation, Electromagnetic fields and public health: mobile phone (June 2010), http://www.who.int/mediacentre/factsheets/fs296/en/
11. Blake, H.: Innovation in practice: mobile phone technology in patient care, Professional Issues
12. Ubaydli, A., Paton, C.: The Doctors PDA and Smartphone handbook. Journal of The Royal Society of Medicine 99(3), 120–124 (2006)
13. Armstrong, N., Nugent, C.D., Moore, G., Finlay, D.: Developing Smartphone applications for people with Alzheimer's disease. In: 10th IEEE International Conference on Information Technology and Applications in Biomedicine, ITAB 2010, Corfu, November 3-5, pp. 1–5 (2010)
14. Lauriks, et al.: Review of ICT-based services for identified unmet needs of people with dementia. Aging Research Reviews 6(3), 223–246 (2007)
15. Lau, et al.: Supporting Patient Monitoring Using Activity Recognition with a smartphone. In: 7th International Symposium Wireless Communication Systems (ISWCS), York, September 19-22, pp. 810–814 (2010)
16. Zhang, et al.: Activity monitoring using a smart phones accelerometer with hierarchical classification. In: Intelligent Environments Sixth International Conference 2010, Kuala Lumpur, July 19-21, pp. 158–163 (2010)
17. Armstrong, N., Nugent, C.D., Moore, G., Finlay, D.D.: Using smartphones to address the needs of persons with Alzheimer's disease. Annals of Telecommunications 65 (2010)

Performance of a Near-Field Radio-Frequency Pressure Sensing Method in Compression Garment Application

Timo Juhani Salpavaara[*], Jarmo Verho, and Pekka Kumpulainen

Tampere University of Technology,
Korkeakoulunkatu 3, 33101 Tampere, Finland
{timo.salpavaara,jarmo.verho,pekka.kumpulainen}@tut.fi

Abstract. Information on the applied pressure is critical to the pressure garment treatment. The performance of a close-range radio-frequency pressure measurement method is evaluated. The aim of the measurement is to sense the pressure under pressure garments. The hand-held measurement unit is used to inductively read a passive resonance sensor. The response and the repeatability of a new pressure sensor structure are tested. The performance of the telemetry is tested by altering the distance, angle and alignment between the measurement unit and the sensor. The functioning of the read-out method is tested within the useful frequency range of the sensor. The effects of the measurement environment are studied. According to the results, the tested measurement method is acceptable in this application.

Keywords: RF telemetry, passive resonance sensor, pressure measurement.

1   Introduction

The pressure garment treatment has potential to improve the healing process of burns and to reduce the swelling. The use of suitable pressure is critical to the treatment and thus, in order to ensure the proper functioning of the pressure garment, this pressure has to be measured periodically. Close-range wireless sensors have an obvious niche in this application since tubing and electrical wiring used in conventional sensors make the measurement of pressure inconvenient and unreliable. Tubing and electrical wiring are a significant hindrance, especially when tight pressure garments are put on. An alternative wireless method for obtaining the measurements is to use passive resonance sensors. The advantage of this method, when compared to the more common wireless techniques, is the simplicity of the sensor, which promotes the idea of disposable sensors. In our earlier work, we have presented the methods needed for this measurement [1]. The other applications for this type of wireless telemetry in medical settings include intra-ocular pressure sensing [2,3] and ECG-measurements [4,5].

In this paper, we present the performance of the new hand-held measurement device and the new pressure sensors. The sensor and the read-out methods are

---

[*] Corresponding author.

K.S. Nikita et al. (Eds.): MobiHealth 2011, LNICST 83, pp. 322–328, 2012.

introduced in section 2. The pressure response of the sensor is measured in section 3.1. The effects of the positioning between the reader and the sensor are tested in section 3.2. The functioning of the read-out method is tested within the useful frequency range of the sensor in section 3.3. The effects of the environment on measurement are studied section 3.4.

2   Pressure Sensor and Read-Out Methods

The instrumentation in this work (Fig. 1) consists of a hand-held measurement unit (13 cm by 7 cm by 1.5 cm) and the passive resonance sensors (17 mm by 17 mm by 3mm). The measurement unit uses a radio-frequency inductive link to wirelessly read the sensors through non-conductive materials like clothing. The measurement unit sweeps over the specified frequency range, measuring the phase response (phase dip) of the sensor. The sensor consists of a pressure dependent capacitor and a resonance coil, which also doubles as a link coil. Thus, the pressure on the sensor alters the capacitance, which in turn alters the phase response. The power consumption of the used short-range inductive link is relatively small: the excitation power used by the measurement unit is 600μW or less.

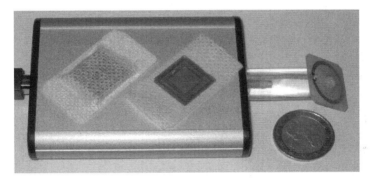

Fig. 1. The hand-held measurement unit and passive resonance sensors (on top of device). The sensors are attached to the measurement target with an adhesive bandage.

The measurement unit transmits the measured data to a PC via a USB port for post processing. First, the phase dip features (uncompensated frequency and the height of the phase dip) are extracted. The value of the height of the phase dip is used to select valid data points (dip height from 3 to 25 degrees). Next the PC post-processing software calculates an estimate of the compensated resonance frequency. The compensated frequency is then compared to the reference value and converted to pressure. The reference value is acquired by the tuning procedure [5]. The preselection of the data points is required, because the data points with a small phase dip are noisy and the data points with a large phase dip do not fit the regression model which is used in the compensation.The more exhaustive description of the used methods can be found in [1,5].

3   Performance Tests

In this section, the response of the sensor is measured and the effects of the positioning between the sensor coil and the reader coil are studied. The functioning of the read-out method is tested within the useful frequency range. Finally, the effects of the measurement environment are tested.

3.1   Pressure Response

In order to evaluate the response and repeatability of the new sensor structure, the resonance frequency of the sensor is measured as a function of pressure in a test setup, where the applied pressure can be controlled.

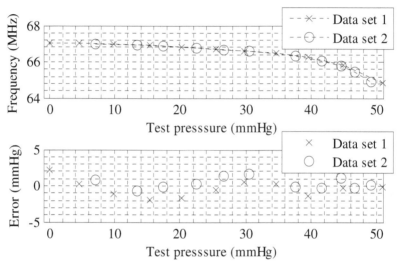

Fig. 2. (a:upper) The response of the sensor was measured with test pressures ranging from 0 to 50 mmHg twice. (b:lower) Identification error.

The resonance frequency of the sensor is measured at each calibration pressure wirelessly with the hand-held reader unit. Each data point is an average of about 100 samples. The pressure response curve measurement is repeated twice. The response curves are shown in Fig. 2a. According to results, the sensor has sufficient repeatability. The maximum measured frequency shift is -2.2 MHz at 50.8 mmHg. The response of the sensor is nonlinear. At around 20 mmHg pressure, 1 mmHg pressure shift equals roughly 80 kHz frequency shift. The model for the response is identified with an ANFIS model [6]. The identification error is calculated (Fig. 2b). These errors are acceptable in this application.

3.2   Effects of Positioning

Since the used coils are smaller than in the previous studies and the reading distance is dependent on the coil dimensions, the maximum reading distance is measured. The effects of positioning (distance, angle and alignment) are also studied because these factors are unknown in real measurement situation and may cause errors to the reading.

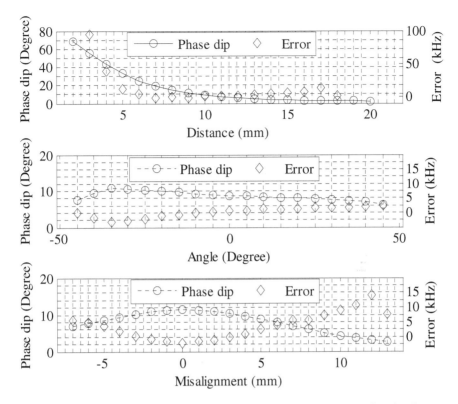

Fig. 3. The performance of the used compensation method is tested by changing the distance (a:upper) and angle (b:middle) between the reader and the sensor coils. The effect of misalignment is also tested (c:lower).

In our measurement method, the sensor causes a dip in the measured phase response curve. The height of the phase dip can be considered as a quantity that indicates the coupling between the coils. The height of the phase dip also shifts the detected resonance frequency but we have a method for compensating this effect. The height of the phase dip and the error in the compensated frequency readings are shown in Fig. 3a as a function of the distance between the coils. These values are an average of about 100 samples. In this measurement, the algorithm is allowed to calculate compensated readings at the extended range (2.02 degrees to 70 degrees). The used read-out method is able to detect the resonance frequency of the sensor between the distances from 2 mm to 20 mm. However, by using regular limits (3 to 25 degrees) the range is limited between 6 mm to 16 mm.

The errors caused by the angle (Fig 3b) and the misalignment (Fig 3c) between the coils are also measured. These measurements are made at the distance of 10 mm. According to these results, neither the angle nor the misalignment cause significant errors compared to the distance.

3.3   Effects of Frequency Range

In this section, the effects of the frequency range of the sensor are studied. Performance of the read-out method is tested when the resonance frequency of the sensor varied within a large frequency range.

First, the uncompensated resonance frequency and the height of the phase dip of a test sensor circuit are measured at the distance varying from 1 mm to 25 mm. This test circuit is similar to used in pressure sensors and it has a 22 pF bulk capacitor. Then 0.5 pF capacitors are added to the circuit to decrease the resonance frequency and the measurement is repeated. The additional capacitors decrease the resonance frequency by 1.75 MHz from 67.64 MHz to 65.89 MHz when all the three 0.5 pF capacitors are added. The frequency decrease is roughly 580 kHz per capacitor. This range of the frequency shifts covers most of the range of the tested resonance sensor. These measured datasets are shown in Fig 4. The corresponding compensated frequencies for each dataset are calculated and drawn at the phase dip value of zero. This drawn value is an average of the valid data points. The standard deviations of the compensated frequencies for the 22 pF and 23.5 pF datasets are 3.4 kHz and 3.7 kHz. In addition, the regression models of each dataset are also shown. These regression models are created according to valid data points which have a phase dip value from 3 degree to 25 degrees.

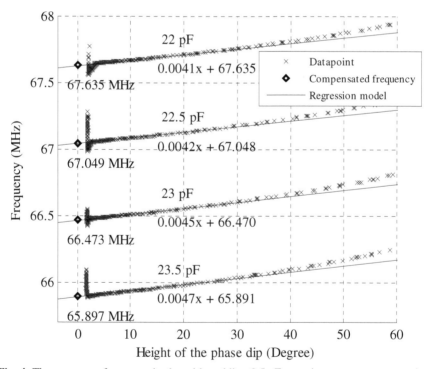

Fig. 4. The resonance frequency is altered by adding 0.5 pF capacitors to a test sensor circuit. The measurements are made from the distance ranging from 1 to 25 mm. The detected uncompensated frequencies are shown with their corresponding heights of phase dips.

The regression model lines almost meet the compensated frequency estimate at the phase dip value of zero. This indicates that the read-out method is valid at tested frequency range. The tuning procedure used for the compensation was done before any 0.5 pF capacitors are added. However, the slope of the regression models seems to increase slightly.

3.4   Effects of Environment

The factors of the environment, for example the changes in the permittivity or permeability, can affect the resonance frequency of an unshielded LC-resonator. Thus, we study the effects of the situations that may occur in the pressure garment application. In this measurement, there is a layer of pressure garment between the sensor and the reader. The garment is prone to absorb sweat and moisture. In addition to that, the effects of the metal objects are tested since a person may carry coins or have snap fasteners in his/her clothing.

The effect of the moisture level in the garment is measured in a test setup in which there is a 10 mm distance between the sensor and the reader. The garment is placed between the sensor and the reader with plastic fasteners. The moisture in the garment is varied by adding salt water. The moisture content in the pressure garment is stated as a percentage of the total mass. The effect of coins and snap fasteners is also tested. The used reading method does not function well if metal objects are placed directly between the sensor and the reader. Instead, we test how the coins or snap fasteners affect the reading if they are placed in the same plane as the sensor. The tests are made by placing the objects right next to and 1 cm away from the sensor.

The errors in the readings are shown in the Table 1. These values are an average of about 100 samples. The errors in pressure are calculated by using the ANFIS-model. There was no external load on the sensor which is the worst case scenario since the sensitivity of the sensor increases with increasing pressure. Results show that, the metallic objects placed in the same plane with the sensor make the measurement underestimate the pressure, while moisture increases the reading. Note that the used ANFIS-model causes a slight error at very small frequency shifts in addition to the environment.

Table 1. The effects of the environmental factors to the reading of a sensor without external load

Disturbance	Error (kHz)	Error (mmHg)
Test setup	-9.7	2.1
Pressure garment (dry)	-20.1	3.0
Pressure garment (10%)	-47.3	5.1
Pressure garment (18%)	-60.2	6.1
Pressure garment (31%)	-121.4	10.8
Pressure garment (47%)	-160.0	13.6
Snap fastener	82.3	-5.1
Snap fastener 1 cm away	-5.3	1.8
Coin	123.9	-8.5
Coin 1 cm away	-5.6	1.9

4   Discussion

The repeatability of the tested sensor is sufficient for this application. The unknown positioning between the reader device and the sensor is not a problem according to the tests. The height of the phase dip indicates when the sensor is within range. This can be indicated to the user. The reading distance that can be achieved is acceptable in this application since the pressure garments are usually made of thin fabric.

The tests made by varying the resonance frequency of the sensor by adding capacitors show that the used compensation method is functioning within the range that is required to measure the resonance frequency of the tested pressure sensor. However, the slight increase in the slopes of the made regression models indicate that the performance of the compensation will decline when the shift in the resonance frequency of sensor increases.

According to the tests made by varying the environment around the sensor, the moisture in the pressure garment has a notable effect on measurement if the garment is soaked. This is a problem especially at the beginning of the pressure range. However, this pressure range is not very important to the application. A dry pressure garment does not seem to disturb the measurement significantly. The metallic objects right next to the sensor affect the measurement, but this effect is insignificant if the objects are moved at least 1 cm away from the sensor.

The overall performance of the used method is good according to the tests. The hand-held measurement device allows much more convenient and unobtrusive measurement of pressures compared to the wired methods. The convenience of the measurement promotes the more regular pressure measurements, which will lead to better pressure garment treatment.

References

1. Salpavaara, T., Verho, J., Kumpulainen, P., Lekkala, J.: Readout methods for an inductively coupled resonance sensor used in pressure garment application. Sens. Actuators, A. In Press, Corrected Proof (March 5, 2011)
2. Chen, P.-J., Saati, S., Varma, R., Humayun, M.S., Tai, Y.-C.: Wireless Intraocular Pressure Sensing Using Microfabricated Minimally Invasive Flexible-Coiled LC Sensor Implant. J. Microelectromech. Syst. 19, 721–734 (2010)
3. Collins, C.C.: Miniature Passive Pressure Transensor for Implanting in the Eye. IEEE Trans. Biomed. Eng. 14, 74–83 (1967)
4. Riistama, J., Aittokallio, E., Verho, J., Lekkala, J.: Totally passive wireless biopotential measurement sensor by utilizing inductively coupled resonance circuits. Sens. Actuators, A. 157, 313–321 (2010)
5. Salpavaara, T., Verho, J., Kumpulainen, P., Lekkala, J.: Wireless interrogation techniques for sensors utilizing inductively coupled resonance circuits. Procedia. Eng. 5, 216–219 (2010)
6. Jang, J.-S.R.: ANFIS: Adaptive-Network-based Fuzzy Inference Systems. IEEE Transactions on Systems, Man, and Cybernetics 23(3), 665–685 (1993)

Hybrid Vital Sensor of Health Monitoring System for the Elderly

Dong Ik Shin[1], Ji Hoon Song[2], Se Kyeong Joo[3], and Soo Jin Huh[3]

[1] Dept. of Biomedical Engineering, Asan Medical Center, Seoul, Korea
[2] SooEe Electronics Co. SeongNam, Korea
[3] Dept. of Biomedical Engineering, University of Ulsan College of Medicine, Korea
{kbread,skjoo,sjhuh}@amc.seoul.kr,
jhsong@sooee.co.kr

Abstract. There are many sensors to monitor vital signs in u-Healthcare system. These vital sensors including ECG, PPG, blood pressure sensor spend heavy processing resource and costs. We propose and developing a new type of hybrid vital sensor. We combine accelerometer and PPG module and control two basic sensors with classified situations. So, we can monitor vital signs more compactly, inexpensively and conveniently using our hybrid sensor. We measured the activity using 3-axis accelerometer and measured the heart rate and oxygen saturation using pulse oxymeter. The major problem of pulse oxymeter is motion artifact. But we suggested a new method using the combination of these two sensors. In case of active motion, we used and analyzed the accelerometer signal and withdraw the pulse oxymeter signal. In case of no activity, we adopt pulse oxymeter signal which has no motion artifacts. The important thing is to categorize activity patterns such as normal or abnormal activity. We categorized activities to 4 patterns which are normal activity, no activity(resting), sleeping and abnormal state. When the device detects abnormal condition, it sends a short message to server and then connected to the u-Healthcare center or emergency center.

Keywords: Hybrid, Vital, Sensor, Health, Monitoring.

1   Introduction

Rapid transition to aging society becomes very important problem day by day. Especially for the single elderly, it is critical problem that whose vital situation. According to the data from Statistics Korea, the aging index will increase rapidly from 9.5%(2006) to 14.3%(2018) and 20.8%(2026). With this trend, the number of single elderly is increases too. Knowing the emergency status of these single elderly is a critical problem in the emergency monitoring system. But the more exact status we want the more expensive and complex sensors will be necessary for this system.

In this research, we proposed a new hybrid method to extract vital sign using the accelerometer and the PPG sensor. Generally, PPG sensor needs more processing resource and power than the accelerometer. If we can classify a person's status to

K.S. Nikita et al. (Eds.): MobiHealth 2011, LNICST 83, pp. 329–334, 2012.

normal or abnormal, we can make more powerful investment in case of abnormal status. As a result, we may reduce processing resource, power and finally physical size of the sensor. The more compact size and reduced processing power will be helpful of wearing it.

2   Methods

2.1   System Block

Fig. 1 shows the illustration of our hybrid vital sensor. Touch sensor and temperature sensor decide the wearing status. Our sensor module will be worn on the wrist. The basic hybrid components are the PPG module and the accelerometer module. The processor classifies a person's status with the data from accelerometer and control the PPG module. We adopt 3-axis accelerometer which has maximum 8g gravity range and 50Hz acquisition rate. PPG sensor module has dual optical sensor which are arranged separately for more robust working. The basic idea is, categorize the human state using accelerometer and then switch on & off PPG sensor according to the categorized state. Then we can reduce the overall power consumption and increase the accuracy. Other sensors such as temperature and touch sensor are used to identifying wearing the module or not, and are used to the wearer's response.

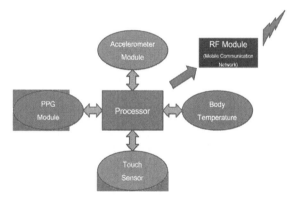

Fig. 1. System block of the hybrid vital sensor

We categorized the human state using the accelerometer. According to these states, our hybrid sensor works in appropriate operation mode. In normal activity, there exist motion artifacts, so pulse oxymeter is useless so we put the PPG in sleep state and activate the accelerometer. In resting state we can get the heart rate using pulse oxymeter. In sleep state further information that is O2 saturation can be obtained. In abnormal state, all sensors will be activated to monitor and judge the accurate status. These operation modes are illustrated in Fig. 2.

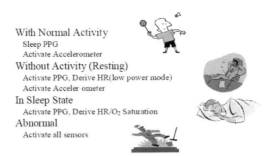

With Normal Activity
   Sleep PPG
   Activate Accelerometer
Without Activity (Resting)
   Activate PPG, Derive HR(low power mode)
   Activate Accelerometer
In Sleep State
   Activate PPG, Derive HR/O₂ Saturation
Abnormal
   Activate all sensors

Fig. 2. Operation modes of hybrid vital sensor

2.2   Algorithm

Fig. 3 is an algorithm of hybrid vital sensor. When the algorithm starts, system power on and initialize each sensors. Then, first the activity data is acquired from accelerometer and classifies activities to appropriate patterns. Periodical activities such as working, running and other regular activities are classified to "Normal Activity". When activity is classified to "Abnormal" the PPG module is powered on for more classification. The PPG module derives heart rate using simple algorithm which uses less power than deriving oxygen saturation. Derived heart rate is analyzed with activity to determine a person's accurate status. If it is classified simple resting status, the algorithm continues heart rate monitoring. In the case of abnormal heart rate, the PPG module measures the oxygen saturation to determine emergency status.

In real world, situations are more complex and ambiguous. So, the classification algorithm is difficult. But as refine more accurately the algorithm, the result will be more realistic.

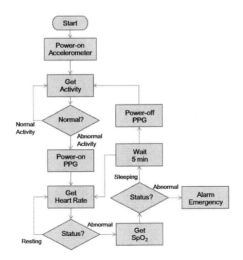

Fig. 3. Hybrid vital sensor algorithm

3   Results

Fig. 4 ~ Fig. 5 show classified results according to our algorithm. Data from sensors are transmitted to personal computer and are processed with Labview software to verify our algorithm. Transplant to micro-controller is being performed.

Once we classify the elderly activity to abnormal we can further investigates the accurate status with the reaction button or pulse oximeter which will be adopted our monitoring device.

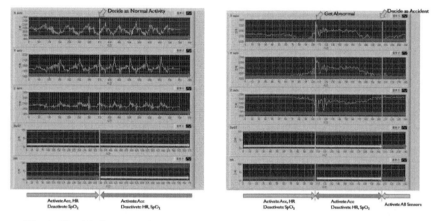

Fig. 4. Classified to normal Activity   **Fig. 5.** Classified to Accident

If we can classify a person's status to normal or abnormal, we can make more powerful investment in case of abnormal status. As a result, we may reduce processing resource, power and finally physical size of the sensor. The more compact size and reduced processing power will be helpful of wearing it.

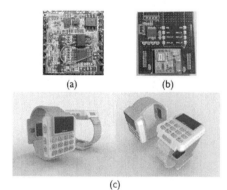

(a)   (b)

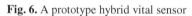

(c)

Fig. 6. A prototype hybrid vital sensor

(a) Hybrid sensor   (b) Main module
(c) An assembled watch-type mobile device

Fig. 6 is our prototype. Our watch-type wearable device is composed of two main parts. Battery part and main module parts. We use the RF module which is commercially released by Seoul Mobile Telecom company. The data rate is six thousands four hundreds bps. Frequency band is separated to up and down.

4   Conclusion

The main idea of our hybrid vital sensor is control each sensor properly. What we want to say is, in each status, by controlling and activating each sensor properly we can raise the overall efficiency and accuracy. So we can compose the vital sensor more inexpensive, power saving and compact. As a result, this cost effective and easily wearable hybrid vital sensor will raise the use of u-Healthcare devices.

References

1. Hafez, A.A., et al.: Design of a low-power ZigBee receiver front-end for wireless sensors. Micro Electronics Journal (in Progress)
2. Anastasi, G., Conti, M., Di Francesco, M., Passarell, A.: Energy conservation in wireless sensor networks: A survey. Ad Hoc Networks 7, 537–568 (2009)
3. Hautefeuille, M., et al.: Miniaturized multi-MEMS sensor development. Microelectronics Reliability 49, 621–626 (2009)
4. Mendoza, P., Gonzalez, P., Villanueva, B., Haltiwanger, E., Nazeran, H.: A Web-based Vital Sign Telemonitor and Recorder for Telemedicine Applications. In: Proceedings of the 26th Annual International Conference of the IEEE EMBS, San Francisco, CA, USA, September 1-5, pp. 2196–2199 (2004)
5. Yang, B.-H., et al.: Development of the ring sensor for healthcare automation. Robotics and Autonomous Systems 30, 273–281 (2000)
6. Kulkarni, P., et al.: mPHASiS: Mobile patient healthcare and sensor information system. Journal of Network and Computer Applications (2010)
7. Letterstål, A., Larsson, F.: Assessment of vital signs on admission to short time emergency wards improves patient safety and cost-effectiveness. Australasian Emergency Nursing Journal 10(4), 191 (2007)
8. Yönt, G.H., et al.: Comparison of oxygen saturation values and measurement times by pulse oximetry in various parts of the body. Applied Nursing Research (2010)
9. Barker, S.J., Morgan, S., et al.: A Laboratory Comparison of the Newest "Motion-Resistant" Pulse Oximeters During Motion and Hypoxemia. Anesthesia and Analgesia 98(55), S2:A6 (2004)
10. Ubeyli, E.D., et al.: Analysis of human PPG, ECG and EEG signals by eigenvector methods. Digital Signal Processing 20, 956–963 (2010)
11. Shin, H.S., Lee, C., Lee, M.: Adaptive threshold method for the peak detection of photoplethysmographic waveform. Computers in Biology and Medicine 39, 1145–1152 (2009)
12. Liu, S.-H., et al.: Heart rate extraction from photoplethysmogram on fuzzy logic discriminator. Engineering Applications of Artificial Intelligence 23, 968–977 (2010)

13. Lee, J., et al.: Design of filter to reject motion artifact of pulse oximetry. Computer Standards & Interfaces 26, 241–249 (2004)
14. Tay, F.E.H., et al.: MEMSWear-biomonitoring system for remote vital signs monitoring. Journal of the Franklin Institute 346, 531–542 (2009)
15. Rialle, V., et al.: Telemonitoring of patients at home: a software agent approach. Computer Methods and Programs in Biomedicine 72, 257–268 (2003)
16. Vergados, D.D., et al.: Service Personalisation for assistive Living in a Mobile Ambient – Healthcare Networked Environment. Personal and Ubiquitous Computing 14(6), 575–590 (2010)
17. Ngo, V., et al.: A detailed review of energy-efficient medium access control protocols for mobile sensor networks. Computers and Electrical Engineering 36(2), 383–396 (2010)
18. Anastasi, G., et al.: Energy conservation in wireless sensor networks: A survey. Ad. Hoc. Networks 7, 537–568 (2009)

Low-Cost Blood Pressure Monitor Device for Developing Countries

Carlos Arteta[1,2,*], João S. Domingos[1,2,*], Marco A.F. Pimentel[1,2,*],
Mauro D. Santos[1,2,*], Corentin Chiffot[2], David Springer[2],
Arvind Raghu[2], and Gari D. Clifford[2]

[1] Centre for Doctoral Training in Healthcare Innovation, Dept. of Engineering Science,
University of Oxford, Oxford, UK
[2] Institute of Biomedical Engineering, Dept. of Engineering Science,
University of Oxford, Oxford, UK
{carlos.arteta,domingos.domingos,marco.pimentel,mauro.santos,
corentin.chiffot,david.springer,arvind.raghu,
gari.clifford}@eng.ox.ac.uk

Abstract. Taking the Blood Pressure (BP) with a traditional sphygmomanometer requires a trained user. In developed countries, patients who need to monitor their BP at home usually acquire an electronic BP device with an automatic inflate/deflate cycle that determines the BP through the oscillometric method. For patients in resource constrained regions automated BP measurement devices are scarce because supply channels are limited and relative costs are high. Consequently, routine screening for and monitoring of hypertension is not common place. In this project we aim to offer an alternative strategy to measure BP and Heart Rate (HR) in developing countries. Given that mobile phones are becoming increasingly available and affordable in these regions, we designed a system that comprises low-cost peripherals with minimal electronics, offloading the main processing to the phone. A simple pressure sensor passes information to the mobile phone and the oscillometric method is used to determine BP and HR. Data are then transmitted to a central medical record to reduce errors in time stamping and information loss.

Keywords: Blood pressure, developing countries, electronic medical records, hypertension, mHealth, resource-constrained healthcare, low-cost devices.

1 Introduction

Globally, hypertension is a major chronic, non-communicable disease and a leading cause of death and disability in economically developing countries [1]. Hypertension, a sustained elevated blood pressure[1] (BP), is a dangerous medical condition that

---

* These authors had equal contributions for this paper.
[1] A systolic BP consistently above 140mmHg and/or a diastolic BP consistently over 85mmHg, although definitions vary with age, gender, disease factors and measurement location.

K.S. Nikita et al. (Eds.): MobiHealth 2011, LNICST 83, pp. 335–342, 2012.
© Institute for Computer Sciences, Social Informatics and Telecommunications Engineering 2012

stresses the heart and promotes vascular weakness and scaring, making blood vessels more prone to rupture [2]. Uncontrolled and untreated hypertension increases the risk of coronary arteries damage, heart attack, stroke, kidney disease, eye damage and is responsible for other conditions such as pre-eclampsia. Most of these problems can be found in developing countries and are a serious economic burden [1,3,4].

Kearney et al. [5] predicted that by 2025 the number of hypertensive adult individuals will approximately be 1.56 billion, 1.15 billion of which are from developing nations [5]. Kearney et al.'s findings also indicate a higher prevalence of hypertension in developed countries (37.3%) than in developing ones (22.9%). However, given the much larger population of developing countries, the absolute number of patients affected by hypertension is considerably larger and is likely to grow [5].

Barriers to the treatment of hypertension include i) lack of detailed data on an individual's BP readings over time, ii) poor infrastructure to deliver medication and information, iii) lack of training in taking reliable BP readings, and iv) lack of financial resources [6,7]. Although manual readings may be suggested to be sufficient, manual recording of data is error-prone and leads to blood pressure overestimation or critical BP-related event misses [8,9]. In particular, lack of data leads to low hypertension awareness rates [10,11] in developing nations and therefore screening for hypertension and the measurement of BP are of critical importance in developing countries.

Nevertheless, current BP electronic devices that enable automatic BP measurements are expensive for the developing countries. In developed countries, one button click, clinically validated electronic BP devices can be found for the upper arm with several features: automatic cuff inflation/deflation cycle, alarm reminders, readings storage, irregular heartbeat or hypertension indication and download of readings to a Personal Computer (PC). These are very useful to monitor BP and heart rate (HR) at home and doctor office visits. Such devices range in price from €20 to €150 depending on the number of features, accuracy of the measurement and quality of the hardware. Therefore they may never be widely available at a reasonable price for the average person in a developing country. However, much of the hardware needed for an automated BP measurement is already in most people's pocket and the extra hardware needed could be manufactured for less than €5. According to the International Telecommunication Union, mobile cellular subscriptions have been increasing significantly over the last decade, particularly in developing countries where the growth is much bigger[12]. Since poor supply-chains are one of the biggest issues in medical device distribution and training, mobile phone offer an enormous potential.

The mobile phone provides the hardware and software required to extract the features from vital signals, calculate and display the results and automatically save them to either a local or remote database. Moreover, it can provide the user with advice on usage, quality and decision support. By off-loading the analysis to a software application running on the phone, software support can be provided without the need for specialist knowledge, thus allowing continual system improvement and optimization for specific communities, such as the hypertense. The mobile platform would also allow for the automatic synchronisation of the BP readings in a remote electronic medical record (EMR). Such data could then be shared with authorized specialists, providing a permanent monitoring service, and an auditing path to improve the system.

In this paper we present and pre-validate a low-cost easy-to-use BP monitor device for developing countries running on a mobile phone through the use of a typical cuff. The solution evades the presence of trained personnel to take BP measurements, and allows for medical data storage, avoiding human mistakes such as transcription errors.

2   Methods

In the device described in this paper, the manometer is removed from a traditional sphygmomanometer and its air tube is connected to a pressure sensor integrated in a low-cost hardware. The cuff and tubing taken from the traditional sphygmomanometer costs around €3 and is widely available in many settings. It is also easily replaced by comparable materials. The hardware is connected to a mobile phone using an USB cable so that power is supplied from the phone, no wireless transmission chips are needed (which would quadruple the system cost) and data pairing between the hardware and the phone is instant, secure and trivial. Pumping is performed manually so that there is no significant power drain, or heavy pumps. The application in the mobile phone allows the user to measure BP and HR and then save it to the device database and synchronize it with the patient EMR.

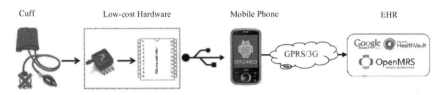

Cuff   Low-cost Hardware   Mobile Phone   EHR

Fig. 1. Overview of the system. Note that the manometer is removed as the phone now calculates the pressure from the output of the pressure transducer.

2.1   Low-Cost Hardware

To minimise the size and cost of the device, the peripheral hardware contains just enough elements to transmit the pressure signal to a mobile phone for the digital signal processing. The final circuit board is shown in Fig. 2.

As is common for electronic BP devices, the air pressure in the inflatable cuff is converted into an analogue electric signal by a pressure sensor (located on the top of Fig. 2). The transducer used in the device presented in this work is the MPXV5050GP from Freescale Semiconductors [13], an on-chip integrated and temperature-compensated pressure sensor. The analogue output signal from the sensor is sampled at 100Hz with a 10-bit resolution (0.29 mmHg per bit) using a PIC18F14K50 from Microchip [14]. The data containing discrete pressure values is sent to the mobile phone through USB, made simple by the USB 2.0 module integrated in the chosen microcontroller. In order to communicate with the peripheral, the mobile phone must support the USB 'On-the-Go' supplement of the USB 2.0 standard, which allows the phone to act as a USB Host to the external device. Furthermore, when using this feature, the peripheral can be powered from the mobile phone.

The resulting peripheral device is a 39x30mm PCB potted inside a small PVC box (recycled from a typical confectionary box at no cost), with a connector for the cuff air tube and a single full size female USB socket. Since most smart phones come with a USB cable of varying connector sizes at one end, which fit the phone, and a full size male USB cable at the other, choosing a female USB socket for this board means that no extra cables need to be purchased.

2.2   Cuff Pressure Signal Processing

The oscillometric method has been widely used in BP monitoring [15-20]. In our algorithm, the source signal is initially pre-filtered using a 5 point median filter to reduce the noise and possible motion artefacts from the signal. A 6th order Butterworth band-pass filter with cut-off frequencies of 0.5 and 5 Hz is then applied in order to obtain the oscillation waveform. This allows the determination of the mean arterial pressure (MAP), systolic BP (SBP) and diastolic BP (DBP) by using a height-based approach [15-17]. Ratios of 50% and 70% out of the maximum amplitude (MAP) were chosen for the SBP and DBP respectively [15]. HR is calculated from the frequency component with highest magnitude of the frequency spectrum generated via Fast Fourier Transform (FFT) of the oscillation waveform.

All signal processing routines were firstly written in MATLAB, and then ported to the Java version compliant with the Dalvik Virtual Machine [21] for Android.

2.3   Software Architecture

The main requirements for the software are: i) the interface must be easy-to-use and guide the user to take high quality BP measurements; ii) the application must support different languages, especially those spoken in developing countries; and iii) the device must save BP measurements to a local database to provide review capabilities, and to a remote system to allow the back-up of data, device-independence, and sharing with a healthcare providers.

The main elements of our system are: 1) the *help page* has pictograms that allow the user to learn easily how to interact with the device and take the BP in a proper way even if they are partially literate; 2) the *measure page* guides the user during the measure and shows his BP at the end of recording; 3) the *measure list page* shows the list of measurements saved by the user and also provides a way to synchronise them with a specific EMR; 4) the *measure view page* displays the information of a specific measure as well as a field to attach useful notes to the data; and 5) the database contains all the saved measurement information. Finally, the application allows the user to save the entire signal to a file in the mobile phone. This information can then be synchronised with a database of BP signals to improve signal processing strategies.

2.4   Prototype

Fig. 2 shows the experimental setup of the mobile BP measure device. The manometer is removed from a traditional sphygmomanometer and its tube connected to the pressure sensor on the PCB. The latter is encased in a free confectionary enclosure and connected to the mobile phone using a standard USB cable.

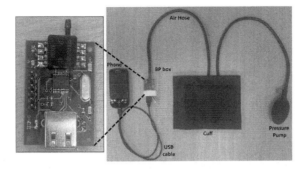

Fig. 2. Experimental setup of the mobile BP measure device. Note that the sphygmomanometer is not used. The cuff and pump were purchased with the sphygmomanometer for less than $10.

The application was implemented using Android 2.2 Software Development Kit (SDK) on an Android device that supported USB host mode. The application comprises 5 pages: *main page, help page, measure page, measure view page* and *measure list page*, that were implemented as Activity classes from the Android SDK [22]. To guide the BP measurement, the *measure page* has a chart that renders the real time pressure values coming from the cuff while the measure is being taken, suggesting that the user to pump until a pressure of 200 mmHg (or another defined limit based on historical or user-selected values) is reached and stop pumping. The device is then allowed to slowly deflate. The application also identifies if the deflation was too fast to extract the oscillations from the recording. In such a case the application requests the user to repeat the measurement with a slower deflation rate. Otherwise, after the pressure drops to 30 mmHg the measurement stops, the SBP, DBP and HR are determined from the processed pressure recording and the result is shown to the user. The user can then save the measurement to the database and also create a CSV file with all the information used to derive the measurement (for debugging and development). This file is only created on the mobile phone if the micro SD card is available with sufficient free space. The application has been translated to 7 different languages. The application, and PCB design are open-source and available in the project website [23].

3   Results and Validation

The prototype was used in 5 different healthy volunteers (1 female, mean age 25.0, range 24-27 years, 1 Hispanic, 3 Caucasian and 1 Afro-European) to measure SBP, DBP and HR. The automatic Boots Arm Cuff BP Monitor 5690447 (The Boots Company PLC, Nottingham, England), which is well known CE marked device, was used in order to pre-validate the BP and HR measurements.

Three measurements with both devices were taken for each subject in the following conditions: i) after lying down for 5 minutes; ii) after running for 5 minutes and iii) after inserting the right hand in a bucket of ice water for 1 minute. These measures try to simulate daily events like, a person at rest, a person after physical activity and a person undergoing a stressful condition, respectively. The measurements were performed on the left arm with the subject sitting down with the cuff at the same height as the heart, except for the first condition where the subject remained lay down. A 5-minute period was given between the readings from the two devices.

The percent difference (mean ± standard deviation) between our device and the Boots device were compared for each individual over all three types of tests and all individuals for each type of test, and the mean percent differences for all individuals over all readings were 14.5 ± 9.8 %, 5.0 ± 5.3 % and 8.1 ± 6.9 % for SBP, DBP and HR, respectively.

4   Conclusion and Discussion

A low-cost and open source BP monitor has been developed to function as a mHealth device for developing countries (see [23]). The preliminary data showed a good agreement between the values of SBP, DBP and HR measured with a known device, with an average error of less than 20%, which is within the accuracy levels of the Boots device itself. A standard inflatable cuff and an inexpensive peripheral make use of the processing power in mobile phones to digitally process the pressure signal and compute the BP and HR values using the oscillometric method. The software developed for Android powered devices allows the user to keep records of the measurements in their mobile phone and/or upload them to EMR systems using the connectivity provided by the phone. Such an implementation would serve as an efficient platform for sampling and analysing BP data from large populations. This could subsequently be used to monitor treatments and in epidemiological studies.

Further work includes completing the validation of the device on larger populations, software improvements to support different cuff sizes and integration with other applications such as SANA Mobile [24]. Calibration for different age groups hypertensive patients is also required, since standard oscillometric methods are known to lead to clinical error through under- and over-estimation of blood pressure [25-28]. We also hope to include warnings and guidelines for bradycardia, tachycardia, hypertension, hypotension and arrhythmia, as well as integration with other decision support tools such as cardiovascular risk indicators. However, we do not envision this system as a replacement for trained medical oversight at this stage. Therefore, integration with a medical record system shared or supported by a medically-trained worker will be important.

Earlier in this paper we noted that barriers to the treatment of hypertension, and in fact healthcare delivery in general, include i) lack of detailed temporal data ii) poor infrastructure, iii) lack of training, and iv) lack of financial resources. The first of these barriers can be addressed by frequent measurement-taking (if a suitable device is available), and transmission of the data to a central medical record. This is only likely to happen if the individual or a close care-giver is able to take readings. Poor infrastructure can be addressed by using existing supply channels, such as Coca Cola's delivery trucks, or a more decentralised approach such as that taken by mobile phone companies. Training can be addressed by adding user-feedback and intelligent processing into the device. Our system addresses all these problems (and solutions).

An obvious criticism remains, in that Android only runs on smart phones, which are not widely available in developing countries. However, Android represents the best choice of development platform for several reasons. First, Android is constantly being ported to cheaper devices with the price point dropping from several hundred dollars to under one hundred dollars in just a couple of years. Over the last year, Android adoption has increase 886%, with the closest competitor, the iPhone, increasing only by 86% [29]. Moreover, in contrast to the iPhone, Android's gains have been largely in developing countries.

Android, being adopted by Google, has followed its philosophy of cloud data storage and analysis. An individual's data medical record can be stored in the cloud, and ported across several devices and shared with several users. Even when phone sharing is common, Android allows you to authenticate a single device with several accounts. Finally, we should note that our BP system is initially aimed at healthcare workers, where deployment of the phone is essentially a low overhead compared to the rest of the clinical trial. By the time the system has been proved to be efficacious (or not), the price point for Android phones is highly likely to have tumbled even further.

Our code is basically vanilla Java and/or C, which means that cross-platform portability is technically simple. The issue lies purely in the way the manufacturers decide to lock down the hardware. Unfortunately most phone OS's do not allow good low level interaction with peripheral sensors or devices. In particular, true USB host capacity is currently only available for mobile phones under Android (and on a limited number of iOS devices). In general there has been a move towards using Bluetooth for medical devices to interface with phones which is partly driven by the lack of USB hosting on phones. However, Bluetooth is more expensive, requires more energy, drops packets, and can be easily 'sniffed' raising privacy concerns. USB tethering increases data quality and privacy, allows delivery of power in an efficient manner and prevents the phone from being plugged into the mains during medical device use, thereby enhancing isolation. Of course, if the low cost pressure sensor is integrated into the phone, these issues are mute.

The final point to consider is how to deploy the non-phone hardware. Although the cuff, pump and tubing in a typical manual device cost about €3, and are easily replaced by local equipment, the electronics are not. It is therefore our intention to make the electronics available either as built-in systems for phones or as a low-cost add-on for the phone available through the same supply channels for phone accessories such as the charger, USB cables and batteries.

Acknowledgments. CA, JSD, MAFP and MDS acknowledge the support of the RCUK Digital Economy Programme grant number EP/G036861/1. The authors thank Catarina Figueiras (MD) for supporting the validation process.

References

1. Murray, C., Lopez, A.: Mortality by cause for eight regions of the world: Global Burden of Disease Study. The Lancet 349(9061), 1269–1276 (1997)
2. Goldman, L., Ausiello, D.: Cecil Medicine. Saunders Elsevier, PA (2008)
3. Daar, A., Singer, P., et al.: Grand challenges in chronic non-communicable diseases. Nature 450(7169), 494–496 (2007)

4. Nugent, R.: Chronic diseases in developing countries. Ann. N. Y. Acad. Sci. 1136(7), 70–79 (2008)
5. Kearney, P., Whelton, M., et al.: Global burden of hypertension: analysis of worldwide data. The Lancet 365(9455), 217–223 (2005)
6. Mittal, B., Singh, A.: Hypertension in the developing world: challenges and opportunities. Am. J. Kidney Dis. 55(3), 590–598 (2010)
7. Malkin, R.A.: Design of health care technologies for the developing world. Annu. Rev. Biomed. Eng. 9, 567–587 (2007)
8. Hug, C.W., Clifford, G.D.: An analysis of the errors in recorded heart rate and blood pressure in the ICU using a complex set of signal quality metrics. In: CIC, pp. 641–644 (2007)
9. Hug, C.W., Clifford, G.D., Reisner, A.T.: Clinician blood pressure documentation of stable intensive care patients: An intelligent archiving agent has a higher association with future hypotension. Crit. Care Med. 39(5), 1006 (2011)
10. Damasceno, A., Azevedo, A., et al.: Hypertension Prevalence, Awareness, Treatment, and Control in Mozambique. Hypertension 54(1), 77–83 (2009)
11. Li, H., Meng, Q., Sun, X., et al.: Prevalence, awareness, treatment, and control of hypertension in rural China: results from Shandong Province. J. Hypertens 28(3), 432 (2010)
12. ICU, http://www.itu.int/ITU-D/ict/statistics/index.html
13. Freescale product code: MPXx5050
14. Microchip product code: en533924
15. Ball-Llovera, A., et al.: An experience in implementing the oscillometric algorithm for the noninvasive determination of human blood pressure. In: Proceedings of the 25th Annual International Conference of the IEEE, vol. 4, pp. 3173–3175 (2003)
16. Lin, C., Liu, S., Wang, J., Wen, Z.: Reduction of interference in oscillometric arterial blood pressure measurement using fuzzy logic. IEEE T. Biomed. Eng. 50(4), 432–441 (2003)
17. Wang, J., Lin, C., Liu, S., et al.: Model-based synthetic fuzzy logic controller for indirect BP measurement. IEEE T. Sys. Man Cy. B 32(3), 306–315 (2002)
18. Baker, P., Westenskow, D., Kück, K.: Theoretical analysis of non-invasive oscillometric maximum amplitude algorithm for estimating mean blood pressure. Medical and Biological Engineering and Computing 35(3), 271–278 (1997)
19. Drzewiecki, G., Hood, R., Apple, H.: Theory of the oscillometric maximum and the systolic and diastolic detection ratios. Ann. Biomed. Eng. 22(1), 88–96 (1994)
20. Geddes, L.: Handbook of blood pressure measurement. Humana Press, Clifton (1991)
21. Dalvik Virtual Machine, http://code.google.com/p/dalvik/
22. Android SDK, http://developer.android.com/reference/android/app/Activity.html
23. EWH-BP-Project, http://code.google.com/p/ewh-bp-project/
24. Sana Mobile, http://www.sanamobile.org/
25. Goonasekera, C.D., Dillon, M.J.: Random zero sphygmomanometer versus automatic oscillometric blood pressure monitor; is either the instrument of choice? J. Hum. Hypertens 9(11), 885–889 (1995)
26. Gupta, M., Shennan, A.H., et al.: Accuracy of oscillometric blood pressure monitoring in pregnancy and pre-eclampsia. BJOG 104(3), 350–355 (1997)
27. Natarajan, P., Shennan, A.H., et al.: Comparison of auscultatory and oscillometric automated blood pressure monitors in the setting of preeclampsia. AJOG 181(5), 1203–1210 (2008)
28. Umana, E., Ahmed, W., et al.: Comparison of oscillometric and intraarterial systolic and diastolic BPs in lean, overweight, and obese patients. Angiology 57(1), 41–45 (2008)
29. Canalys, Android smart phone shipments grow 886% year-on-year in Q2 (2010), http://www.canalys.com/newsroom/android-smart-phone-shipments-grow-886-year-year-q2-2010

Telemedical System for Diagnosis
and Therapy of Stress Related Disorders

Stefan Hey

Karlsruhe Institute of Technology,
House of Competence, hiper.campus,
Fritz-Erler-Str. 1-3, 76133 Karlsruhe, Germany
stefan.hey@kit.edu

Abstract. The consequences of stress on the health of a person as well as the impact on the health system attracted more and more the public's attention during the last years. In contrast to this, only a few instruments for an objective assessment of stress for the use in diagnosis and therapy of stress related disorders exist. In this article, a model for the phenomenon "stress" is introduced to understand the complex correlations between psychological, physiological and behavioral level. According to this model, a telemedical system that allows a comprehensive assessment of stress-related parameters is presented. This system consists of a psycho-physiological measurement device based on a chest-strep with dry electrodes in order to measure physiological parameters like ECG, physical activity, breathing and photoplethysmogramm (PPG) and a wirelessly coupled smartphone used as an e-diary and e-questionnaire to assess subjective and behavioral parameters.

Keywords: stress measurement, ambulatory assessment, physical activity, pulse transit time.

1   Introduction

Over the last years, stress and its negative consequences have become one of the most important health problems in the industrialized countries. Stress is one cause for a number of the most serious widespread diseases and responsible for the increase of costs in the health system.

According to the Federal Statistical Office of Germany, in 2004 Germany spent EUR 22.8 billion on therapy of mental diseases where stress is one of the main causes. The high burden of stress is also shown in a survey conducted by a health insurance (Techniker-Krankenkasse) in 2009 (see Fig. 1). After this study, 80 % of Germans are suffering from stress, one third is always stressed.

K.S. Nikita et al. (Eds.): MobiHealth 2011, LNICST 83, pp. 343–350, 2012.
© Institute for Computer Sciences, Social Informatics and Telecommunications Engineering 2012

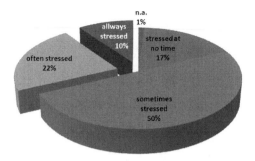

Fig. 1. The incidence of stress in German population (Techniker Krankenkasse, F.A.Z.-Institut, 2009)

Beside this, there has been an increasing interest over the last years in finding new, technology based methods for capturing physiological, subjective and behavioral data in "real time" in the field. Especially in the area of telemonitoring of cardiovascular diseases, a large number of products appeared on the market [1]. The use of these systems in diagnosis and therapy support of stress has some restrictions, because of specific requirements for a stress monitoring system that is able to measure all the relevant aspects of a stress reaction.

A high burden of stress releases reactions on different levels. After a cognitive appraisal of the situation, there could be changes in the emotional state, in the vegetative and hormonal system as well as in the muscular tonus [2]. The reactions of the vegetative and hormonal system could be measured by different methods [3]. It can be differentiated between invasive and non-invasive methods on the one hand and between stationary explorations in a lab environment and mobile explorations in the field on the other hand. There have been a large number of systems that measure some of these aspects, like electrodermal reaction caused by hormonal system or reactions of heart rate or heart rate variability (HRV). Systems for measuring psycho-physiological parameters or stress could be divided in two groups. On one side, you have special devices for stress measurement and biofeedback. Normally these systems are used in relaxation and stress prevention training. They could not be used for mobile and continuous monitoring (e.g. connection to PC). On the other side, you can find mobile monitoring devices. These systems are very small and are normally used for monitoring of one vital signal, but they have disadvantages in establishing long-term electrical contact that do not irritate the skin of the users. For the assessment of subjective data you can find mobile solutions for mood monitoring, which allows the user to fill in basic information on current circumstances and experience.

The goal of a reliable measurement of stress is to assess a holistic view of the whole stress reaction of a person. This has to be done with a simple monitoring system that allows a non-invasive, mobile, continuous and unobtrusive measurement. For this, we need a system with very small size, light weight and electrodes that do not cause discomfort. To understand the complex relationships in the stress process in order to design an appropriate system for the assessment of the different levels of reaction, we use a model based design approach.

1.1   The Stress Model

An integrated stress model that contains aspects from different disciplines was developed by Ice and James [4], see Fig. 2. According to this model stress can be defined as a process that induces emotional, behavioral and physiological reactions. These reactions are dependent on the personal, biological and cultural context of the individual and affect the physical and mental health of that person.

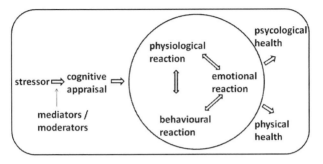

Fig. 2. Stress Model based on Ice and James (2007)

The physiological reactions are caused by an activation of the sympathetic nervous system and by the release of hormones (adrenalin, noradrenalin, testosterone and cortisol). These hormones are released as a consequence of the activation of the hypothalamic-pituitary-adrenal axis (HPA or HTPA axis), the neuroendocrine system that controls reactions to stress and regulates many body processes, i.e. mood and emotions, the immune system, as well as many others. Beside this we can observe behavioral and emotional reactions (like changes in physical activity or time spent for sleeping) which lead to negative affects in physical and psychological health (e.g. heart attack, depression).

2   System for Stress Measurement

The different parts of this stress process need different measurement methods. To get a broad and valid assessment of the whole process, you have to acquire as much parameters as possible and take them into account for the analysis of someone's stress level. During the last years, many hardware and software solutions for multiparametric assessment of physiological signals and experience sampling have been developed [6]. But none of them is able to provide an integrated platform for the assessment of all the parameters defined above.

For the measurement of stressors and the assessment of the cognitive appraisal some approved questionnaires like SAM (Stress Appraisal Measure) [9] and PSS (Perceived Stress Scale) [10] could be used as mobile application on a smartphone. On the smartphone we use the open source software MyExperience as an e-Diary and e-Questionnaire system with the possibility of special extension needed for stress measurement. Therefore we developed some own assessment formats for special parameters like time budget analysis, vigilance or localization of a person (see Fig. 3).

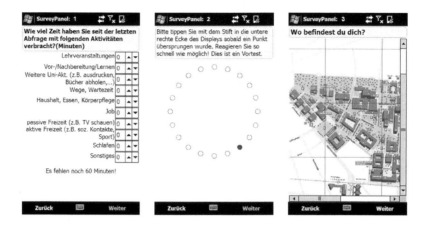

Fig. 3. Special assessment formats for time budget analysis (left), vigilance (middle) and localization (right)

The measurement of physiological reactions in stress process could be done by mobile and unobtrusive physiological sensors. To achieve this goal, we developed a sensor platform for data logging, data processing and wireless data transmission, see Fig. 4. The platform consists of an ultra-low-power microcontroller (MSP430F1611) with 12 bit AD/DA converter, 2 UART interfaces and 48 kB flash and 10 kB RAM. For the storage of the data we use a 2 GB microSD card. The physical activity frontend is built by an ultra low power MEMS 3D acceleration sensor (adxl345) with a range of ± 8 g and a resolution of 4 mg. The acceleration signal is sampled with 64 Hz. The ECG module is a single channel ECG with 12 bit and a sampling rate up to 1024 Hz. With this system we are able to store raw data of the physiological signals up to 12 days. Besides the physiological signals we measure temperature and air pressure to calculate changes in the altitude (BMP085, Bosch GmBH with resolution 15 cm and sampling frequency 1 Hz). These signals are used as context information (e.g. weather) and combined with the acceleration signal for the classification of different types of physical activity and the estimation of energy expenditure. The transmission of raw data after a complete measurement and the power supply is realized by a USB 2.0 interface. The user interface should be very simple, provide an easy to use system and avoid interaction of the subject to the measurement. This is built by a LED with three colors and a vibration motor to give a feedback to the user. User input could be done by tapping on the sensor. This could be detected by the acceleration sensor. Based on this platform we achieved a wearable sensor strap with dry electrodes that is light, small and comfortable for long-term ambulatory measurement of the electrocardiogram and the physical activity.

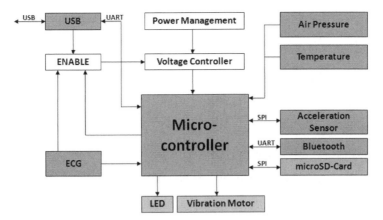

Fig. 4. Hardware Architecture of the Sensor Platform

From the sensor signals physiological features and contextual information is extracted in real-time and can be transmitted to a smartphone via Bluetooth connection, see Fig. 5. For online communication between the sensor and the smartphone we use Bluetooth connection (WML46, Bluetooth-Version 2.0+EDR). Online communication is necessary for interactive assessments where subjective data are collected according to changes or events in physiological signals. The mobile data collecting software is designed to run on smartphones. Smartphones are not only a very good platform for mobile data acquisition but they also provide multimedia features such as digital photo or video capture.

Fig. 5. Integrated measurement system for assessment of stress

Besides the synchronous recording of the physiological data received from the mobile sensors MyExperience also collects self-reports from the participant [7]. These self-reports can be triggered by the user or automatically by physiological or contextual events. With this system a context-aware experience sampling can be achieved [8]. The different levels of stress reactions could also be measured by this system. We use Positive Affect - Negative Affect Schedule (PANAS) [11] and Profile of Mood States (POMS) [12] for assessment of emotional reaction and Ways of

Coping Questionnaire (WCQ) to get information about behavioral stress reaction and the use of coping strategies [13]. An objective measurement of behavior is realized with acceleration sensors, which measure physical activity, energy expenditure and other parameters like step counts [14]. The physiological level of stress reaction can be measured with different sensors based on the platform described before. Besides the measurement of ECG there are mobile sensors for photoplethysmogramm and electrodermal activity that can also be included in the system. With these sensors it is possible to measure the most important stress-related parameters like Heart Rate, Heart Rate Variability (HRV), Additional Heart Rate (AHR), Pulse Transit Time (PTT) [15] [16] and Galvanic Skin Response (GSR).

3   Results

To validate the physiological measurement of stress related vital parameters, we designed a detailed and complex procedure for a study, where we used the „Trier Social Stress Test" (TSST) for induction of moderate psycho-biological stress responses in laboratory environment. To control the evidence of the physiological signals during the different phases of the study, we calculated a Poincare-plot of the R-R-intervals in the baseline phase and the anticipation phase. As we can see in figure 6, there is a significant difference in HRV in phases with stress (anticipation) and phases of relaxation (baseline).

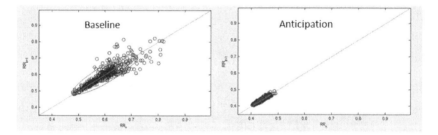

Fig. 6. Poincare Plot for a test person on stress experiment

One of the main aspects in mobile monitoring is the operating time of the system when transmitting data via Bluetooth connection. We measured this time for the sensor chest strap and the smartphone (HTC touch diamond II) for different data transmission rates. The difference between continuous transmission of ECG data and the transmission of some events (if the devices stay connected) is negligible for the sensor, only the life time of the smart phone increases significantly. We also calculated power consumption if the devices are only connected for the transmission of an event. In this case we achieve an operating time for monitoring and online analyzing of ECG data of more than 24 hours.

Bluetooth continuous connection	Chest Strap with logging	Smartphone
High continuous data (256Hz ECG)	3:10 h	6 h
Low continuous data (1Hz HR)	3:15 h	9 h
Connection only (e.g. events)	3:20 h	24 h
Connecting on events only	>24 h, depending on the amount of events	>48 h

Fig. 7. Power Consumption for data transmission for different modi on Sensor and Smartphone

4   Conclusion

The presented telemedical system for diagnosis and therapy of stress related disorders is able to assess synchronously a large number of parameters of the whole process of stress reaction. Besides the measurement of physiological data the system is able to get information about emotional and behavioral reaction. The basis of the system is built by a sensor platform for different physiological parameters which is connected wireless to a smartphone. On this smartphone, subjective data is collected with e-questionnaires and it is also used to get context information. Through the coupling of physiological sensors and mobile device we achieved a system for interactive ambulatory assessment of stress. With this system, questionnaires for emotional and behavioral reactions can be executed as a function of the appearance of specified (physiological) events. With this we get new possibilities in diagnosis and treatment of stress that can lead to an improvement of the present situation.

References

1. Stork, W., Kunze, C., Hey, S.: Mobile Intelligenz für die medizinische Überwachung. In: Niederlag, W., Lemke, U. (eds.) Telekardiologie. Health Academy 01/2004, Dresden (2004)
2. Pan, R.L.-C., Chen, B., Li, J.K.-J.: Noninvasive monitoring of autonomic cardiovascular control during stress. In: Proceedings of the 1993 IEEE Nineteenth Annual Northeast Bioengineering Conference, pp. 187–188 (1993)
3. Dickerson, S.S., Kemeny, M.E.: Acute stressors and cortisol responses: a theoretical integration and synthesis of laboratory research. Psychological Bulletin 130(3), 355–391 (2004)
4. Ice, G.H., James, G.D. (eds.): Measuring Stress in Humans: A Practical Guide for the Field. Cambridge Studies in Biological and Evolutionary Anthropology (2007)
5. Solomon, G.F., Benton, D.: Psychoneuroimmunologic aspects of aging. In: Glaser, R., Kiecolt-Glaser, J. (eds.) Handbook of Human Stress and Immunity, pp. 341–363. Academic Press, New York (1994)
6. Ebner-Priemer, U.W., Kubiak, T.: Psychological and Psychophysiological Ambulatory Monitoring. EJPA 23, 214–226 (2007)

7. Froehlich, J., Chen, M.Y., Consolvo, S., Harrison, B., Landay, J.A.: MyExperience: a system for in situ tracing and capturing of user feedback on mobile phones. In: Proceedings of the 5th International Conference on Mobile Systems, Applications and Services, p. 70 (2007)
8. Stumpp, J., Anastasopoulou, P., Hey, S.: Platform for ambulatory assessment of psycho-physiological signals and online data capture. In: 7th International Conference on Methods and Techniques in Behavioral Research "Measuring Behavior 2010", Eindhoven (2010)
9. Peacock, E.J., Wong, P.T.P.: The Stress Appraisal Measure (SAM): A multidimensional approach to cognitive appraisal. Stress Medicine 6, 227–236 (1990)
10. Cohen, S., Williamson, G.: Perceived stress in a probability sample of the U.S. In: Spacapam, S., Oskamp, S. (eds.) The Social Psychology of Health: Claremont Symposium on Applied Social Psychology, pp. 31–67. Sage, Newbury Park (1988)
11. Watson, D., Clark, L.A., Tellegen, A.: Development and validation of brief measures of Positive and Negative Affect: The PANAS Scales. JPSP 54, 1063–1070 (1988)
12. McNair, D.M., Lorr, M., Droppleman, L.F.: Manual for the Profile of Mood States. Educational and Industrial Testing Service, San Diego (1971)
13. Folkman, S., Lazarus, R.S.: An analysis of coping in a middle-aged community sample. JHSB 21, 219–239 (1980)
14. Härtel, S., Gnam, J.-P., Löffler, S., Bös, K.: Estimation of Energy Expenditure using Accelerometers and activity-based Energy-models – Validation of a new device. In: EURAPA. Springer, Heidelberg (2010)
15. Hey, S., Gharbi, A., von Haaren, B., Walter, K., König, N., Löffler, S.: Continuous non-invasive Pulse Transit Time Measurement for Psycho-physiological Stress Monitoring. In: Proceeding of the International Conference on eHealth, Telemedicine, and Social Medicine eTELEMED 2009, Cancun. IEEE Computer Society Conference Proceedings, pp. 113–116 (2009)
16. Hey, S., Sghir, H.: Psycho-physiological Stress Monitoring using Mobile and Continous Pulse Transit Time Measurement. In: Proceeding of the International Conference on eHealth, Telemedicine, and Social Medicine eTELEMED 2011, Gosier (2011)

An Integrated Approach towards Functional Brain Imaging Using Simultaneous Focused Microwave Radiometry, Near-Infrared Spectroscopy and Electroencephalography Measurements

Panagiotis Farantatos, Irene Karanasiou*, and Nikolaos Uzunoglu

School of Electrical and Computer Engineering, National Technical University of Athens
9, Iroon Polytechniou, Zografou Campus, 15773, Athens, Greece
{pfaran,ikaran}@esd.ece.ntua.gr, nuzu@cc.ece.ntua.gr

Abstract. The scope of our ongoing research is the development of a multi-modal, multi-spectral methodology using non-ionizing radiation to study brain function. With this view, a novel microwave radiometry imaging and a near-infrared spectroscopy device have been integrated with an electroencephalogram (EEG) for concurrent measurements of blood flow, neural activity, temperature and conductivity changes in the brain. In this paper, a simulation study to identify whether the focusing properties of the microwave radiometry monitoring system are affected when used in conjunction with concurrent EEG and near-infrared spectroscopy measurements is presented. The simulations are performed using two head models and a phased array system as radiometric antenna receiver, which ensures scanning of the brain area of interest without the need of moving the subject or the monitoring configuration. The results of the electric field distributions inside the entire proposed imaging system illustrate the potential of integrating the three techniques into a single non-invasive monitoring intracranial system.

Keywords: functional imaging, microwave radiometry, functional near-infrared spectroscopy, electroencephalography.

1 Introduction

Functional neuroimaging is broadly defined as the imaging techniques that provide measures of brain activity [1]. These imaging modalities measure correlates of brain activity, and aim at linking the relationship between neural activity in certain brain areas to specific cognitive functions. The activation of specific brain regions is related to increased local neural activity and/or increased regional cerebral blood flow, blood volume, blood oxygen content, and changes in tissue metabolite concentration [2]. It is only in the past few decades that significant advancements have been made in both basic and clinical neuroscience towards the understanding of the subtle mechanisms

---

* Corresponding author.

K.S. Nikita et al. (Eds.): MobiHealth 2011, LNICST 83, pp. 351–357, 2012.

and complex relations of the structure and function of the human brain. This progress is of paramount importance, since the brain is not only the most essential organ in the human body because it ensures the vitality, quality and functionality of the rest of the body, but also because it is the organ of the mind and consciousness, the locus of our sense of selfhood and human existence.

In view of the recent advances in functional neuroimaging, current and future trends focus on synchronous combination of imaging modalities by integrating more than one measures of brain function, e.g., hemodynamic and electrophysiological (EEG and fMRI). These multi-modal approaches aim at achieving sufficient temporal and spatial resolution in order to localize neural activity by providing different expressions of the same phenomenon and identify the functional connectivity between different brain regions, assuming that the multi-modal information represents the same neural networks [3].

In this context, the scope of this research is to provide non-invasive, non-ionizing functional imaging comprising combined blood flow and neural dynamics information, as well as passive measurement of temperature and conductivity fluctuations during activation of specified brain areas in-vivo. The simultaneous acquisition of both hemodynamic and electrophysiological measures is critical for a better understanding of this neurovascular coupling. Moreover, different techniques measure different correlates of neural activity and are characterized by different attributes, e.g. spatial and temporal resolution. In addition, acquiring at the same time measurements of brain temperature and conductivity variations which have been associated with brain activation [4] provides additional insights to the understanding of the underlying mechanisms of brain function. Efforts are being made to achieve the aforementioned aims using a novel Microwave Radiometry Imaging System (MiRaIS) [5]-[7] integrated with a near-infrared spectroscopy system [8] and an electroencephalograph.

It should be noted that, the MiRaIS being a novel technique and non-standardized brain imaging technique, has been used for the past 6 years in various experiments in order to evaluate its potential as an intracranial imaging device [5]-[7]. One of the most important advantages of this method is that it operates in an entirely passive and non-invasive manner. It is able to provide real-time temperature and/or conductivity variation measurements in water phantoms and animals and potentially in subcutaneous biological tissues. Importantly, the system has been used in human experiments in order to explore the possibility of passively measuring brain activation changes that are possibly attributed to local conductivity and/or temperature changes. The results indicate the potential value of using focused microwave radiometry to identify brain activations possibly involved or affected in operations induced by particular psychophysiological tasks [5]. Following this rationale, if such changes can be measured with the proposed method, then it could be used to image brain activity in an entirely passive and non-invasive manner that it is completely harmless and can be repeated as often as necessary without any risk even for sensitive populations.

The work presented herein focuses on certain aspects of the incorporation of the various brain monitoring modules to a single modality and mainly on the effect of concurrent EEG and near-infrared measurements on the focusing properties of the

MiRaIS system. The simulations that were carried out in order to calculate the electric field distributions inside the entire proposed imaging system are performed using two head models, a spherical and a more anatomically detailed one. The radiometric antenna receiver is a phased array system, which ensures scanning of the brain area of interest without the need of moving the subject or the monitoring configuration. In the following sections, the details and the results of the simulation study are fully described and finally discussed in the relevant sections, investigating the potential of integrating three non-invasive brain monitoring techniques in a single system.

2 Materials and Methods

Based on the aforementioned, the systems and methodologies that are in the process of being implemented in one modality to achieve synchronous measurements of neural correlates and blood flow in conjunction with temperature and conductivity variation information, are presented in this section.

2.1 System Description

The operating principle of the Microwave Radiometry Imaging System (MiRaIS) is based on the use of an ellipsoidal conductive wall cavity to achieve beamforming and focusing on the brain areas of interest. The ellipsoidal beamformer is axis-symmetric with an opening aperture to host the human head that is monitored. The cavity has 1.25m length of large axis and 1.20m length of small axis (Fig. 1).

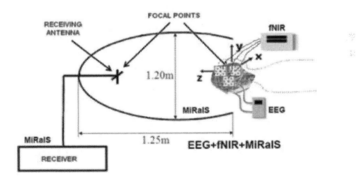

Fig. 1. Block diagram of integrated MiRaIS, Near-Infrared and EEG system

The geometrical properties of the ellipse indicate that rays originating from one focal point will merge on the other focal point. Exploiting this characteristic, when the system is used for microwave radiometry the medium of interest is placed at one focal point, whereas a receiving antenna is placed at the other one. In this way, the chaotic electromagnetic energy emitted by the medium of interest is received by the antenna and driven to a multiband radiometer (operating at 1-4GHz) for detection (Fig. 1). The receiving antenna used in this case is a phased array setup comprising patch antennas operating at 1.53GHz to achieve scanning of the areas under measurement. In the simulations presented herein the reciprocal problem is solved in order to reduce the computational cost imposed by the solution of the initial "forward" problem.

The MiRaIS is intended to be combined with a newly-developed CW pulsed fNIR system operating in the range of 650-850 nm that is compatible with both MiRaIS and EEG [8]. Finally, the electroencephalograph that will be used is a commercial system. All fNIR source and detector components as well as the EEG electrodes will be soon integrated and mounted on a customized head cap, according to the 10-20 system. This will allow concurrent EEG/fNIR and MiRaIS recordings. Therefore, the resulting multi-modal measurements will comprise electrical activity (EEG), blood flow and volume (fNIR), as well as temperature and conductivity variations (MiRaIS) in activated areas. The setup of the entire system is depicted in Fig. 1. The subject enters the ellipsoidal cavity wearing a headcap on which the EEG electrodes and the optical fibres of the near-infrared system are mounted (Fig. 3). The area to be scanned is placed on one focal area of the reflector while the microwave receiving antenna of the MiRaIS is placed on the other focus.

2.2   Simulation Setups

The analysis of the electromagnetic problem is approached numerically using commercial FEM solver (High Frequency Structure Simulator, HFSS, Ansoft Corporation). The simulations have been performed using a two element microstrip patch emitting antenna operating at 1.53GHz placed at the ellipsoidal focus with each element forming an angle of 60° with the horizontal axis (Fig. 2). It has been used as a phased array system and the system's focusing properties have been investigated for phase difference value $\Delta\varphi=60°$ and $\Delta\varphi=90°$.

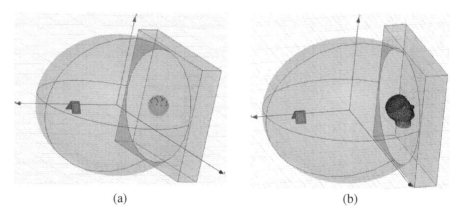

(a)   (b)

Fig. 2. MiRaIS Configuration with a) spherical head model b) SAM model placed at the ellipsoid's reflector focus while the phased two-element antenna is placed on the other focal area

The analysis is performed for two types of head models; a spherical head model and a more detailed anatomic one (SAM -Standard Anthropomorphic Mannequin) whose shape and dimension are specified in a CAD (computer aided design) file included with EN 50361-2001 and IEEE 1528-2003. Both models are single layered having dielectric permittivity and conductivity mean values for brain grey matter at 1.53GHz. Both head models are surrounded by a matching lossless dielectric material of εr= 6.15 and 1cm thickness which, as previously shown [9], significantly improves the system focusing properties minimizing the electromagnetic wave scattering due to the more stepped change of the refraction index on the head–air interface.

The EEG electrodes and the optic fibers and photodiode detectors of the fNIR system have been placed on the head models. The EEG system comprises 16 electrodes modeled by aluminum cylinders of 11mm diameter and 3mm height. Three optic fibers for each one of the operating wavelengths of the fNIR system are placed on the frontocentral electrodes (FP1, FP2, Fz) and the three detectors in between having a distance of 5cm from each corresponding optic fiber. The diameters of the latter are 0.5mm and are modeled as glass fibers while the detectors are indium cylinders of 10mm diameter and height. The EEG and fNIR sensors have been integrated on a dielectric head cap and are depicted in Fig. 3.

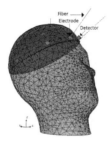

Fig. 3. Dielectric headcap with EEG electrodes and fNIR sources and detectors on SAM model

3   Results

In order to use a diagnostic device such as the proposed one, it is of great importance to have the ability to image any arbitrary area inside the human head, placed on the ellipsoid's focal point where the maximum peak of radiation is achieved. The main drawback that may occur when using the MiRaIs and the EEG, fNIR at the same time is the shielding mesh that may be created by the metallic parts of the EEG electrodes and fNIR system placed on the surface of the human head.

Therefore, the electromagnetic field inside the cavity volume and especially inside the spherical head model with the presence of the EEG electrodes has been calculated. As it is observed in Fig. 3a, sixteen electrodes (eight approximately at each hemisphere) maybe used at the frequency of 1.53GHz (and at all higher operation frequencies up to 4GHz) without raising any electromagnetic compatibility issues; the EEG electrode metallic parts do not affect the systems focusing properties and the electric field converges at the centre of the human head model when it is placed at the ellipsoidal focus. It should be noted as well that the electrode wiring should be cross-polarized to the MiRaIS receiving antenna in order to minimize any additional electromagnetic compatibility issues.

The fNIR system is not expected to particularly interfere with the MiRaIS operation since the three glass optical fibers have low dielectric property values and the only partially metallic parts (detectors) are placed on the frontocentral electrodes. In Fig. 3b, all system components have been mounted on the SAM head model with the latter's center placed on the ellipsoid's focus point. The electric field distribution inside the head model is depicted in the same figure and the simulation results fully verify theoretical expectations.

In all simulation cases with the integration of the three brain monitoring modalities, clear focusing inside the head models is achieved. Additionally, because of the two-element phased antenna, the focusing area inside the head model moves, performing a linear scan of 20 mm, approximately starting at a distance of varying from 5mm to 10mm away from the ellipsoidal focal point. The phase difference of the two microstrip antennas significantly affects the way the scanning inside the head model is achieved and thus with the appropriate selection of phase difference the scanning area may be successfully manipulated. This way scanning of the area of interest can be performed without moving the subject.

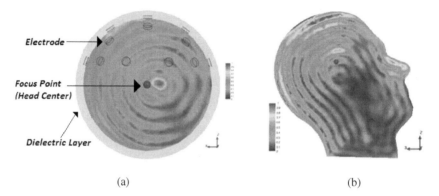

(a)   (b)

Fig. 4. Electric field distribution in a) the spherical head model and b) SAM model when concurrent EEG, fNIR and MiRaIS measurements are performed

4   Discussion and Conclusion

Ongoing research during the past few years, envisions the development of an integrated functional imaging methodology to study brain function in-vivo through a multi-modal, multi-spectral approach using non-ionizing radiation to provide combined brain functional information. The proposed system comprises a microwave radiometry imaging module that provides the measurements of the local temperature and conductivity tissue variations, a near-infrared spectroscopy and EEG system that provide the blood flow and neural activity information respectively.

The present paper focuses on certain aspects of the incorporation of the various brain monitoring modules to a single modality and mainly on the effect of concurrent EEG and near-infrared measurements on the focusing properties of the MiRaIS system. Simulations were carried out to calculate the electric field distributions inside the entire proposed imaging system using two head models, a spherical and a more anatomically detailed one. The radiometric antenna receiver is a phased array system, which ensures scanning of the brain area of interest without the need of moving the subject to acquire tomographic data. The results show that the modules can be successfully integrated in a single imaging modality without any compatibility issues. Nevertheless, extensive experimentation of the whole system especially including human volunteers is essential in order to validate it as a potentially useful imaging tool in practice.

More precise assessment of the underlying biophysics of the measured signals will be extremely important for the better understanding of the healthy and diseased brain. The ultimate goal of the proposed methodology is to make joint inferences about the neural activity, hemodynamics, and other biophysical parameters of activated brain areas and the correlations between them, by exploiting complementary information from multimodal and multispectral information, through implementation of non-invasive and non-ionizing imaging techniques. The proposed unprecedented concurrent assessment of multiple biophysical correlates of neural activation (blood flow, volume and oxygenation, conductivity and temperature) will allow detailed correlation studies that will potentially further elucidate the nature of brain activation measurement and the neurovascular coupling.

References

1. D'Esposito, M.: Functional Neuroimaging of Cognition. Semin. Neurol. 20, 487–498 (2000)
2. Villringer, A., Dirnagl, U.: Coupling of Brain Activity and Cerebral Blood Flow: Basis of Functional Neuroimaging. Cerebrovasc. Brain Metab. Rev. 7, 240–276 (1995)
3. Desmond, J.E., Chen, S.H.A.: Ethical Issues in the Clinical Application of fMRI: Factors affecting the Validity and Interpretation of Activations. Brain Cogn. 50, 482–497 (2002)
4. Kiyatkin, E.A.: Brain Temperature Fluctuations during Physiological and Pathological Conditions. Euro. J. Appl. Physiol. 101, 3–17 (2007)
5. Karanasiou, I.S., Uzunoglu, N.K., Papageorgiou, C.: Towards Functional Non-invasive Imaging of Excitable Tissues inside the Human Body using Focused Microwave Radiometry. IEEE Trans. Microwave Theory and Tech. 52, 1898–1908 (2004)
6. Gouzouazis, I.A., Karathanasis, K.T., Karanasiou, I.S., Uzunoglu, N.K.: Contactless Passive Diagnosis for Brain Intracranial Applications: a Study using Dielectric Matching Materials. Bioelectromagnetics 31, 335–349 (2010)
7. Karathanasis, K.T., Gouzouazis, I.A., Karanasiou, I.S., Giamalaki, M., Stratakos, G., Uzunoglu, N.K.: Non-invasive Focused Monitoring and Irradiation of Head Tissue Phantoms at Microwave Frequencies. IEEE T Inf. Technol. B 14, 657–663 (2010)
8. Oikonomou, A., Korini, P., Karanasiou, I., Klinkenberg, B., Makropoulou, M., Uzunoglu, N., Serafetinidis, A.: Study of Brain Oxygenation with Near-Infrared Spectroscopy. In: Proc. of the 3rd National Conference of Electrical and Comp. Engineering Schools (2009)
9. Gouzouasis, I., Karathanasis, K., Karanasiou, I., Uzunoglu, N.: Passive multi-frequency brain imaging and hyperthermia irradiation apparatus: the use of dielectric matching materials in phantom experiments. Meas. Sci. Technol. 20, 104022 (2009)

A New Home-Based Training System for Cardiac Rehabilitation

Christian Menard and Raimund Antonitsch

Carinthia University of Applied Sciences, Department of Medical Information Technology
Primoschgasse 10, 9020 Klagenfurt, Austria
{C.Menard,R.Antonitsch}@cuas.at

Abstract. Epidemiological studies show that regular exercise sessions are the most important pillars in the field of cardiologic rehabilitation. For many diseases a correctly implemented strength training therapy can improve physical fitness and health. In this work a novel pictorial strength measurement method for home-based training is presented. In order to reduce the risk of injury during the training session, simple elastic bands can be used as training devices. Such systems can make therapeutical processes more effective while simultaneously reduce the overall costs of therapy.

Keywords: Ambient Assisted Living, Home Monitoring, Home based Training.

1 Introduction and Motivation

Based upon epidemiological studies we know today that structured, regular exercise sessions are the most important pillars to the prevention and rehabilitation of society's most widespread disease, the metabolic syndrome [1]. Pedersen and Saltin describe in great detail how sport and exercise can be used as therapies and treatments for diseases that are associated with the metabolic syndrome [2]. In all of these diseases specialized, correctly implemented strength training therapy improved the individuals' psycho-socially and bodily well-being as well as their physical fitness, reduced the number of medical complaints and stopped or even reversed the advancement of these diseases. Due to these reasons new innovative technologies for home-based training need to be developed that not only keep people healthier longer but also make the therapeutical processes in treating health problems more effective while simultaneously reducing the costs of therapy.

Especially for the elderly as well as people with coronary heart disease carefully monitored weight lifting programs are becoming very important therapeutical methods [3]. A precise validation of the progress of the training is currently only available in outpatient settings and with expensive strength measurement devices. Costs compared to normal stationary stays at therapy centers can be greatly reduced on the one hand while on the other hand it is easier for the elderly to participate in medically directed health training programs.

K.S. Nikita et al. (Eds.): MobiHealth 2011, LNICST 83, pp. 358–365, 2012.
© Institute for Computer Sciences, Social Informatics and Telecommunications Engineering 2012

This work will give a short overview to cardiac rehabilitation and the developed home based training system and will then focus on the methods for the pictorial strength acquisition.

2   Strength Training for Cardiac Rehabilitation

Heart diseases, also known as cardiac diseases, are the major cause of death in Austria. Four deaths of ten were due to heart diseases in 2011. The most prevalent heart disease is the heart attack, which belongs to the group of coronary diseases, caused mainly by the calcification of coronary vessels.

According to the World Health Organization (WHO) the aim of cardiac rehabilitation is the stabilization and prevention of heart diseases [4]. In addition to the preservation of life cardiac rehabilitation also provides health economical benefit such as decreases in hospital stays and reduction of blood pressure lowering medication.

For many years endurance training was the key aspect in cardiac rehabilitation. But recently results of several studies have shown that moderate strength training is becoming more and more important. Strength training helps to improve stability and coordinative skills, which are essential for elderly to avoid falls. Especially bedridden patients after a surgery are understrength and therefore strength training is an important step to regain independency and sense of well-being. Furthermore, bone density increases due to well-directed strength exercises. For these reasons strength training should be an integral part of cardiac rehabilitation [5].

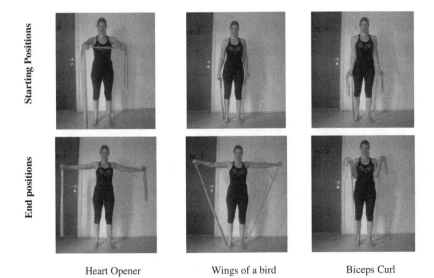

Heart Opener   Wings of a bird   Biceps Curl

Fig. 1. Three training exercises used for home based training

The Thera-Band® is an established training device for cardiac rehabilitation. The multitude of possible exercises and the low price make it a popular choice. These bands are available with different strength levels, indicated by the color. Patterson et al. describe the material properties depending on color and strain in great detail [2]. In this work three typically exercises, which are used in cardiac rehabilitation, were used to test the proposed method (see Fig. 1).

Training exercises have a detailed description with exact starting and end positions. For each patient a specific trainings profile can be specified with all necessary parameters like type of the exercise, duration, set count and the color of the used band. Using these parameters the therapist is able to adapt the training program to the actual situation of the patient. In the field of home-based training exercise monitoring and immediate feedback to the patient is essential for the training success. Especially for people who have no training experience the developed measurement system guarantees that the spent force is adequate and does not stress the patient.

3   System Overview: Home-Based Training

In the framework of the research project "Health@Home" a home-based training system is developed. It is designed as a simple user friendly system to be used directly in the user's living room [6]. In Fig. 2 an overview of the home-based training system is depicted. A *Training-Coach* downloads the user-specific training profile and starts with the training instructions. For all exercises standard elastic bands are used. The measured pulse and the strength exerted on the band are continuously measured and transmitted via Bluetooth to the *Training-Coach*. The pulse is recorded by means of a standard pulse sensor (e.g. Polar).

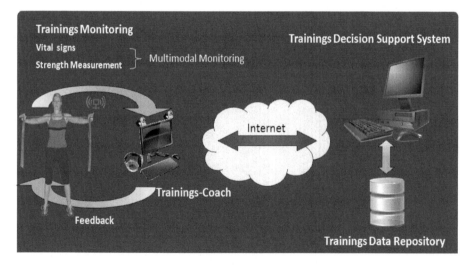

Fig. 2. System overview of the home-based training system

During the training the *Training-Coach* provides the patient with audio feedback about the current training status. Time series data of both sensors are chronologically synchronized for the entire training session, transmitted and stored in the *Training Data Repository* at the therapy center. Thus, therapists have the possibility via the *Decision Support System* (DSS) to monitor and control each training session individually. Using the *DSS* the therapists have the possibility to visualize and analyze all recorded training sessions in detail.

In order to have a flexible and simple training setup a pictorial strength measurement method was developed. This pictorial strength measurement reduces the risk of injury tremendously, compared to physically mounted sensors that may hurt the patient during the exercise.

4   Pictorial Strength Measurement

The strength measurement is realized via a visual analysis of the training session using a calibrated stereo camera setup (see Fig. 3). A simple coding technique using predefined ball markers on the band is used to calculate the strength during the training session. So the elastic band itself is the measurement device and physically mounted force sensors are obsolete. The ball markers are detected and the extension of the band can be used to calculate the strength.

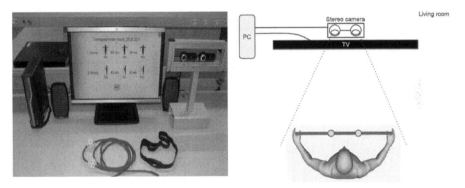

Fig. 3. Equipment of the home-based training system and a schematic overview

The larger the distance between the markers the higher is the strength. By using the stereo setup the 3D position of the band and the markers can be reconstructed, thus making the strength measurement more robust. Due to the fact that exclusively standard hardware is used, the system is designed as a low cost system.

The workflow of the pictorial measurement algorithm is depicted in Fig. 4. For most of the process tasks standard image processing methods could be used [7][9].

Fig. 4. Process of the pictorial strength measurement

First the image area is segmented according to the color of the mounted ball markers so that pixels with the same color as the balls remain [10]. In order to avoid fail detections only moving pixels are filtered out on the basis of motion information [8][9]. A blob detection based on a certain blob size is then used to determine the position of the ball markers. Corresponding balls in the left and right image together with the stereo calibration parameters are used for 3D reconstruction. The 3D positions of the balls are transformed to real-world distances which are then used for strength calculation.

In order to determine the strength out of the measured marker distances, a strength calculation model was determined depending on the different bands. Fig. 5 shows force values for different bands depending on the elongation.

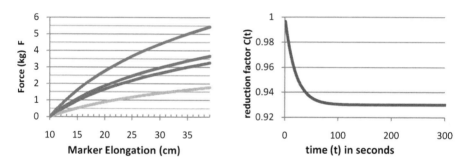

Fig. 5. Strength calculation model with time-dependent force characteristic

The force

$$F(t) = C(t)\, B_{\text{color}} \ln(\varepsilon),\tag{1}$$

needed to stretch the band by the elongation ΔD is time-dependent and it is a nonlinear function of the current strain $\varepsilon = \Delta D/D_0$, where D_0 denotes the marker distance in the zero-force configuration. The matter constant B_{color} accounts for the mechanical properties of the materials used for the different band types. Below the constants B_{color} of the four different, color coded, band types are listed.

$$
\begin{aligned}
B_{\text{blue}} &= 4, \\
B_{\text{green}} &= 2.7, \\
B_{\text{red}} &= 2.6 \text{ and} \\
B_{\text{yellow}} &= 1.3.
\end{aligned}
\tag{2}
$$

Due to the fact that the strength of the band reduces over time exponentially a force reduction factor $C(t)$ is modeled by

$$C(t) = (1 - C_s)e^{-t/\tau} + C_s ,$$

(3)

whereby τ is the time constant of the relaxation process in minutes and C_s defines the steady state factor at the end of the relaxation process. Several tests on different bands using the force sensor have shown to use the values $C_s = 0.93$ and $\tau = 20$.

5   Results

The home based training system was tested for all three exercises with several patients. In Fig. 6 the user interface of the home based trainings system is depicted. The *Training-Coach* application can be installed on the standard PC and was designed as low cost system, thus the components are available for approximately 400 € (excl. TV).

The patient can see the video together with the actual pulse and the calculated strength on the screen. During the training all relevant data are transferred to the *Trainings Data Repository*. So the therapists are able to analyze the training and if necessary adapt the training profile of the patient. The progress bar indicates the actual progress during the exercises. In addition the *Training-Coach* permanently supports the user with audio feedback about the actual trainings status like the tempo or how many exercises are missing.

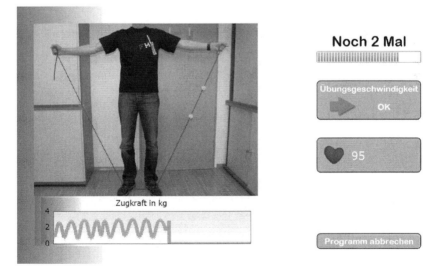

Fig. 6. User interface showing a user during the training. The strength together with the pulse is visualized.

In order to evaluate this new pictorial strength measurement method a physically mounted force sensor was used on the band to collect ground truth strength data. The visual strength measurement method was tested for the three different exercises using the measured ground truth data. With the visual sensed data on the one hand and ground truth data on the other hand, the underlying calculation model could be calibrated.

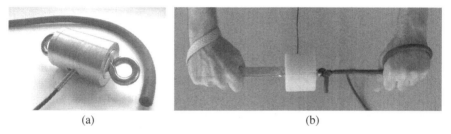

(a)   (b)

Fig. 7. Force sensor (a) force sensor with case and (b) mounted on the band

For each band a so-called rating curve could be determined in order to fit best to the physical behavior of the specific band. During the training the strength could be calculated at 20fps with an accuracy of ±5% to the physical force sensor. Fig. 8 shows both the strength obtained by the force sensor and the calculated strength with the presented method.

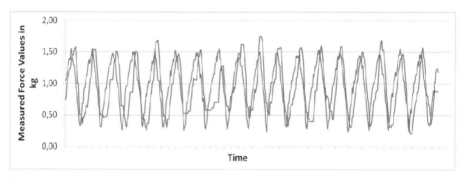

Fig. 8. This chart shows a comparison between force data obtained by the physically mounted sensor (blue) and the calculated strength (red)

6   Conclusion

A new pictorial strength measurement method was presented to be used in the field of cardiac rehabilitation. Using this system home based training can be performed very easily in the patient's home environment, thus the proposed method offers a new methodology in the field of home based training. The motivation effect on the one hand and the facility to track the training progress on the other hand are the major advantages. Furthermore the risk of injury is reduced tremendously compared to other measurement methods that depend on mounted sensors. Future steps are directed towards an infrared setup to be more flexible and robust against varying light conditions. In addition the integration of this method into an easy to use training application and a field study with over 80 patients is planned in the near future.

References

1. Maxwell Mark, S., et al.: Metabolic Syndrome and Related Disorders, vol. 6(1), pp. 8–14. Springer, Heidelberg (2008)
2. Pedersen, B.K., Saltin, B.: Evidence for prescribing exercise as therapy in chronic disease. Scandinavian Journal of Medicine & Science in Sports 16(s1), 3–63 (2006)
3. O'Connor, G.T., Buring, J.E., Yusuf, S., Goldhaber, S.Z., Olmstead, E.M., Paffenbarger Jr., R.S., Hennekens, C.H.: An overview of randomized trials of rehabilitation with exercise after myocardial infarction. American Heart Association. Circulation 80, 234–244 (1989)
4. Balady Gary, J., et al.: A Scientific Statement From the American Heart Association Exercise. Cardiac Rehabilitation, and Prevention Committee, Circulation 115, 2675–2682 (2007)
5. Jonas, S., Phillips, E.: ACSM's Exercise is Medicine™ A Clinician's Guide to Exercise Prescription Softbound
6. Menard, C., Hayn, D., Traninger, H.: Gesundes Altern durch Heimbasiertes Training. In: Forschungsforum der Oesterreichischen Fachhochschulen (2011)
7. Bradski, G., Kaehler, A.: Learning OpenCV - Computer Vision with the OpenCV Library. O'REILLY (2008)
8. Cheung, S., Kamath, C.: Robust techniques for background subtraction in urban traffic video. Visual Communications and Image Processing 5308(1), 881–892 (2004)
9. Sonka, M., Hlavac, V., Boyle, R.: Image Processing, Analysis, and Machine Vision. Thomson (2008)
10. Schanda, J.: Colorimetry - Understanding the CIE System. Wiley (2007)

Context-Aware System for Neurology Hospital Wards

Ingrid Flinsenberg, Roel Cuppen, Evert van Loenen,
Elke Daemen, and Roos Rajae-Joordens

Philips Research, High Tech Campus 34, 5656AE Eindhoven, The Netherlands
{Ingrid.Flinsenberg,Roel.Cuppen,Evert.van.Loenen,
Elke.Daemen,Roos.Rajae}@philips.com

Abstract. In this paper we describe the context-aware Adaptive Daily Rhythm Atmosphere (ADRA) system. The ADRA system is designed to stimulate the healing process of hospital patients, neurology patients in particular. We first report on the needs and issues of neurology patients identified by an observation study in a hospital neurology ward. Based on these needs, we define several concepts to promote the healing process. Finally, the context-aware system we have designed to realize these concepts is described.

Keywords: Context-aware system, agent-based system, hospital ward, ambient healing, neurology.

1   Introduction

We explore possibilities of enhancing the patient healing process in future (single) hospital patient rooms by means of a context-related adaptation of the environment while addressing needs of both patients and staff. We focus specifically on neurology patients with the emphasis on stroke and the inpatient environments these patients find themselves in during the post event recovery process. We spent a full week in two leading hospitals to make observations on-site.

Four relevant key conclusions from workflow observations and interviews emerge. First of all, the *amount and intensity of stimuli* that a patient can handle in this environment is very dependent on his or her condition: too many stimuli could lead to aggression and restlessness. Too few stimuli could lead to boredom. Secondly, the right *balance* between a clinical environment and a personal environment needs to be achieved for all stakeholders in the neurology department. A patient room that *adapts the environment* of the patient without interfering with the activities of other stakeholders is therefore expected to be very beneficial to the patients, family and staff. Thirdly, a clear *structure* of the day is important for stroke patients to decrease the risk of disorientation and confusion, achieve a healthy sleeping pattern, avoid delirium, better handle rehabilitation therapy, and to consolidate their memories. Fourthly, stroke patients have a large *risk of falling and accidents*. The main reason for falling accidents is the patients' limited insight in their disease. The incidence of falling can be reduced by reducing clutter and improving lighting conditions.

K.S. Nikita et al. (Eds.): MobiHealth 2011, LNICST 83, pp. 366–373, 2012.
© Institute for Computer Sciences, Social Informatics and Telecommunications Engineering 2012

2   Literature

Many research efforts have focused on providing staff members with the right information at the right time. What constitutes 'right' information is determined based on the activity the staff member is carrying out. For example, the CISESE institute in Mexico has carried out a number of workplace studies in a public hospital, an overview of which is given by Favela et al. [5]. Based on the workplace studies, Sanchez et al. [8, 9] describe different methods for classifying activities of the hospital workers to enable context-aware communication between staff members. Bardram [1,2,3] discuss a context aware hospital bed that displays relevant information for the nurse when administering medication by e.g. displaying the medication scheme, patient record, lighting the proper medication container, when the nurse and the medication container are close to the bed. Siewe et al. [10] show how this application can be formulated using context-aware calculus. Kjeldskov et al. [7] describe a prototype to support morning procedure tasks in a hospital ward by showing patients lists and patient information based on the location of the nurse and time of day. Weal et al. [11], and Cassens et al. [4] discuss the annotation of staff activities on a patient ward to facilitate further development of context-aware systems. Our approach is unique compared to these previously proposed systems; our system focuses on providing the patient with the proper healing environment, while the focus of the previously proposed systems is on providing the staff with the proper information.

3   System Description

Based on our observations, we propose the Adaptive Daily Rhythm Atmosphere (ADRA) system, which generates a dynamic atmosphere that supports the daily rhythm of the patient. Where needed the atmosphere adapts to specific interrupts and visits, e.g. when doctor or cleaner is visiting. By using ADRA, the mentioned negative effects of the rigid environmental conditions are alleviated because the system provides a daily rhythm atmosphere in sync with, and optimized for, patient needs and the care agenda and intelligently adapts to deviations thereof.

The ADRA system starts with a series of pre-defined *day schedules* per patient. Each day schedule consists of a sequence of phases of the day with fixed start and end times, an example of which is given in Table 1.

Each phase within the day schedule is also described by at least one pair of start and end triggers. These triggers indicate the start and end events of the phase. Triggers can be used for *expected* events, such as breakfast, for which the timing is typically uncertain, but the occurrence is not. However, they can also be used for *unexpected* events such as a physical examination in the afternoon. An example of these triggers is given in Table 2.

Table 1. An example of a pre-defined day schedule

Ambiance	Start time	End time
Sleep	00:00	7:30
Wakeup	7:30	8:00
Breakfast	8:00	8:30
Morning	8:30	11:30
Lunch	11:30	12:00
Rest	12:00	14:00
Wakeup	14:00	15:00
Visitors	15:00	17:00
Dinner	17:00	18:00
Rest	18:00	24:00

Table 2. An example of start and end triggers of different phases.

Phase	Start trigger	End trigger
Wakeup	Time>7:00 & Time<7:30 & sleepState = awake	Caterer enters department OR caterer enters room OR Time=8:00
Meal	Caterer enters department OR caterer enters room	DurationMeal = 0:45
Sleep	(Lights off & Time>20:00) OR (sleepState = asleep & Time>20:00)	Time = 7:30
Nurse Call	Call button pressed	Nurse call completed button pressed

Each trigger schema is stored in a database and retrieved by the ADRA system based on the date and patient identifier. The status of all defined phases and their triggers are continuously monitored in the system. Next to the triggers the system reacts to the entering and leaving of the staff into the patient room. Based on an identifier of the current phase, the corresponding atmosphere for that phase is retrieved and used by an ambience system to start the predefined corresponding e.g. lighting and audio settings.

We envision that the ADRA system runs for multiple patient rooms (n > 100) at the same time. Scalability and maintainability are therefore the most important non-functional requirements. We therefore created a distributed system architecture, in which each room is an independent system that can be maintained separately without interrupting the ADRA systems in other patient rooms. The only resource that all patient rooms share is a centralized database server that contains the patient information and ambience settings. All processing resides in controllers within the separate room that are connected to one or more sensors and that communicate over a TCP/IP network. Concurrency only occurs in the database calls to the centralized database. However, these calls happen at low resolution making the ADRA system very scalable.

For each room the system consists of a separate room controller, bedside controller, and ambience controller. The bedside controller communicates its observed bed state to the corresponding room controller over TCP/IP, which utilizes this information together with the room state to determine the current phase of the day for the patient. Based on this phase, the room controller retrieves and communicates the corresponding ambience settings to the ambience controller over TCP/IP that generates the correct lighting, audio, and video settings. The system architecture is illustrated in Figure 1.

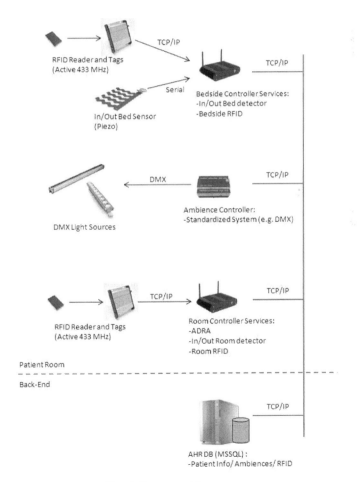

Fig. 1. System architecture

The development and execution of the different services within the different controllers and the communication between the services is based on the agent paradigm to execute peer-to-peer applications that can seamless work and interoperate in a network environment. The ADRA system makes use of standard runtime services that enable the execution, search, and discovery of agents and the communication between them. These standard agent services that run on different machines automatically connect with each other, resulting in one distributed agent support layer. Each agent is identified by a unique name and provides a set of services that it registers within this agent support layer. Agents communicate by exchanging asynchronous messages. The properties of an agent-based system make this technology very suitable to realize a distributed ADRA system for multiple rooms. The services within the controllers are represented by agents that communicate with other agents in order to provide the intended functionality. The agent architecture is illustrated in Figure 2.

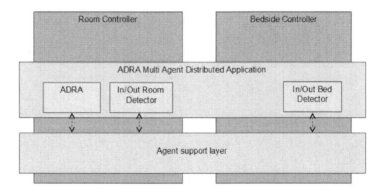

Fig. 2. Agent architecture

To be able to evaluate this concept and its performance, we are developing a prototype ADRA system integrated in a concept patient room. The system contains the commercially available Emfit monitor device (D-1070-2G) and the under-mattress bed sensor (L-4060SL) to detect if the patient is in or out of bed, and the Wavetrend system with active tags to detect that a person is entering a room or a department and what that person's role is.

The actuators of the ADRA system consist of several lighting devices. All light sources are connected to the main ambience controller, which is a DMX controller supporting static, dynamic and interactive scenes. Scene creation is done by programming timelines for the different lighting devices. These timeline specify the DMX values of the lighting devices over a predefined time period.

Figure 3 to Figure 6 illustrate how a patient room can be adapted over the day to the patient's needs and the clinical activities taking place in the patient room. These environments differ per patient and depend on for example the stage in the recovery process, and planned or unplanned treatment sessions. Furthermore, the timing of the different phases can differ per patient, based on the planned and unplanned actions of patient and staff.

When it is time for the patient to wake up, lights are slowly turned on, giving the patient sufficient time to wake up in a relaxed manner, and avoiding too much stimuli at once. A nature video is visible on the center panel, with calming nature sounds. A clock is displayed on the left panel at all times to provide the required structure. Figure 3 shows the patient room at the end of the wake up phase. During clinical care, see Figure 4, bright white light is used to create a clinical environment, and the nature view in the center panel is removed to prevent distraction during the clinical care.

Fig. 3. Wake up

Fig. 4. Clinical care

During the resting hour after lunch, the curtains are closed half, lights are dimmed, and a nature video with calming sounds is displayed as illustrated in Figure 5. This stimulates relaxation, yet maintains the connection with the outside world by not fully closing the curtains. To stimulate the feeling of being connected to the outside world, the right panel shows photos and/or drawings of family and friends. When going to sleep, the lighting becomes more reddish as illustrated in Figure 6, because blue light keeps us awake and alert, while the absence of it makes us sleepy, as shown by Gooley et al. [6].

Fig. 5. Rest

Fig. 6. Going to sleep

4   Conclusion

There are needs in hospital contexts which may be addressed by intelligent, context-aware systems in order to realize an optimal healing environment for patients. We have shown that neurology patients need a personalized environment to optimally stimulate their recovery process. Patients have limited abilities to cope with stimuli, are in need of clear structure, and are at risk of falling and other accidents. Their capabilities, needs and risks differ per individual, per time of day, and depend on their stage in the recovery process. Therefore the environment needs to be continuously adapted to each individuals needs at every moment in the recovery process. While doing so, the right balance between the clinical environment needed by the staff, and a personal environment for the patient and his visitors, is needed as well. To realize these objectives, we propose the ADRA system. The distributed nature of the ADRA system is versatile, and is particularly suited for easily extending the number of rooms, patients, sensors, and actuators in the system. Although our solutions have been tailored to neurology patients, many of the identified issues generalize to other diseases. Hence, the proposed ADRA system can be applied, possibly after some modifications, to other patient groups. As next steps, we will evaluate the system in a concept patient room lab setting and in an actual hospital environment.

Acknowledgments. We thank all patients, medical staff, researchers and designers that have contributed to defining and validating the issues, concepts, and solutions proposed in this work.

References

1. Bardram, J.E.: A Novel Approach for Creating Activity-Aware Applications in a Hospital Environment. In: Gross, T., Gulliksen, J., Kotzé, P., Oestreicher, L., Palanque, P., Prates, R.O., Winckler, M. (eds.) INTERACT 2009, Part II. LNCS, vol. 5727, pp. 731–744. Springer, Heidelberg (2009)
2. Bardram, J.E.: Applications of Context-Aware Computing in Hospital Work – Examples and Design Principles. In: Proc. ACM Symposium on Applied Computing, pp. 1574–1579 (2004)
3. Bardram, J.E.: Hospitals of the Future – Ubiquitous Computing support for Medical Work in Hospitals. In: Proc. Hospitals Workshop Ubihealth (2003)
4. Kofod-Petersen, A., Cassens, J.: Using Activity Theory to Model Context Awareness. In: Roth-Berghofer, T.R., Schulz, S., Leake, D.B. (eds.) MRC 2005. LNCS (LNAI), vol. 3946, pp. 1–17. Springer, Heidelberg (2006)
5. Favela, J., Martinez, A.I., Rodriguez, M.D., Gonzalez, V.M.: Ambient Computing Research for Healthcare: Challenges, Opportunities and Experiences. Computacion y Sistemas 12(1), 109–127 (2008)
6. Gooley, J.J., Lu, J., Fischer, D., Saper, C.B.: A broad role for melanopsin in nonvisual photoreception. Journal of Neuroscience 23, 7093–7106 (2003)
7. Kjeldskov, J., Skov, M.B.: Supporting Work Activities in Healthcare by Mobile Electronic Patient Records. In: Masoodian, M., Jones, S., Rogers, B. (eds.) APCHI 2004. LNCS, vol. 3101, pp. 191–200. Springer, Heidelberg (2004)
8. Sanchez, D., Tentori, M., Favela, J.: Hidden Markov Models for Activity Recognition in Ambient Intelligence Environments. In: Proc. Eighth Mexican International Conference on Current Trends in Computer Science, pp. 33–40. IEEE Computer Society, Washington (2007)
9. Sanchez, D., Tentori, M., Favela, J.: Activity Recognition for the Smart Hospital. IEEE Intelligent Systems 23(2), 50–57 (2008)
10. Siewe, F., Zedan, H., Cau, A.: The Calculus of Context-aware Ambients. Journal of Computer and System Sciences 77(4), 597–620 (2011)
11. Weal, M.J., Michaelides, D.T., Page, K.R., De Roure, D.C., Gobbi, M., Monger, E., Martinez, F.: Location based semantic annotation for ward analysis. In: Proc. 3rd International Conference on Pervasive Computing Technologies for Healthcare, University of Southampton (2009)

Towards Utilizing Tcpcrypt in Mobile Healthcare Applications

Stefanos A. Nikolidakis, Vasileios Giotsas, Emmanouil Georgakakis,
Dimitrios D. Vergados, and Christos Douligeris

Department of Informatics. University of Piraeus
80, Karaoli & Dimitriou St., GR-185 34, Piraeus, Greece
{snikol,egeo,vergados,cdoulig}@unipi.gr,
giotsas@ieee.org

Abstract. The evolution and growth of networks has made the personal data of the users available to many applications. In this direction, one of the main concerns is to protect the sensitive personal information, while at the same time avoid delays in the provision of services like healthcare to the general public. An extension of TCP, the Tcpcrypt, is a promising technology that can be used on this field. Tcpcrypt is designed to provide end-to-end encryption in the transport layer with low overhead, rendering it a very promising solution in order to protect medical data that are often handled by devices with limited resources. In this paper Tcpcrypt performance is evaluated against TCP, in terms of additional overhead incurred in the total size of the transmitted data and the total number of CPU instructions that are executed. Moreover, a solution for reducing overhead through fine-grained packet handling is proposed.

Keywords: Tcpcrypt, SSL, Healthcare.

1   Introduction

Preserving the privacy and the integrity of data transmitted over networks is a well established requirement and many solutions have been proposed and implemented towards this direction. Security and encryption mechanisms can be deployed in the upper layers of the network stack. Some of the most widely used solutions to provide authentication and encryption mechanisms are SSH (Secure Shell) and Https [1, 2] which are deployed in the application layer, TLS (Transport Layer Security) / SSL (Secure Sockets Layer) [3] are deployed in the transport layer and IPsec [4], is deployed in the Internet Layer. Tcpcrypt has emerged as an alternate solution in the transport layer that will address some of the shortcomings of the existing technologies [5].

Tcpcrypt enhances TCP by adding cryptographic capabilities. One of the key benefits of Tcpcrypt is transparency as it requires no configuration, no changes to applications and the network connections will continue to work even if the remote end does not support Tcpcrypt. In the latter case the connections will gracefully fall back

K.S. Nikita et al. (Eds.): MobiHealth 2011, LNICST 83, pp. 374–379, 2012.
© Institute for Computer Sciences, Social Informatics and Telecommunications Engineering 2012

to standard clear-text TCP. Tcpcrypt operates in the transport layer. In [6] a comparison of the performance of Tcpcrypt, TLS and SSL in terms of the number of connections a server can handle per second and the possible transfer rate was provided. In both metrics Tcpcrypt appears to have superior performance. One of the reasons Tcpcrypt is less demanding comparing to TLS/SSL is the fact that it does not utilize asymmetric cryptography mechanisms, which are computationally demanding operations. Moreover, the need of digital certificates and some form of PKI (Public Key Infrastructure) and CA (Certificate Authority) is essential for the use of SSL rendering its deployment cumbersome. The use of digital certificates enables SSL to perform strong authentication of the involved entities. Tcpcrypt authentication mechanisms cannot defend against active attacks. However Tcpcrypt can rely on application level authentication to ensure proper authentication and does not specify the means of the authentication e.g. certificates, passwords, tokens etc. Tcpcrypt is vulnerable to active attacks such as Man In the Middle Attacks (MIMA). For example, an attacker can modify a server's response to claim that Tcpcrypt is not supported (when in fact it is) so that all subsequent traffic will be transmitted in clear text and be susceptible to eavesdropping.

Given the promising capabilities of Tcpcrypt it is worth-investigating the performance of Tcpcrypt in mobile healthcare applications. A widely adopted paradigm entails a WBAN (Wireless Body Area Network) which collects and transmits data to a mobile sink attached to the patient. The sink is usually a IEEE802.11 capable mobile device which handles the communication with a healthcare server or other sinks in an ad-hoc manner. The data transmitted by the sink should be protected from malicious attackers, but at the same time the sink has to maintain low power consumption to achieve the longest possible availability. Tcpcrypt can become a severe handicap for the expected battery lifetime. Until now there is no published attempt to characterize the overhead incurred by Tcpcrypt at the client side. In this paper, such an attempt is presented and the realistic conditions under which Tcpcrypt can be deployed on mobile resource-limited devices are provided.

The paper is organized as follows: In Section 2.1, the comparison of transmitted bytes with the use of TCP and Tcpcrypt as the file size increases is presented. In Section 2.2, the CPU utilization and total duration for the transmission of data is depicted. In Section 2.3, the results of the previous sections are discussed. In Section 2.4, the reducing overhead through fine-grained packet handling is suggested. Finally, in 3, conclusions are given and future work is discussed.

2   Tcpcrypt Overhead Evaluation

The performance of Tcpcrypt against TCP, in terms of additional overhead incurred in the total size of the transmitted data and the total number of CPU instructions executed, are evaluated and compared in this section. Currently there is no Tcpcrypt implementation for ARM architecture (Advanced RISC Machine), thus it is not possible to evaluate it on handheld devices (e.g. android or iphone smart phones).

Therefore, the user-level implementation of Tcpcrypt protocol on a single-core Intel U3500 CPU (1.4 GHz) netbook with 2GB of RAM that operated over Ubuntu 10.04 Linux distribution is adopted. The network measurements were collected through a Wireshark Network Protocol Analyzer. Hardware measurements were obtained by instrumenting Tcpcrypt using Intel's VTune Performance Analyzer.

User-level Tcpcrypt exhibits slower performance than its kernel-level implementation [6] but it can be used in for hardware performance events measurement. Since native TCP is in the kernel-level, the performance comparisons are biased in favor of TCP. To remove this bias the case where both client and server communicate over Tcpcrypt against the case where the server communicates over TCP is compared. Also it is considered that the client communicates over Tcpcrypt having all the security functionality deactivated.

2.1   Overhead on Transmitted Data

A metric of particular interest for battery-powered mobile devices, whose energy consumption depends on the amount of transmitted and received data, is the overhead incurred by Tcpcrypt on the total volume of transmitted data. This overhead is due to two factors, extra bytes in the header of the packets for Tcpcrypt options, or extra bytes as a result of the encryption. In this evaluation the total transmission size in bytes of a file uploaded from a client to a server is measured, using six different file sizes. The data used for this evaluation were encapsulated in CDA (Clinical Document Architecture) format [7]. For each file transmission 50 iterations were executed and the average transmission size was calculated. The results are presented in Fig. 1. The overhead varies between 12.8 – 17.5% with the exception of the case where the file size is 897KB, for which the overhead is less than 1%. This overhead may become larger in a noisy channel due to the increased number of retransmissions.

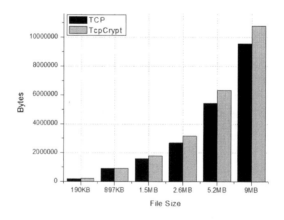

Fig. 1. Comparison of *transmitted bytes* with TCP and Tcpcrypt as the *File Size* increases

2.2   Overhead in CPU Instructions

A second metric related to the performance of mobile devices is the number of CPU instructions required. Given the limited resources of mobile clients, CPU utilization ideally should remain low. When encryption is disabled, the transmission of a 9MB file requires 1,036,000,000 instructions and spends 0.9 seconds of full CPU utilization. When encryption is enabled the number of executed instructions increases to 10,590,000,000, i.e. it requires an order of magnitude more instructions. The total time of full utilization increases to 8.3 seconds[1].

As shown in Fig. 2, the public-key connection initiation (the bursts during the first seconds) incurs the biggest cost. Tcpcrypt performs this operation in the client-side to reduce the stretch of the server's performance. After this initial phase the keys are cached and reused during further TCP communications, even for different TCP sessions. Thus, for a long-lived communication it is preferable to study the CPU utilization after the establishment of keys (second bursts in Fig. 2). When decryption is deactivated the data transmission requires 154,000,000 instructions, while when encryption is supported by both sides the number of instructions is increased to 1,036,000,000 (6.7 times more instructions). When a client communicates with only a limited number of servers, this initial phase does not impose a significant overhead. On the contrary, when the clients have to establish multiple connections with numerous different machines, the key generation incurs a prohibited large overhead. However, in a realistic scenario where a mobile device is engaged, in order to exchange health care data, a limited number of connections is required, rendering the use of Tcpcrypt an efficient solution. [8-10].

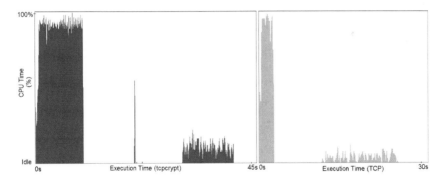

Fig. 2. CPU utilization and total duration for the transmission of data

2.3   Reducing Overhead through Fine-Grained Packet Handling

Availability is a critical performance metric for healthcare applications thus a system design that will enable a fine-grained handling of the TCP packets based on whether

---

[1] The CPU time depends on the CPU architecture and type and it is expected to differ for different processors.

the carried information is sensitive or not is proposed. Currently Tcpcrypt works as an on/off switch. If both ends support Tcpcrypt, encryption is activated for all the packets regardless of whether the information is confidential. This is unimportant for clients with spare CPU cycles and energy; however it would be more desirable to defer from encrypting packets with trivial information in resource-limited devices. Fig. 3 describes this functionality.

An application-specific module characterizes the packets criticality depending on the origin and type of the data (e.g. physiological sensors are marked as a sensitive source, temperature sensors as trivial). Then it passes the packets to the dual-stack TCP (Tcpcrypt/TCP) which operates simultaneously. The "Characterizer" module assigns five different levels of criticality to packets, 0-4, where 0 is trivial and 4 is highest confidentiality. If there is adequate energy all data regardless to the level assigned to them are passed to Tcpcrypt. As the battery discharges, only packets of a higher criticality level are passed to Tcpcrypt. Although this approach does not mitigate the initiation overhead it ensures low-cost CPU operations for long-lived sessions.

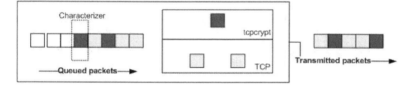

Fig. 3. The Characterizer addon to TCP dual stack. Only critical *(red/dark) packets* are passed to tcpcrypt. Two separate TCP sessions are maintained for each stack.

3   Discussion and Conclusions

When employing security mechanisms it is expected that certain overhead will be introduced. Although privacy and security requirements are of high importance it is imperative that they are used in a sensible manner especially in environments where resources are limited in terms of processing power, bandwidth, battery life etc. In such cases it may be preferable to downgrade security requirements in favor of the performance or the lifetime of the network.

In this paper, a comparison of Tcpcrypt against TCP was presented with a focus on health care applications which are sensitive in terms of integrity and confidentiality. In particular the overhead introduced was examined in terms of:

- increase in the volume of data to be transmitted
- CPU instructions

In the first case the increase occurred in different CDA file sizes was measured and was found roughly to be 12.8 – 17.5%. In the later case we have measured the CPU instructions and processing time overhead incurred. Enabling encryption results in an increase of CPU processing time approximately 9 times up. While the instructions required for full encryption increase in a magnitude of almost 7 times.

As expected, Tcpcrypt is more demanding in both metrics and especially when it comes to CPU instructions. The increase in CPU operations results in increased power consumption that may have undesirable side effects in environments with limited resources. Furthermore non critical data may utilize Tcpcrypt depending on the availability of resources, and if for example a mobile device that is transmitting healthcare information of a patient and suffers from low battery situation TCP could be the most preferable solution.

Towards this direction, a methodology for classifying data in terms of criticality has been proposed. The goal is to minimize the overhead introduced by consuming valuable resources only for information that is considered critical. The implementation of Tcpcrypt, either standalone or in combination with the proposed classification scheme, in mobile devices would be valuable as the benefits provided can be maximized in such environments.

References

1. Ylonen, T., Lonvick, C.: The Secure Shell (SSH) Authentication Protocol, Network Working Group of the IETF, RFC 4252 (2006)
2. Rescorla, E.: HTTP Over TLS, Network Working Group of the IETF, RFC 2818 (2000)
3. Dierks, T., Rescorla, E.: The Transport Layer Security (TLS) Protocol, Network Working Group of the IETF RFC 5246 (2008)
4. Kent, S., Seo, K.: Security Architecture for the Internet Protocol Network Working Group of the IETF, RFC 4301 (2005)
5. Bittau, A., Boneh, D., Hamburg, M., Handley, M., Mazieres, D., Slack, Q.: Cryptographic Protection of TCP Streams (Tcpcrypt) draft-bittau-tcp-crypt-00.txt (2011)
6. Bittau, A., Hamburg, M., Handley, M., Mazieres, D., Boneh, D.: The Case for Ubiquitous Transport-Level Encryption. In: USENIX Security Symposium, Washington, DC (2010)
7. Alschuler, L., Dolin, R.H., Boyer, S., Beebe, C.: HL7 Clinical Document Architecture Framework, Release 1.0.ANSI-approved HL7 Standard (2000)
8. Widya, I., van Halteren, A., Jones, V., Bults, R., Konstantas, D., Vierhout, P., Peuscher, J.: Telematic Requirements for a Mobile and Wireless Healthcare System Derived from Enterprise Models. In: The Proceedings of 7th International Conference on Telecommunications, Croatia, pp. 527–534 (2003)
9. Boukerche, A., Yonglin, R.: A Secure Mobile Healthcare System using Trust-Based Multicast Scheme. IEEE Journal on Selected Areas in Communications, 387–399 (2009)
10. Nikolidakis, S., Georgakakis, E., Giotsas, V., Vergados, D.D., Douligeris, C.: A Secure Ubiquitous Healthcare System Based on IMS and the HL7 Standards. In: The Proceedings of the 3rd International Conference on Pervasive Technologies Related to Assistive Environments, Samos (2010)

An Access Control Framework for Pervasive Mobile Healthcare Systems Utilizing Cloud Services

Mikaela Poulymenopoulou, Flora Malamateniou, and George Vassilacopoulos

Department of Digital Systems, University of Piraeus, Piraeus 185 34, Greece
{mpouly,flora,gvass}@unipi.gr

Abstract. Mobile in conjunction with cloud computing can fulfil the vision of "Pervasive Healthcare" by enabling authorized healthcare participants to access services and required patient information without locational, time and other restraints. Of particular importance on such healthcare systems that incorporate mobile devices and cloud services is protecting the confidentiality of patient information. On these grounds, this paper proposes an access control framework for providing role-based context-aware authorization services with regard services invocations and patient information accesses. According to this, authorization decisions are taken according to contextual constraints that result from the domain ontology inferring that is used to represent context information.

Keywords: context-aware, access control, mobile and cloud computing, healthcare.

1 Introduction

Pervasive computing with the use of appropriate technologies like mobile technology, wireless networks and cloud computing has received considerable attention in the healthcare field recently for providing anytime and anywhere access to appropriate patient information and services to users during healthcare delivery process according to their changing environment [1], [2]. Healthcare delivery is inherently a decentralized process with participating users crossing many institutional boundaries. With the use of a pervasive healthcare system, participating users without locational and time restraints can access context-aware services existing on cloud by their mobile devices in order to view, update and share patient information that is usually structured in the form of XML documents and are also stored to the cloud [2], [3]. In such a pervasive healthcare system, it is important to meet the global security requirements of the healthcare organization(s) involved in the healthcare processes in order to protect the confidentiality of patient information, whilst at the same time allow authorized users to access it conveniently. This is a crucial requirement in the unpredictable environment of healthcare where context is often changing and users change roles at runtime. Thus, there is a need for an access control policy that is adaptable according to the active context, in a way that when the context changes the access control policy must reflect this change [4], [5], [6], [7].

K.S. Nikita et al. (Eds.): MobiHealth 2011, LNICST 83, pp. 380–385, 2012.

On these grounds, in this paper emphasis is on a security framework proposed to provide discretionary role-based and context-aware access control services with regard to services executions and patient XML documents (existing on the cloud) accesses according to an access control model developed. This work is motivated by our involvement in an emergency healthcare project that is still under development and concerns the implementation of a pervasive emergency healthcare system with the use of mobile and cloud computing. To illustrate the feasibility and applicability of the proposed security framework a simplified version of the emergency healthcare process is described.

2   Methods

In Figure 1 the proposed context-aware access control framework is presented. At cloud servers there exists an application server that hosts the web and cloud (provided by the cloud vendor) services and the context manager (CM) that acts as a mediator among users' environment and the semantic knowledge base. The CM uses an inferring engine for context reasoning. In addition, there exists the access control mechanism (ACM) that takes context-based authorization decisions for (cloud and web) services invocations and XML documents accesses existing on cloud. Moreover, at cloud servers there exists a database server where the access control policy and the knowledge base are stored. It is assumed that during healthcare delivery, authorized staff uses a mobile application on their mobile devices that sends context information from users environment to the CM. On users' request, the mobile application calls services existing on cloud that are orchestrated into workflows in order to create, view and update the XML documents with patient medical data.

Context information might be domain-dependent like the subjects (users) and objects (patients) involved in the healthcare processes as well as the relationships between them or might be domain-independent such as this related to the environment (e.g. time) [2], [5], [6]. In order to allow the ACM to interpret context accurately and to have a common understanding among the participating healthcare organizations, context information was organized in the form of a domain ontology using Ontology Web Language (OWL) which also enables context sharing in a semantic way and context reasoning. Those OWL files are stored to the knowledge base and additionally Semantic Web Rule Language (SWRL) rules were written in order to capture the relationships between subjects and objects [4], [5]. In particular, after services execution appropriate context information is sent to the CM that creates any individuals to the domain ontology. Then, the inferring engine infers the domain ontology and the relevant SWRL rules that may result to a user role change (that creates a relationship among subjects and objects). User roles are divided to permanent roles (e.g. ambulance nurse) and temporary roles (e.g. attending ambulance nurse). Users are alleviated from the burden to change roles at run time by automatic role changes (e.g. from permanent roles to temporary roles and vice versa) [6]. For example, in emergency healthcare on ambulance selection for an emergency case, the ambulance staff with permanent role "ambulance physician" is granted the temporary role "attending ambulance physician". This is revoked on ambulance arrival at the hospital emergency department ED, that is realized by another service execution. In particular, role change rules were described as follows [6]:

Definition (Role change): A role change is a 4-tuple (u, ri, rj, {pk}) stating that a user u holding the role ri receives the role rj subject to contextual constraints {pk}.

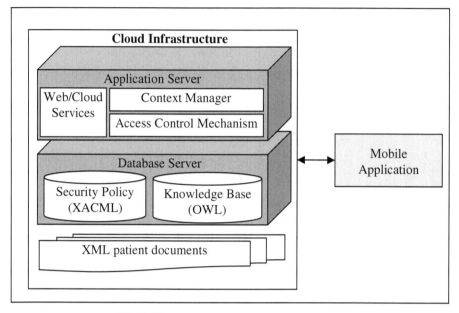

Fig. 1. The proposed access control framework

The role-based access control policy rules specified are evaluated by the ACM according to contextual constraints resulting from the ontology inferring. Those contextual constraints might include authorization delegations that are required when a service S2 invocation is requested after a service S1 execution. Then, the authorization for invoking the S1 should be delegated for the same user role to the S2. This is achieved by sending appropriate context information after S1 execution to the CM. Then, the domain ontology is inferred that result to authorization delegation for S2 invocation. For example the authorization for a service that produces a XML document is delegated to the same user role to enable cloud service invocation for storing this document to the cloud. The access control rules for services invocations and XML document accesses are described as follows [6]:

Definition (web/cloud services invocation): An access control rule for web/cloud service invocation is a 4-tuple (r, "invoke", S, {pk}) stating that a user holding the role r is allowed to invoke web/cloud service S subject to contextual constraints {pk} including authorization delegations.

Definition (XML document access): Given a rule for invoking web/cloud service S by a user holding the role r, an access control rule for XML document access is a 5-tuple (r, "access", S, XML, {pk}) stating that a user holding the role r is allowed to access (read/write) XML document during S execution subject to contextual constraints {pk}.

3   Results

For testing the proposed security framework, a small part of the emergency healthcare process is considered that involve the prehospital care provided by the ambulance service [3]. Figure 3 shows a simplified view of the emergency healthcare process. The activities "Create CDA doc" and "Update CDA doc" have been implemented by two web services developed based on RESTful technology and the other two activities "Store CDA doc" are implemented by the cloud service Amazon S3 for storing the CDA documents created to the cloud servers. The first REST service (CreateCDA) involves the creation of XML Clinical Document Architecture (CDA) based documents with initial emergency case data. The second REST service (UpdateCDA) involves updating the CDA-based documents with emergency case medical data.

The Protégé editor was used for creating the domain ontology and the Jess rule engine for inferring the ontology and SWRL rules. For the access control mechanism the XML Access Control Language (XACML) implementation by Sun Microsystems was used to implement the authorization services on the cloud infrastructure [7]. Authorizations for CDA documents accesses are specified at the level of XML schema. In order to insert the contextual constraints including the authorization delegations to the XACML rules of the XACML policy the attributes of the XACML subjects and resources were used [4].

According to the prototype implementation, on ambulance request context information "new emergency case" is sent to the CM that consults the ontology to retrieve any contextual constraints that are used to form the XACML request send to the ACM. This results to a decision to allow or not the telephone operator of ambulance service to invoke the CreateCDA service to create a CDA document with case data including the selected ambulance and hospital for the case. After service execution the context information "ambulance and hospital selected" is send to the CM that consults the ontology to trigger appropriate (role change) rules for changing the permanent roles (e.g. ambulance nurse) to appropriate temporary roles (e.g. attending ambulance nurse). The same context information results to an authorization delegation in order to allow telephone operator (who executed the CreateCDA service) to invoke Amazon S3 for storing the CDA document.

While at the place of incident or en-route, "attending ambulance nurse" through the mobile application can invoke the UpdateCDA service for updating the CDA document with the medications administered and/or the procedures performed to the case. Hence, context information "update medical data" is sent to the CM that consults the ontology to retrieve any contextual constraints used to form a XACML request send to the ACM in order to allow or not the service invocation. After service execution, the context information "CDA document updated" is sent to the CM that results to an authorization delegation in order to allow attending ambulance nurse (who executed the UpdateCDA service) to invoke Amazon S3 for storing the updated CDA document. Due to lack of space, a more detailed description of this prototype system implementation will be presented elsewhere.

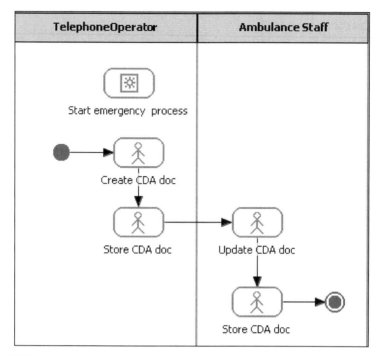

Fig. 2. The emergency healthcare process model designed using Oracle BPM Studio

4   Concluding Remarks

In this paper, is presented an access control framework for providing role-based context-aware authorization services in pervasive healthcare systems that involve the provision of context-aware cloud services to users through their mobile devices. Those authorization services take into consideration the contextual constraints as result from the domain ontology inferring and involve services invocations and XML document accesses. The OWL language is used for representing the domain ontology and SWRL rules were written for expressing role change rules for automatic runtime role changes to users and delegation authorization rules that enables delegating authorizations from a service to another. The prototype system implemented for evaluating the feasibility of the proposed security framework. The rationale behind the envisaged access control framework was to provide users secure access to appropriate services and patient information anytime and anywhere according to context information that represent their changing environment and in a way that the complexity is hidden by the users that is the ultimate goal of pervasive healthcare. Hence, users are alleviated from the burden to change roles manually and authorizations for service executions and patient document accesses are granted to users automatically. However, the proposed framework presented here needs further evaluation in a real world before accepted in a real healthcare environment with adverse circumstances.

References

1. Arnrich, B., Mayora, O., Bardram, J., Troster, G.: Pervasive Healthcare: Paving the Way for a Pervasive, User-centered and Preventive Healthcare Model. Methods Inf. Med. 49(1), 67–73 (2010)
2. Bouzid, Y., Harroud, H., Boulmalf, M., Karmouch, A.: Context-Based Services Discovery in Mobile Environments. In: 16th International Conference on Telecommunications, NJ, USA, pp. 13–18 (2009)
3. Poulymenopoulou, M., Malamateniou, F., Vassilacopoulos, G.: E-EPR: A Cloud-based Architecture of an Electronic Emergency Patient Record. In: 4th International Conference on Pervasive Technologies Related to Assisted Environments, Crete, Greece (2011)
4. Dersingh, A., Liscano, R., Jost, A.: Context-aware Access Control Using Semantic Policies. UbiCC J., Special Issue Autonomous Computing Systems and Applications, 19–32 (2008)
5. Wrona, K., Gomez, L.: Context-aware Security and Secure Context-awareness in Ubiquitous Computing Environments. In: XXI Autumn Meeting of Polish Information Processing Society Conference, Wisla, Poland, pp. 255–265 (2005)
6. Koufi, V., Malamateniou, F., Vassilacopoulos, G., Papakonstantinou, D.: Healthcare System Evolution towards SOA: A Security Perspective. In: 13th World Congress on Medical and Health Informatics, Cape Town, South Africa. Stud. Health Technol. Inform., pp. 874–878 (2010)
7. Zhang, R., Liu, L.: Security Models and Requirements for Healthcare Application Clouds. In: 3rd International Conference on Cloud Computing, Miami, Florida, USA, pp. 268–275 (2010)

Distributed Management of Pervasive Healthcare Data through Cloud Computing

Charalampos Doukas[1,4], Thomas Pliakas[2],
Panayiotis Tsanakas[3], and Ilias Maglogiannis[4]

[1] University of the Aegean, Greece
[2] Nokia Siemens Networks, Greece
[3] National Technical University and Greek Research and Education Network, Greece
[4] University of Central Greece, Greece
doukas@aegean.gr, thomas.pliakas@nsn.com,
tsanakas@admin.grnet.gr, imaglo@ucg.gr

Abstract. This paper presents a distributed platform based on Cloud Computing for management of pervasive healthcare data. Pervasive applications through continuous monitoring of patients and their context generate a vast amount of data that need to be managed and stored for processing and future usage. Cloud computing and service-oriented applications are the new trends for efficient managing and processing data online. The *Cumulocity* cloud platform utilized in this work is especially developed for the support of sensors and machine-to-machine (M2M) communication infrastructures. This paper presents an integrated system for managing sensor data related to the detection of disabled or elderly citizens falls. Wearable sensors collect the fall related data, which are then handled by the Cumulocity Cloud Platform.

Keywords: Cloud Computing, Wearable Sensors, Sensor data management, Fall Detection, Distributed programming, web services.

1 Introduction

The proper delivery of healthcare services among with patient monitoring are considered key issues for improving the quality of life and ensuring efficient health and social care. Mobile pervasive healthcare technologies can support a wide range of applications and services, including mobile telemedicine, patient monitoring, location-based medical services, emergency response and management, personalized monitoring and pervasive access to healthcare information, providing great benefits to both patients and medical personnel ([1], [2]). The realization, however, of health information management through mobile devices introduces several challenges, like data storage and management (e.g., physical storage issues, availability and maintenance), interoperability and availability of heterogeneous resources, security and privacy (e.g., permission control, data anonymity, etc.), unified and ubiquitous access. One potential solution for addressing all aforementioned issues is the introduction of Cloud Computing concept in electronic healthcare systems. Cloud

K.S. Nikita et al. (Eds.): MobiHealth 2011, LNICST 83, pp. 386–393, 2012.

Computing provides the facility to access shared resources and common infrastructure in a ubiquitous and pervasive manner, offering services on-demand, over the network, to perform operations that meet changing needs in electronic healthcare application.

In this context, a distributed platform based on Cloud Computing for management of pervasive healthcare data has been developed. The platform contains the appropriate mechanisms for collecting sensor data. It is based on *Cumulocity*, a horizontal Machine-to-Machine (M2M) Cloud Solution platform provided by Nokia Siemens Networks (NSN). It contains a comprehensive set of tools for managing meters and sensors, collecting and validating data and providing it to enterprises back-office applications. A use case regarding the collection and management of pervasive motion data for fall detection is demonstrated. The rest of the paper is organized as follows; Section 2 discusses related work in distributed pervasive healthcare data management. Section 3 introduces briefly the Cloud computing framework, while Section 4 presents the proposed architecture that utilizes the Cumulocity platform. Section 5 describes the developed use case and, finally, Section 6 concludes the paper.

2   Related Work

The application of mobile devices for pervasive healthcare information management has already been acknowledged and well established ([1], [8]). Authors in [3] present the benefits of using virtual health records for mobile care of elderly citizens. The main purpose of the work is to provide seamless and consistent communication flow between home health care and primary care providers using devices like PDAs and Tablet PCs. Smart cards and web interfaces have been used in [4] for storing patient records electronically. The MADIP system [5] is a distributed information platform allowing wide-area health information exchange based on mobile agents. In [2] authors present a mobile platform for exchanging medical images and patient records over wireless networks using advanced compression schemes.

The majority of the aforementioned works is based on proprietary architectures and communication schemes and requires the deployment of specific software components. Furthermore, these works focus mostly on delivering data to healthcare applications and do not address issues of data management and interoperability issues introduced by the heterogeneous data resources found in modern healthcare systems. The usage of Cloud Computing provides data management and access functionality overcoming the aforementioned issues as discussed in previous sections. The concept of utilizing Cloud Computing in the context of healthcare information management is relatively new but is considered to have great potential [9].

3   Cloud Computing Utilization

Cloud Computing is a model for enabling convenient, on-demand network access to a shared group of configurable computing resources (e.g., networks, servers, storage, applications, and services) that can be rapidly provisioned and released with minimal management effort or service provider interaction. Resources are available over the network and accessed through standard mechanisms that promote use by heterogeneous thin or thick client platforms (e.g., smart phones). Examples of resources include storage, processing, memory, network bandwidth, and virtual machines. Given the characteristics of Cloud Computing and the flexibility of the services that can be developed, a major benefit is the agility that improves with users being able to rapidly and inexpensively re-provision technological infrastructure resources. Device and location independence enable users to access systems using a web browser, regardless of their location or what device they are using (e.g., mobile phones). Multi-tenancy enables sharing of resources and costs across a large pool of users, thus allowing for centralization of infrastructure in locations with lower costs. Reliability improves through the use of multiple redundant sites, which makes Cloud Computing suitable for business continuity and disaster recovery. Security typically can be improved, due to centralization of data and increased availability of security-focused resources. Sustainability comes about through improved resource utilization, resulting in more efficient systems.

A number of Cloud Computing platforms are already available for pervasive management of user data, either free (e.g., iCloud [28], ¬Okeanos [31], Pithos [32] and DropBox [30]) or commercial (e.g., GoGrid [27], Amazon AWS [29] and Rackspace [33]). Most of them, however, do not provide substantial developer support, to create custom applications and incorporate Cloud Computing functionality, apart from Amazon AWS. None of them is optimized for the provision of services to sensor-based applications.

3.1   The Cumulocity Cloud Computing Platform

Cumulocity is a horizontal Machine-to-Machine (M2M) Cloud Solution platform provided by Nokia Siemens Networks (NSN). It contains a comprehensive set of tools for managing meters and sensors, collecting and validating data and providing it to enterprise back-office applications. In addition to this, Cumulocity contains a set of tools for building sensor-based and M2M applications. The platform is used both for integrating sensors and meters into enterprises back-office applications and processes, as well as a stand-alone environment for deploying and running a number of innovative M2M applications. The primary benefit of this integration is increased visibility into the real assets of enterprises and thus improved performance of business processes as well cost reduction.

The Cumulocity based solution consists of three layers: (1) Connected meters and sensors, (2) the management platform, and (3) the integrated vertical applications and enterprise processes. Any meter or sensor can be integrated to the platform through its open smart device integration API. The platform itself consists of device and sensor management functionalities like data collection and validation, fulfillment, monitoring, performance management, configuration management, inventory, identity service, tenant management and open northbound interfaces for application integration. Users can manage and monitor all of these components and features through the embedded management dashboard.

Cumulocity has mainly three different exposure Application Programming Interfaces (APIs): Functional REST, batch data and near-real time publish/subscribe. The first one is RESTful exposure API for northbound applications to use its functionalities. The batch interface is used for exporting large datasets. It is used for example in billing integration, where meter readings are transferred to a billing system. The Event API is a Publish/Subscribe interface that allows for receiving event information from a device or set of devices in near real time. This allows for the creation of independent event driven applications. Through the latter APIs, the interconnection and interoperability with pervasive healthcare applications is direct and straightforward. The sensors can be connected directly through their wireless interfaces to the platform and use simple REST calls for sending and retrieving data. Alternatively, appropriate s/w gateways with similar functionalities can be developed for the sensors that cannot connect directly to this platform. Regarding the caregivers, treatment experts and monitoring personnel, appropriate web applications will be developed giving them access to collected data and events.

The following section presents the proposed architecture for utilizing the M2M platform as a means for distributed management of pervasive healthcare data.

4   The Proposed Architecture Utilizing the Cumulocity Cloud Computing Platform

Fig. 1 presents an illustration of the proposed architecture for managing pervasive healthcare data over the Cumulocity Cloud platform. A variety of pervasive sensors can be utilized for monitoring the patient status and context. The latter can be wearable and textile sensors that monitor vital biosignals and patient motion and generate alerts in cases of stroke or fall detection. Contextual sensors like overhead cameras and microphone arrays can provide more information about the patient condition, context and location and assist with the better assessment of an emergency situation. All sensors are equipped with appropriate networking interfaces (e.g., WiFi, Bluetooth or ZigBee) for communicating directly with the Cloud platform or through intermediate nodes, e.g., like a smartphone. Software interfaces are developed that can act as the intermediate nodes for forwarding the data to the Cloud using REST web service calls. Web applications have been developed that are also hosted by Cumulocity and visualize the data to the caregivers providing them the ability to retrieve information anywhere and anytime. Mobile applications can also be developed especially for alert management, in cases of fall event detections (utilizing the Event API).

An example of a REST web service call for storing a sensor value to the Cloud is the of the following form: *https://<tenantname>.cumulocity.com: port/ webapplication/storevalue?=sensorvalue&key=xxx*

'Sensorvalue' represents the reading from the sensor and 'key' is a secret key for authenticating the sensor to the system. Sensors that communicate directly with the platform can make the REST call which can also very easily be embedded to the intermediate nodes and/or mobile applications.

The communication between the sensors or the intermediate nodes and the Cloud is performed over the SSL protocol providing the essential encryption of the data over transmission. The Cumulocity platform, is deployed on top of the Amazon AWS infrastructure, and is HIPAA compliant. The latter means that all appropriate security techniques and technologies have been adopted in order to store data safely and at the same time maintain the appropriate data anonymity.

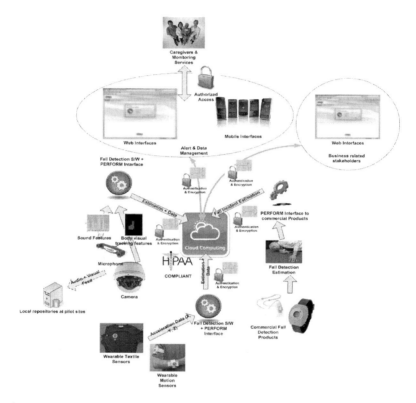

Fig. 1. The proposed architecture for managing pervasive healthcare data in the Cloud

5   Managing Fall Detection Data through Cumulocity Cloud Platform

A use case of pervasive healthcare data management through the Cumulocity platform is presented in this Section. Fall-related injuries are among the most common, morbid, and expensive health conditions involving older adults ([1] – [13]). Falls account for 10% of emergency department visits and 6% of hospitalizations among persons over the age of 65 years and are major determinants of functional decline, nursing-home placement, and restricted activity (14 - 17). The most common and way to monitor patients for fall detection and emergency management is through wearable motion sensors – accelerometers.

In previous works ([21] – [25]) several sensors have been used for collecting motion data. The Arduino microcontroller ([18]) equipped with 3-axis accelerometer and tilt sensor has also been widely utilized. Arduino is an open-source electronics prototyping platform based on flexible, easy-to-use hardware and software. It supports a variety of extensions (shields) that provide additional functionality (e.g., collecting motion data) and networking capabilities (ZigBee, Bluetooth, WiFi, 3G/UMTS, etc.). It exists in various forms with different sizes. It also exists as wearable solution (LilyPad Arduino [19]) that can be sewn to fabric and similarly mounted power supplies, sensors and actuators with conductive thread.

Fig. 2. Arduino sensor board equipped with WiFi module, accelerometer and tilt sensor the LilyPad Arduino sewed on cloth along with accelerometer textile sensors

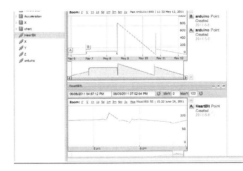

Fig. 3. Screenshot of the web-based application hosted on Cumulocity for monitoring the output of sensors

By using the appropriate network interface (e.g., WiFi and/or 3G/UMTS), Arduino can collect and transmit motion data, wirelessly in both indoor and outdoor environments maximizing this way the availability of the platform. Additionally, the recently introduced Google's Android Open Accessory Development Kit (ADK) [20] provides an implementation of Android USB accessories that are based on the Arduino open source electronics prototyping platform. This allows Arduino to easily interface with android-enabled mobile phones, providing better means of data communication between the sensors and the cloud platform especially in cases where user is located outdoors.

Arduino with the appropriate libraries can make directly calls to the REST API of Cumulocity. An appropriate web-based application has been developed on the platform (see Fig. 3) that receives and displays the sensor data. Through the same

REST API, external applications like in [22] can retrieve data for further analysis and fall detection.

During the initial experimentation with the system, a drop packet rate of 20-30% has been detected. This fact is either due to the Arduino low resources for high rate sampling of sensors and transmitting the data at the same time, or due to network congestion because of the repetitive REST calls at such a high sampling rate (i.e. 10 acceleration samples per second). In order to address this issue, a memory buffer has been introduced on the Arduino side that collects motion data during a 10 second time frame and then transmits the latter to the Cloud. This way the drop rate has been minimized between 2-5%, which is quite acceptable for the application.

6   Conclusions

Pervasive healthcare applications generate a vast amount of sensor data that need to be managed properly for further analysis and processing. Cloud computing through its elasticity and facility to access shared resources and common infrastructure in a ubiquitous and pervasive manner is a promising solution for efficient management of pervasive healthcare data. In this paper we have presented a system for managing fall detection data from wearable sensors using the Cumulocity M2M Cloud Platform. Future work includes the deployment of the service in a wide range of sensors and the further evaluation of the system in terms of sustainability, availability and energy efficiency.

References

1. Varshney, U.: Pervasive Healthcare. IEEE Computer Magazine 36(12), 138–140 (2003)
2. Maglogiannis, I., Doukas, C., Kormentzas, G., Pliakas, T.: Wavelet-Based Compression With ROI Coding Support for Mobile Access to DICOM Images Over Heterogeneous Radio Networks. IEEE Transactions on Information Technology in Biomedicine 13(4), 458–466 (2009)
3. Koch, S., Hägglund, M., Scandurra, I., Moström, D.: Towards a virtual health record for mobile home care of elderly citizens. Presented in MEDINFO 2004, Amsterdam (2004)
4. Chan, A.T.S.: WWW_smart card: towards a mobile health care management system. International Journal of Medical Informatics 57, 127–137 (2000)
5. Su, C.J.: Mobile multi-agent based, distributed information platform (MADIP) for wide-area e-health monitoring. Computers in Industry 59, 55–68 (2008)
6. Hameed, K.: The application of mobile computing and technology to health care services. Telematics and Informatics 20, 99–106 (2003)
7. Doukas, C., Maglogiannis, I.: Managing Wearable Sensor Data through Cloud Computing. In: 3rd IEEE International Conference on Cloud Computing Technology and Science, CloudCom 2011, Athens, Greece (2011)
8. Mendonça, E.A., Chen, E.S., Stetson, P.D., McKnight, L.K., Lei, J., Cimino, J.J.: Approach to mobile information and communication for health care. International Journal of Medical Informatics 73, 631–638 (2004)
9. Shimrat, O.: Cloud Computing and Healthcare. San Diego Physician, 26–29 (2009)

10. Sattin, R.W.: Falls among older persons: a public health perspective. Annu. Rev. Public Health 13, 489–508 (1992); Medical expenditures attributable to injuries — United States. MMWR Morb. Mortal Wkly. Rep. 53, 1–4 (2004)

11. Nevitt, M.C., Cummings, S.R., Hudes, E.S.: Risk factors for injurious falls: a prospective study. J. Gerontol. 46, 164–170 (1991)

12. Tinetti, M.E., Doucette, J., Claus, E., Marottoli, R.: Risk factors for serious injury during falls by older persons in the community. J. Am. Geriatr. Soc. 43, 1214–1221 (1995)

13. Englander, F., Hodson, T.J., Terregrossa, R.A.: Economic dimensions of slip and fall injuries. J. Forensic Sci. 41, 733–746 (1996)

14. Tinetti, M.E., Williams, C.S.: Falls, injuries due to falls, and the risk of admission to a nursing home. N. Engl. J. Med. 337, 1279–1284 (1997)

15. Idem: The effect of falls and fall injuries on functioning in community-dwelling older persons. J. Gerontol. A. Biol. Med. Sci. 53A, M112–M119 (1998)

16. Gill, T.M., Desai, M.M., Gahbauer, E.A., Holford, T.R., Williams, C.S.: Restricted activities among community-living older persons: incidence, precipitants, and health care utilization. Ann. Intern. Med. 135, 313–321 (2001)

17. Gillespie, L.D., Gillespie, W.J., Robertson, M.C., Lamb, S.E., Cumming, R.G., Rowe, B.H.: Interventions for preventing falls in elderly people. Cochrane Database Syst. Rev. 4 (2003)

18. The Arduino Open Source microcontroller platform, http://www.arduino.cc

19. Wearable microcontroller solution, LilyPad Arduino, http://arduino.cc/en/Main/ArduinoBoardLilyPad

20. The Android Open Accessory Development Kit, http://developer.android.com/guide/topics/usb/adk.html

21. Doukas, C., Maglogiannis, I.: Emergency Fall Incidents Detection in Assisted Living Environments Utilizing Motion, Sound and Visual Perceptual Components. IEEE Transactions on Information Technology in Biomedicine 15(2), 277–289 (2011), doi:10.1109/TITB.2010.2091140

22. Doukas, C., Maglogiannis, I.: An Assistive Environment for Improving Human Safety Utilizing Advanced Sound and Motion Data Classification. Accepted for publication in Universal Access in the Information Society. Springer, Heidelberg

23. Doukas, C., Maglogiannis, I.: Advanced Classification and Rules-Based Evaluation of Motion, Visual and Biosignal Data for Patient Fall Incident Detection. Artificial Intelligence Techniques for Pervasive Computing. International Journal on AI Tools (IJAIT) 19(2), 175–191 (2010)

24. Doukas, C., Maglogiannis, I.: Advanced Patient or Elder Fall Detection based on Movement and Sound Data. In: 2nd International Conference on Pervasive Computing Technologies for Healthcare (2008)

25. Doukas, C., Maglogiannis, I., Tragkas, P., Liapis, D., Yovanof, G.: Patient Fall Detection using Support Vector Machines. In: Proc. of the 4th IFIP Conference on Artificial Intelligence Applications & Innovations (AIAI), Athens, Greece, September 19-21 (2007)

26. Reese, G.: Cloud Application Architectures: Building Applications and Infrastructure in the Cloud. O'Reilly Media, Paperback (April 17, 2009), ISBN: 0596156367

27. GoGrid Storage Services, http://www.gogrid.com

28. iCloud, http://www.icloud.com

29. Amazon Web Services (AWS), http://aws.amazon.com/

30. DropBox, https://www.dropbox.com

31. Okeanos cloud services for the Greek academic community, http://okeanos.grnet.gr

32. Pithos network storage service for the Greek academic community, http://pithos.grnet.gr

33. Rackspace cloud computing provider, http://www.rackspacke.com

Efficient Exploitation of Parallel Computing on the Server-Side of Health Organizations' Intranet for Distributing Medical Images to Smart Devices

Athanasios Kakarountas and Ilias Mavridis

Computer Science and Biomedical Informatics, University of Central Greece,
Papasiopoulou 2-4, 35100 Lamia, Greece
{kakarountas,emavridis}@ieee.org

Abstract. The distribution of high-resolution medical images to smart devices, in a health organization premises, is considered in this work. The aim is to reduce network traffic and computation load (of the smart devices). Security issues are also considered. The approach takes full advantage of the parallel processing on medical images performed on the server side, exploiting GPGPUs processing power. Data are pre-calculated and modified to best fit the targeted smart device's display. The evaluation results show minimization on processing requirements at the smart device, while network traffic is reduced significantly when few actions are performed on the image.

Keywords: Parallel computing, medical data processing, medical image, safe access.

1 Introduction

The distribution of medical content from a patient's Electronic Health Record (EHR) has been under major discussion due to privacy and life critical issues. The most common approach until now was aiming to share the original medical content stored in a datacenter to registered clients, to which someone may download data and perform any kind of processing at the client-side [1], [2]. This approach is mainly targeting workstations (connected to medical equipment), desktop PCs or bulky mobile devices (e.g. laptops). However, this introduces significant increase to the network's traffic load; it offers a variety of security vulnerabilities for malicious attacks, while it doesn't take into consideration future trends for health care including the mobile smart devices revolution [3], which may be a critical tool for paramedics or professionals offering health services away from the health organization. Furthermore, this approach doesn't take into consideration the computing power that is required for proper processing and display of medical data (i.e. a medical image of several tens of MB). Finally, considering that most of the available mobile medical solutions are limited in devices with poor displaying capabilities, it is questionable if this approach actually contributes to the needs of a wireless monitoring application.

K.S. Nikita et al. (Eds.): MobiHealth 2011, LNICST 83, pp. 394–398, 2012.
© Institute for Computer Sciences, Social Informatics and Telecommunications Engineering 2012

This work aims in getting together several technologies, from various scientific fields, and suggest a new approach for distributing medical images on mobile smart devices. With the term smart device, we refer to a device offering wireless communication, embedding a descent processor and memory capacity, a reasonable display resolution, and finally offering an easy to use User Interface (UI) through an extensible and robust Operating System (OS). The scientific fields that get under the umbrella of the proposed approach are: Parallel Computing, Distributed Computing, Security, Image Processing, Wireless Communication and Internet Technologies.

2   Exploiting Parallel Computing on the Server-Side

The proposed approach was derived from a simple concept. The majority of the smart mobile devices lack computing power and display capabilities. Thus, there is no need to distribute medical information as is, under the constraint of no resilience in information loss, since the available hardware is inherently imposing this loss in order to display it. Issues concerning latency for downloading data and the security of the storage media (client) are also critical, introducing extra threats and vulnerabilities.

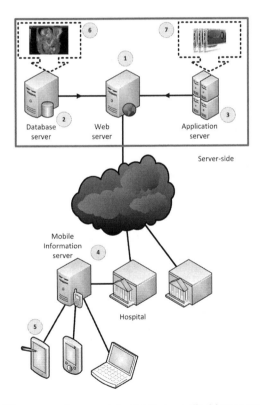

Fig. 1. The proposed approach to distribute medical image content

Thus, in terms of security, it may be said that the distribution of critical parts of data, pre-processed on the server and modified accordingly to fit the characteristics of the targeted smart device, may protect the EHR's data and expose in threat only part of it (which is a result of processing and not the originally stored data). In Fig. 1 a graphical representation of the approach is presented.

The latter mentioned framework consists of four servers, one for distributing the web content through an https connection (circled 1), a second one for serving queries and responses to and from the EHR database (circled 2), a third one for processing data in near-real-time (circled 3) and a fourth one (circled 4) for serving the modified content to the smart devices (circled 5). The first three servers are located to the server-side, while the fourth server corresponds abstractly to the hospital's telecommunication equipment. Although, this configuration has been identified, an alternative configuration may apply for various reasons (i.e. one server for all services to reduce cost).

The main contribution of this work is the migration of heavy processing load to the server side instead of the client side. This allows a wider range of smart devices to be used on medical image distribution (and other similar applications), but also moves complicated calculations, codecs and technology issues (i.e. compatibility, memory capacity etc.) from the client to the server.

2.1   Display of the Information

The smart device is registered during the establishment of a TCP connection, and from a Devices Database, the pre-characterized display resolution is set as the target of the application. The user is authenticated and then a dynamic webpage is created from which the user gets access to the medical image. Data remain in the database server and only a manipulated projection (suitable for the targeted resolution) is offered to the user. This approach makes possible the connection to the system of older smart devices with low processing characteristics.

2.2   Data Processing

The available data (circled 6 in Fig. 1) in the EHR is fetched from the database server to the processing servers. In the proposed framework, those servers are capable for parallel processing in General Purpose Graphic Processing Units (GPGPUs), since medical information consists, apart from text, of multimedia content such as image, video, sound in their appropriate container formats. The solution of GPGPUs allow multiple processing using several kernels, allowing thus pre-processing of information and appropriate modification to fit the targeted display. The GPGPU architecture that is considered in the proposed Framework is the CUDA NVidia [4],[5]. An example of pre-processing is illustrated in Fig. 2, where the possible actions (move around, zoom in-out and lighten-darken) of the user are pre-calculated and are available before the action takes place.

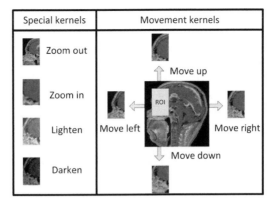

Fig. 2. An example of kernels executed on GPGPUs allowing pre-calculation of actions and modification for the targeted display

2.3   Security Issues

Security is always a hot topic of discussion and debate in such applications. The nature of the proposed framework guarantees that the actual data (except text) never gets out of the server boundaries. In contrast, a projection of data is offered to the user, to allow best viewing of the Region of Interest (RoI), but never acquire the data in whole. Another security mechanism is the authentication of the user and the time to live of the offered information [6].

During the authentication phase, the user's and the patient's authentication information are requested and a temporal communication key is generated from a hash function [7]. Also the MAC address of the device is considered and checked if it is registered in the hospital's registered devices. The derived message digest is used to create a temporal communication session. If the session remains inactive for a pre-defined time period, it ends and further access to the data requires the establishment of a new session. The communication in application level is exploiting the HTTPS protocol.

3   Evaluation of the Proposed Approach

Although the proposed approach is considered for a Web-based application, it may be installed as a wireless service at a governmental building, i.e. hospital. Such a scenario was considered, implementing a case of continuous access requests to a large collection of DICOM files [8]. An Apache server (on a Linux Ubuntu 10.10 computer with 2 GB RAM), embedding mySQL and PHP was responsible for distributing content through the web and a handheld mobile phone (Samsung Nexus S embedding Android 2.3) was used as the client device. This scenario was considered to evaluate under hard conditions the traffic load, the data processing time and the display latency. The network was based on WiFi 802.11g and a wireless 54 Mbps dedicated connection was the communication medium. As it is observed in Table 1, the performance (processing time) achieved with the proposed Framework is significantly increased compared to a typical implementation.

Table 1. Evaluation of the proposed Framework, by accessing DICOM images in real-time over an 802.11g wireless connection

Approach	Network Traffic (Mbps)	Processing Time (s in average)	Time to display after action (s)	Initial download latency (s in average)
Typical	50,0	2,3	0,3	11,9
Proposed	1,0	0,4	0,2	0,3

4   Conclusions

The proposed approach takes full advantage of the parallel processing to data performed on the server side, exploiting the GPGPUs processing power. Data are pre-calculated and modified to best fit to the targeted display. The network traffic is decreased, and in cases of in-hospital installations, the performance is significantly increased. Critical issues of security are also considered and addressed, offering a good degree of security in communication and a high degree of security for the EHR. Finally, it is the first time in the technical literature that a parallel processing architecture is becoming part of a EHR system.

References

1. Zhang, X.M., Zhang, N.: An Open, Secure and Flexible Platform Based on Internet of Things and Cloud Computing for Ambient Aiding Living and Telemedicine. In: Int. Conf. Computer and Management 2011, pp. 1–4 (2011)
2. Kannoju, P.K., Sridhar, K.V., Prasad, K.S.R.: A New Paradigm of Electronic Health Record for Efficient Implementation of Health Care Delivery. In: Second International Conference on Intelligent Systems, Modelling and Simulation 2011, pp. 118–120 (2011)
3. Smartphoning it in. Harvard Medical School, Harvard Health Letter (November 2010)
4. Franco, J., Bernab, G., Fernndez, J., Acacio, M.E.: A parallel implementation of the 2d wavelet transform using cuda. In: Euromicro Conference on Parallel, Distributed, and Network-Based Processing (2009)
5. Matela, J., Rusňák, V., Holub, P.: Efficient JPEG2000 EBCOT Context Modeling for Massively Parallel Architectures. In: Data Compression Conference 2011, March 29-31, pp. 423–432 (2011)
6. Michail, H.E., Kakarountas, A.P., Goutis, C.E.: Server Side Hashing Core Exceeding 3 Gbps of Throughput. Int. J. Net. Sec. 1, 43–53 (2007)
7. Michail, H.E., Kakarountas, A.P., Milidonis, A.S., Goutis, C.E.: A Top-Down Design Methodology for Implementing Ultra High-Speed Hashing Cores. IEEE Trans. Dep. Sec. Comp. 6, 255–268 (2009)
8. OsiriX sample DICOM images, http://pubimage.hcuge.ch:8080/

Body Absorbed Radiation and Design Issues for Wearable Antennas and Sensors

Stavros Koulouridis

Electrotechnics Laboratory, Electrical and Computer Engineering Department
University of Patras, Rio-Patras, Greece
koulouridis@ece.upatras.gr

Abstract. Wearable antennas and sensors are placed inside the body, on the body or in its very close proximity as part of wireless bi-directional communication networks supporting mainly medical applications. Human bodies, having high dielectric permittivity and losses, can greatly affect elements radiation resulting to unwanted power absorption on one hand (raising safety limits questions) and affecting antennas and sensors performance greatly. To deal with these problems, several configurations need to be investigated including optimum radiating element position, use of multiple elements, antennas' special designs. In this work we evaluate several proposed scenarios found in literature to an effort to draw some basic conclusions.

Keywords: flexible antennas, planar antennas, on-body sensors, implanted devices, power absorption, antenna efficiency.

1 Introduction

Implanted and on body radiating devices are being continuously introduced in the late years. Wireless Body Area Networks (WBAN) and Body Area Networks (BAN) [1], biomedical implants [2], "smart" clothes for firefighters [3], Vests loaded with GPS systems [4], UHF applications [5], place several design needs for integrated sensors, conformal antennas, miniaturized designs, low power devices. Various BAN using a variety of communication technologies (MICS, Wi-Fi, GPRS, UMTS) are considered for the several medical application demands in order to obtain the health picture of a patient. Antennas proposed are usually directional microstrip patches [1], [3], [4] but sometimes (depending on the frequency) omnidirectional solutions are being implemented. Further, common configurations for wearable antennas employ textile materials that replace the classical substrates. When implantable designs are pursued, radiating devices are encapsulated in dielectrics for matching purposes but also for avoiding the direct contact of tissues with the metal [2].

Three basic communication scenarios can be identified for the medical applications. (a) First, we consider cases where on body sensors or wearable (integrated in clothes, helmets, etc) antennas will propagate towards open space and will need to transfer information to surrounding environment or to another antenna on the same body; (b) second comes the situation where an antenna or a sensor are

K.S. Nikita et al. (Eds.): MobiHealth 2011, LNICST 83, pp. 399–402, 2012.

placed inside the body and need to transmit information outside the body. Usually the device outside the body is in direct contact with the skin or integrated in a cloth as part of a t-shirt for example; (c) In third case the communication is taking place inside the body. The small sensor or antenna is planted inside deeper in the body and needs to transmit information to a receiver inside the body, usually placed just under the skin.

In any case, antennas need to operate at the minimum power and certainly under safety limits since absorbed power raise safety and health issues for the antenna carrier and/or those around. Energy losses can degrade antenna or sensor efficiency and can drain battery power. Consequently, radiation links or in general communication capabilities are restricted in conjunction with the radiation scenario they fall under (see above). Further, human body, being a large lossy and complex vessel can make things even harder. Optimum designs should be pursued that take into account all these issues posing contradictory goals.

In this work we will examine several exposure scenarios employed in implanted and wearable antennas field. The effort will be devoted to identifying possible exposure patterns. Some basic conclusions will be drawn for the effect of body on the antenna operation and the result of "smart" loading of the antennas substrate, initially intended to decrease back radiation and mutual coupling of wearable antennas.

2   Exposure Scenarios and Design Issues

In the first case (radiator and receiver placed outside the body) the goal is to decrease the energy lost inside the body or increase the antenna efficiency. Depending on the receiver and radiator position, antenna's polarization needs usually be taken into account. Also, antenna's radiation pattern which can be affected by the body and surrounding objects draws specific limitations on antenna type employed. Hence, in this scenario the antenna might need to have a relatively omnidirectional pattern or a directive pattern if radiator and receiver are at fixed positions (placed for example on the same body). Sometimes more than one antenna collaborate to achieve coverage of the surrounding area. It is very interesting also, that as noted in [4], reported wearable antennas have measured gains close to -10 dBi –the radiating elements are of low efficiency.

Unwanted exposure from wearable antennas can also occur from secondary sources. Multiple antennas on the same body for example could lead to increased absorption, or people in the vicinity of the antenna could significantly alter antenna far field radiation behavior.

Except for the few cases where omnidirectional antennas are proposed, wearable antennas usually consist of patches that require a ground plane. Ground planes supposedly decrease back radiation. However, inefficiency of wearable antennas shows that great fraction of antenna forward radiation returns to the body. Further, commonly used textile loading for flexibility and lightness adds to their inefficiency. Therefore till now wearable antennas capabilities are restricted and a lot remain to be done.

In the second scenario a link is sought between the external antenna and the internal microwave device. Problems can arise from the great difference between the dielectric properties of air and human body. Also unwanted surface waves can occur. Apart from safety issues and possible intervention from other sensors or antennas, power is lost and signal integrity might also worsen. Of importance is also the relative position of the two elements since it might not be fixed, because of different body properies or small movements of the wearable sensor for example.

Last scenario covers the connection between two elements inside the body. One device (sensor for example) performs the measurement, tracks physiology parameters, etc and the other device (antenna for instance) receives the information, stores it or sends it to a receiver outside the body (second scenario). Again the relative position between the two can have many uncertainties. If it is a "smart" pill for instance, obviously there many possible places inside the body. Thus the "receiver" will have to occupy an optimum position and have a relatively broadband pattern in order to cover multiple possibilities.

All three cases will be examined in order to shed more light on the various issues that arise.

3   Operational Frequency and Employed Models

Operation frequency can certainly play a role on the design followed in order to have the best result. Medical Implant Communication Service for example occupies the 402-405MHz frequency band. In these frequencies electromagnetic waves can certainly have increased penetration (as compared to higher frequencies). Yet, electromagnetic devices can have relatively large sizes. To minimize them high dielectrics can be employed which will lead to higher sensitivity to design uncertainties. On the other hand UMTS, Wi-Fi, GPRS, etc services that operate at higher frequencies and can be used for BAN networks for example can certainly allow for more stable designs but will demonstrate lower penetration and this can be a problem if we need to cover second scenario as described above. Hence, frequencies employed need to be taken into account when cases are examined and conclusions are drawn.

Second, the unique body characteristics that each human has can increase uncertainties and design failures especially for the sensitive, to uncertainties, designs. Thus it is necessary that in the design process several models will be employed. In the current work, models from Virtual Population [8] will be used in order to include various body characteristics in the studied problems. SEMCAD X from SPEAG [9] will be used to analyze all the examined cases.

Acknowledgments. The author would like to thank Schmid & Partner Engineering, AG, for providing the SEMCAD software.

References

1. Declercq, F., Rogier, H.: Active Integrated Wearable Textile Antenna With Optimized Noise Characteristics. IEEE T. Antennas Propagation 57, 3050–3054 (2009)
2. Mizuno, H., Takahashi, M., Saito, K., Haga, N., Ito, K.: Design of a Helical Folded Dipole Antenna for Biomedical Implants. In: Proc. 5th European Conference on Antennas and Propagation, Eucap, pp. 3640–3643 (2011)

3. Hertleer, C., Rogier, H., Vallozzi, L., Langenhove, L.V.: A Textile Antenna for Off-Body Communication Integrated Into Protective Clothing for Firefighters. IEEE T. Antennas Propagation 57, 919–925 (2009)
4. Vallozzi, L., Vandendriessche, W., Hertleer, C., Scarpello, M.L.: Wearable textile GPS antenna for integration in protective garments. In: 4th European Conference on Antennas and Propagation, EuCAP (2010)
5. Psychoudakis, D., Volakis, J.L.: Conformal Asymmetric Meandered Flare (AMF) Antenna for Body-Worn Applications. IEEE Antennas Wireless Prop. Letters 8, 931–934 (2009)
6. Gallego-Gallego, I., Quevedo-Terue, O., Inclan-Sanchez, L., Rajo-Iglesias, E., Garcia-Vidal, F.J.: On the Use of Soft Surfaces to Reduce Back Radiation in Textile Microstrip Patch Antennas. In: 5th European Conference on Antennas and Propagation, EuCAP (2011)
7. Salonen, P., Rahmat-Samii, Y., Kivikoski, M.: Wearable antennas in the vicinity of human body. In: Proc. IEEE Antennas Propag. Soc. Int. Symp., vol. 1, pp. 467–470 (2004)
8. Christ, A., et al.: The Virtual Family – development of surface-based anatomical models of two adults and two children for dosimetric simulations. Physics in Medicine and Biology 55(2), N23–N38 (2010)
9. Schmid & Partner Engineering AG, http://www.speag.com/speag/

Development of a FDTD Simulator
for the Calculation
of Temperature Rise in Human Heads
from Mobile Phones Operation

Adamos G. Kyriakou[1], Elias Aitides[2], and Michael T. Chryssomallis[2]

[1] IT'IS Foundation, Zeughausstrasse 43, 8004 Zurich, Switzerland
[2] Dept. of Electrical and Computer Engineering
of Democritus University of Thrace, GR-67100 Xanthi, Greece
adamos@itis.ethz.ch, {iaitidis,mchrysso}@ee.duth.gr

Abstract. In this work a complete software tool, which uses the finite-difference time-domain (FDTD) method, was developed from scratch. This tool comprises a full three-dimensional (3D) wave electromagnetic simulator and a bioheat equation solver. The application of this tool, using an anatomically based model of the human head, allows the electromagnetic and thermal analysis of a head exposed to the radiation of a mobile phone, through the determination of the specific absorption rate (SAR). Preliminary results show good agreement with previous published numerical and measurement results taken for similar formulations.

Keywords: Biological effects of electromagnetic radiation, Finite difference time domain (FDTD), bioheat equation, anatomic model of the head, cellular telephones, temperature increase, specific absorption rate (SAR).

1 Introduction

In recent years, the generalized use of mobile radio systems had as a result to increase the public concern regarding the hazards from handheld terminals. These terminals, known as mobile phones, emit electromagnetic fields and operate in close proximity to the human head. Although a lot of work has been done on SAR estimations, only in a few recent studies there have been attempts to include the biological and physical mechanisms of heat transfer in conjunction with the Maxwell equations, which describe the electromagnetic energy propagation around and within the exposed human head. Wang [1] calculated the SAR and temperature increase distributions due to an approximate mobile phone model consisting of a quarter-wavelength monopole over a conducting box. Bernardi [2] calculated SAR and temperature increase for four different approximate mobile phone models irradiating an anatomical model of the human head. Hirata [3], Rodriguez [4] and Kim [5] examined the influence of a half-wavelength dipole near head models. In all of these studies the anatomical models (when used) were simplified consisting of 6-18 tissue types. In addition, most of these

K.S. Nikita et al. (Eds.): MobiHealth 2011, LNICST 83, pp. 403–407, 2012.

studies [1-5] do not agree with each other in their findings, which calls for additional investigations to be performed with more detailed anatomical models. In this work, both electromagnetic and thermal simulations of a highly detailed anatomical head model exposed to the fields of an approximate mobile phone model were performed. The resulting SAR and temperature increase where extracted and compared to results from other studies.

2   Description of the Process

The used human head is based on a segmented anatomical model obtained from [6]. This model, developed from Computer Tomography (CT) image data of a human head, has a resolution of 4×4×4 mm cubic cells. In the form that it was used in this work it includes 26 distinct different tissues and every cubic cell contains only one tissue, which makes it very efficient for FDTD simulation. The dielectric and thermal parameters of the involved tissues for the desired frequency were based on similar studies found in bibliography [7]. The handheld phone is modeled as a monopole antenna on a dielectric covered metal box and it is the same used in previous works [1, 2]. The temperature rise in this work is computed in two steps. During the first step, the electromagnetic problem is solved, where from the solution of Maxwell equations using the FDTD method the electromagnetic fields are calculated [1]. From these fields the distribution of SAR in the human head is evaluated. On the second step, the thermal problem is treated, where the determination of temperature rise is evaluated by solving the bioheat equation. The temperature increase is calculated as the difference between the temperature distribution of the head during thermal equilibrium with and without exposure to the EM fields of the mobile phone. The Pennes bioheat equation [2], which accounts for various heat-exchange mechanisms such as heat conduction, blood flow and electromagnetic heating, is also solved through the FDTD method. Considering the fact that the temperature rise due to handheld phones operation is not sufficient to cause a significant change on the electric parameters of tissues, the steady state SAR distribution is used as the input electromagnetic heating source into the bioheat equation. Validation of the EM solver was performed by calculating the complex input impedance of dipoles for different frequencies and comparing it to analytically derived results provided by textbooks and the numerically calculated results from commercial FDTD software packages. The observed error was less than 5%.

3   Simulation Results

For the purposes of this work, as well as for other similar cases, an electromagnetic and a thermal simulator were developed from scratch. Based on Python, the open source programming libraries NumPy and SciPy for the core computational and numerical needs, and the Visual Tool Kit for visualization and post processing needs were used.

The electromagnetic solver is a full wave 3D FDTD electromagnetic simulator, capable of solving the Maxwell equations, based on the standard E-H paradigm. It was flexibly developed to simulate any properly defined electromagnetic problem, such as waveguide and antenna design cases, or to evaluate biomedical applications, such as wave-tissue interaction to biomedical implant safety assessment. The discretized partial differential equations were solved on a uniform rectangular grid. The truncation of the generally unbounded computational domain was achieved using a combination of 1st and a modified version of the 2nd order Mur absorptive boundary conditions in order to keep the computational resource requirements to a minimum. These were discretized using upwind finite differences to counter the late-time instabilities, that is the spurious values at the domain boundaries which occur when using centered finite differences to discretize the Mur equations, effectively countering this error source. The power radiated from the phone has been computed on the basis of the feed point impedance.

The thermal solver implementing the bioheat equation for the calculation of the absolute temperature and the temperature increase of a 3D electromagnetically irradiated geometry was also developed from the ground-up. Starting from the evaluated SAR distribution, the thermal response as a function of time, until the steady state is reached, has been calculated through an explicit finite difference formulation on the bioheat equation. This equation was formulated to account for the metabolic heat generation as well as the cooling effect of blood according to the model proposed in [8]. Figs. 1-3 present some results from the application of the method. For a 910 MHz mobile phone with antenna output of 250 mW, which is the output of a GSM900 mobile phone, Fig. 1 shows the total electric field distribution in a vertical cross section, Fig. 2 shows the 1-gr-averaged SAR's distributions and Fig. 3 shows the temperature rise distributions, all for a vertical cross section which includes the monopole antenna. The achieved values are compared well with results found in literature for similar cases. Using a mobile phone operating at 900 MHz with antenna output at 600 mW, which radiates at a distance of 16 mm from the head, the 1-gr-averaged SAR maximum value was evaluated to 3.2 W/kg while the maximum temperature rise was 0.43 °C, for the whole head. Table 1 shows the comparison between the results of this study and those from other publications.

Table 1. Comparison between the results of this study and those from other publications

	Head 1g SAR [W/Kg]	Head Temperature Increase [C]	Brain 1g SAR [W/Kg]	Brain Temperature Increase [C]
Current Study	3.20	0.43	2.03	0.16
Wang [1]	1.60	0.16	0.89	0.05
Bernardi [2]	2.74	0.33	1.85	0.19
Hirata [3]	2.10	0.80	1.20	0.8
Kim [5]	2.84	0.50	1.76	0.24

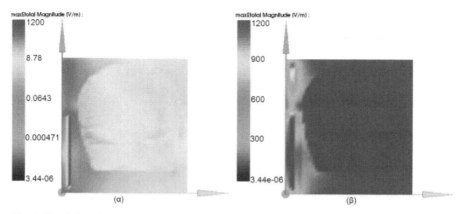

Fig. 1. Total electric field distributions in a vertical cross section for a 910 MHz mobile phone with antenna output of 250 mW (a) linear and (b) chromatic view

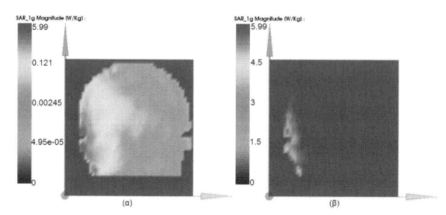

Fig. 2. 1-gr- averaged SAR's distributions in a vertical cross section, for a 910 MHz mobile phone with antenna output of 250 mW (a) linear and (b) chromatic view

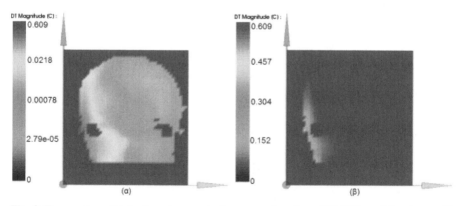

Fig. 3. Temperature distributions in a vertical cross section, for a 910 MHz mobile phone with antenna output of 250 mW (a) log and (b) linear scale

4   Conclusions

The electromagnetic and thermal analysis of a head exposed to the radiation of a mobile phone, through the determination of the specific absorption rate (SAR) was performed, by using a combination of an electromagnetic and thermal solver, both working using the FDTD method. These solvers were developed from the beginning and their application proved their validity. Preliminary results show good agreement with previous published numerical and measurement results [1-5].

References

1. Wang, J., Fujiwara, O.: FDTD Computation of Temperature Rise in the Human Head for Portable Telephones. IEEE Trans. Microwave Theory Tech. 47(8), 1528–1534 (1999)
2. Bernardi, S.P., Paolo, C.M., Piuzzi, E.: Specific Absorption Rate and Temperature Increases in the Head of a Cellular-Phone User. IEEE Trans. Microwave Theory Tech. 48(7), 1118–1126 (2000)
3. Hirata, A., Morita, M., Shiozawa, T.: Temperature Increase in the Human Head Due to a Dipole Antenna at Microwave Frequencies. IEEE Trans. on Electromagnetic Compatibility 45(1), 109–116 (2003)
4. Rodrigues, J.J.V.L.O.R., Ana, O., Malta, L., Ramirez, J.: A Head Model for the Calculation of SAR and temperature Rise Induced by Cellular Phones. IEEE Trans. on Magnetics 44(6), 1446–1449 (2008)
5. Kim, W.-T., Yook, J.-G.: Thermal Steady State in Human Head under Continuous EM Exposure. In: IEEE MTT-S International Symposium Digest (2005), doi:10.1109/MWSYM.2005.1517075
6. Zubal, I.G., Harrell, C.R., Smith, E.O., Rattner, Z., Gindi, G.R., Hoffer, P.B.: Computerized three-dimensional segmented human anatomy. Med. Phys. 21(2), 299–302 (1994)
7. Gabriel, R.W.L.S., Gabriel, C.: The dielectric properties of biological tissues. Physics in Medicine and Biology 41, 2271–2293 (1996)
8. Specific Absorption Rate – Wikipedia, the free encyclopedia, http://en.Wikipedia.org/wiki/Specific_absorption_rate

Numerical Assessment of EEG Electrode Artifacts during EMF Exposure in Human Provocation Studies

Maria Christopoulou*, Orestis Kazasidis, and Konstantina S. Nikita

Biomedical Simulations and Imaging Unit (BIOSIM)
School of Electrical and Computer Engineering, National Technical University of Athens
9 Iroon Polytechniou, Zografou Campus, 15780, Athens, Greece
mchrist@biosim.ntua.gr, orestis.kaza@gmail.com,
knikita@cc.ece.ntua.gr

Abstract. The paper presents the numerical evaluation of the electroencephalogram (EEG) electrode artifacts that are caused during exposure to electromagnetic fields (EMF), in volunteers study. The scope of the study is to differentially present the electromagnetic (EM) power absorption and local Specific Absorption Rate (SAR) distribution, with and without the electrodes. Versions of two basic exposure scenarios are evaluated: flat layered tissue phantom and anatomical head model exposed to plane wave or patch antenna radiation at operating frequency of 1966 MHz. Finite Difference Time Domain (FDTD) method is used in order to model the computational domain. E-field distributions and SAR values are calculated. The electromagnetic power absorption by the brain tissues is correlated with the presence of the EEG electrodes and the relative positioning of their leads. Results conclude in significant alternations in EM power absorption, E-field and SAR distributions, due to the co-polarization between the leads and the E-field. Concerning the realistic scenario, the presence of 32 electrodes and their leads enhances (11% without and 12.3% with electric contact) the $psSAR_{10g}$, comparing to the reference simulation.

Keywords: Specific Absorption Rate (SAR), Finite Difference Time Domain (FDTD) method, numerical dosimetry, electroencephalogram (EEG), electrode, human provocation study.

1   Introduction

In human provocation studies, the electromagnetic (EM) exposure prior the sleep electroencephalogram (EEG) [1] or the Event Related Potentials (ERP) recordings is often performed with the volunteers having the EEG cap already worn, in order to minimize time between exposure and sleep onset or cognitive task initiation. In [2] the Specific Absorption Rate (SAR) enhancement has been numerically evaluated due to simultaneous MRI and EEG recordings. Simulations have been conducted for 128 MHz-3 Tesla and 300 MHz-7 Tesla with 16, 31, 62 and 124 electrodes. Hamblin *et al.*

---

* Corresponding author.

S. Müller Arisona et al. (Eds.): DUMS, CCIS 242, pp. 408–415, 2012.
© Institute for Computer Sciences, Social Informatics and Telecommunications Engineering 2012

[3] have experimentally and numerically assessed the effect of two 64-electrodes EEG caps on the SAR values for 900 MHz. The main outcome of the study included reduction of peak spatial SAR averaged over 10g ($psSAR_{10g}$) due to the presence of electrode leads. Lately [4]-[5], the shielding effect, the SAR alternation and E-field artifacts have been evaluated for UMTS-like exposure, according to the reported exposure scenario.

This paper is part of the numerical dosimetry for a human study, according to published guidelines for provocation studies [6]. The volunteers study is a collaboration between the Biomedical Simulations and Imaging (BIOSIM) Unit and the University Mental Health Research Institute, aiming at the assessment of potential alternations in electroencephalogram (EEG) and event related potentials (ERP) recordings during acoustic stimuli, due to exposure to UMTS-like EM signal. Before EEG and ERP recordings, the subjects are exposed for 30 min to EM radiation, having the EEG cap already worn. In this paper, the EEG electrodes artifacts are numerically evaluated, concerning the power absorption, the local SAR and E-field distribution. The electrodes that will be used during the human study are modeled and the evaluation is carried out with i) flat phantom and ii) realistic head model.

2   Materials and Methods

Two basic exposure scenarios, including modifications, are numerically modeled and simulated: a) flat layered phantom with one electrode attached and b) anatomical head model with an EEG cap of 32 electrodes attached. Both scenarios are comparatively assessed with the corresponding reference ones, without electrodes. The used operating frequency is 1966 MHz, corresponding to UMTS operating frequency band. Apart from a plane wave, the numerical models are exposed to the radiation of the wideband patch antenna SPA 2000/80/8/0/V (Huber & Suhner), placed at x=-180 mm separation distance. Measurements data of the antenna operation characteristics are in agreement with the corresponding simulated ones and they are both presented in [7].

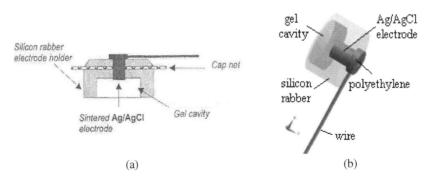

(a)   (b)

Fig. 1. Description of the electrode's structure: a) real and b) numerical model

Fig. 1(a) illustrates the real model of the electrodes that are attached to Softcap (Spes Medica) and will be used in the volunteers study. Fig. 1(b) illustrates the derived numerical model, where the Ag/AgCl electrode (PEC), silicon rabber electrode holder (ε_r=3.2, σ=0.0265 Si/m), gel cavity (air) and the polyethylene connection (ε_r=2.25, σ=0.0005 Si/m), between the electrode and the lead (PEC) are separately denoted. The electrode lead is covered by PVC (ε_r=2.8, σ=0.019 Si/m). In order to simulate the electric contact between the skin and the electrode, gel cavity is selectively characterized as PEC. Computational dosimetry is conducted for flat layered tissue phantom and anatomical head model using the SEMCAD-X v14.2 software (SPEAG, Zurich, Switzerland) and the Finite Difference Time Domain (FDTD) method [8].

2.1   Flat Layered Phantom Exposure Scenario

For preliminary evaluation, a flat phantom with one electrode attached, is used. The flat phantom is structured in eight (8) layers [4] and it is considered to simulate the head biological tissues' sequence, from the external tissue to the inner one, as it is tabulated in Table 1. Table 1 also includes the dielectric properties of the tissues at 1966 MHz [9] and the thickness that is used, according to anatomy information [10].

Table 1. Thickness and dielectric properties (at operating frequency 1966 MHz) of the biological tissues in flat layered phantom

biological tissue	thickness (mm)	ε_r	σ (S/m)	ρ (kg/m^3)
dry skin	2	38.62	1.25	1100
fat (not infiltrated)	1	5.33	0.08	916
muscle	4	53.33	1.43	1041
cortical bone	6	11.67	0.30	1990
dura matter	1	42.67	1.40	1013
CSF	2	66.96	3.05	1007
grey matter	4	49.76	1.49	1039
white matter	150	36.78	0.99	1043

The flat layered phantom exposure scenario is illustrated in Fig. 2(a). One electrode with its lead is attached to the external tissue (dry skin). The electric contact between Ag/AgCl electrode and skin is modeled by altering the gel cavity dielectric properties and height. The electrode lead is set differently in order to be co- and cross-polarized with the incident E-field. A plane wave at 1966 MHz or the patch antenna at distance x=-180 mm are both used, as electromagnetic sources. The computational grid consists of ~24 Mcells and the simulation time is set to 20 periods.

2.2   Anatomical Head Exposure Scenario

The proposed anatomical head exposure scenario is illustrated in Fig. 2(b). As realistic head model, 'Ella' from Virtual Family [11] is used, corresponding to MRI data of a 26 year old adult woman. 'Ella' model has a resolution of 0.5×0.5×0.5mm^3 and consists of 41 head structures. 32 electrodes (with and without their leads) are mounted on the numerical head, according to 10-20 extended system [12]. The leads are placed as horizontally as possibly, in order to minimize the alternations in E-field distribution and the enhancement in SAR values [5]. Only the patch antenna is used, simulating the real experimental scenario, during the human provocation study. The center of the antenna is placed at (x,y,z)=(-180,0,42) mm, considering as (0,0,0), the right ear canal. The computational grid consists of 41-95 Mcells, depending on the exposure scenario ((a) reference, (b) with electrodes, (c) with electrodes and leads), and the simulation time is set to 35 periods.

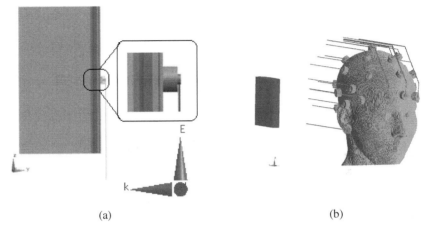

(a)   (b)

Fig. 2. Geometry of the proposed exposure scenarios at 1966 MHz: a) flat layered phantom with one electrode and lead attached, exposed to plane wave (co-polarization), b) realistic head with 32 electrodes and leads attached-according to 10-20 system, exposed to irradiation of the patch antenna.

3   Simulation Results

All the results are differentially presented as compared to the reference simulations, i.e. the corresponding ones without the presence of the electrodes and their leads. All results are normalized to 1 W input power. The gel cavity is generally simulated as air, except the worst case scenario that is presented for anatomical head. In this section, results of both exposure scenarios are presented.

3.1   Flat Layered Phantom Exposure Scenario

For the case of the flat layered phantom exposed to plane wave, the surface E_{rms}-field distribution is illustrated in Fig. 3. The results confirm that the co-polarization of the electrode lead can cause significant distortion in the E-field distribution, as previous studies [4]-[5] emphasize. Additionally, there is an E-field enhanced area, parallel to the electrode lead, between two regions of significantly low values. In case of cross-polarization (Fig. 3(c)), there is a slight amplification in the center of the flat phantom, corresponding to the electrode. This amplification is local and almost superficial and it is restricted to the external layers of the phantom (up to 10 mm). The presence of the lead has no result in the E-field distribution.

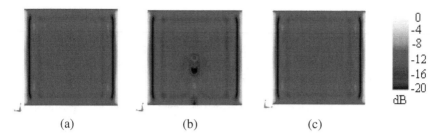

(a)   (b)   (c)

Fig. 3. Surface E_{rms}-field distribution for the a) reference simulation, b) co- and c) cross-polarization. All values are normalized to 14.5 V/m (0 dB).

Additionally, the $psSAR_{1g}$ [13] is calculated for each exposure scenario. Comparing to the reference simulation (0.2459 W/kg), an increase of over 11% in the co-polarization (0.2629 W/kg) and a slight decrease of 1% in the cross-polarization (0.2339 W/kg) are calculated for the $psSAR_{1g}$ values.

For the case of the flat layered phantom exposed to the SPA 2000/80/8/0/V patch antenna irradiation, the local SAR distribution at x=0 mm and z=0 mm, where the electrode has been placed, is comparatively illustrated in Fig. 4. The results confirm that the local SAR distribution presents almost the same pattern for the reference and the cross-polarization scenario. In the co-polarization scenario, the presence of the electrode lead causes SAR attenuation of approximately 15 dB in selected tissues, such as dry skin. Comparing the local SAR calculated for each biological tissue, dry skin, muscle and CSF are the ones that absorb the most of the radiated EM power. This is related to their comparatively large values of electrical conductivity as well as their small distance from the EM source.

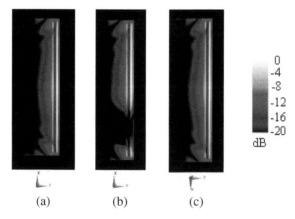

(a)   (b)   (c)

Fig. 4. Local SAR distribution for the a) reference simulation (x=0 mm) b) co- (x=0 mm) and c) cross-polarization (z=0 mm). All values are normalized to 0.6 W/kg (0 dB).

3.2   Anatomical Head Exposure Scenario

In case of the anatomical head model exposed to the radiation of the patch antenna, E-field distribution and SAR values in each brain structure are assessed. Fig. 5

illustrates the E-field distribution at y=0 slice for (a) reference, (b) with electrodes and (c) with electrodes and leads simulations. Enhancement of the E-field values along the leads is obvious at both sides of the head. Comparing the three simulations, the E-field distribution in the head remains almost unchanged.

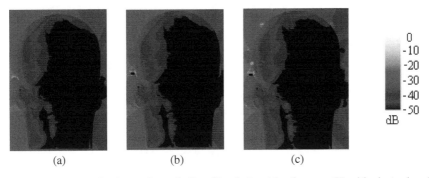

(a)   (b)   (c)

Fig. 5. E-field distribution at the y=0 slice. Simulation (a) reference, (b) with electrodes, (c) with electrodes and leads. All values are normalized to 10^3 V/m (0 dB).

In order to simulate the worst case scenario of electric contact between the electrode and the skin, the gel cavity for each of the 32 electrodes is characterized as PEC. Therefore, Fig. 6 compares the local SAR surface distribution for the following simulations: (a) reference, (c) electrodes and leads, with gel cavity characterized as air and (d) electrodes and leads, with gel cavity characterized as PEC. It is obvious that when there is no electric contact between the skin and the electrode, no significant difference in the local SAR surface distribution is reported. In (d) scenario, the characterization of the gel cavity as PEC leads to attenuation of the SAR values below the cavity and amplification of the EM absorbed power around the electrode.

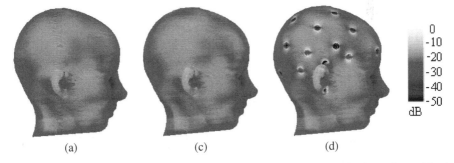

(a)   (c)   (d)

Fig. 6. Local SAR surface distribution. Simulation (a) reference, (c) with electrodes and leads (gel cavity: air), (d) with electrodes and leads (gel cavity: PEC). All values are normalized to 50.37 W/kg (0 dB).

Additionally, the $psSAR_{1g/10g}$ [13] and the averaged SAR over the whole mass of selected brain structures are calculated for all simulations. Comparing to the reference simulation (0.5015 W/kg), an increase of 11% in the (c) scenario simulation (0.5569

W/kg) and 12.3% in the (d) scenario (0.5632 W/kg) are calculated, correspondingly for the head $psSAR_{10g}$ values. Indicatively, Table 2 includes the $psSAR_{1g}$ and averaged SAR calculations in selected brain structures and head for both hemispheres, comparing (a), (c) and (d) simulation scenarios. Concerning the $psSAR_{1g}$, the maximum alternation (45% decrease) is noticed in thalamus for simulation (d), comparing to the reference one. For averaged SAR, the corresponding maximum alternation (44% decrease) is assessed also in thalamus for simulation (d).

Table 2. $psSAR_{1g}$ and averaged SAR values calculated in selected brain structures and head for (a) reference, (c) electrodes with leads and (d) electrodes with leads and gel cavity as PEC simulations.

brain structure or head	$psSAR_{1g}$ (W/kg)			Avg. SAR (W/kg)		
	(a)	**(c)**	**(d)**	**(a)**	**(c)**	**(d)**
grey matter	0.312	0.308	0.360	0.036	0.036	0.030
white matter	0.164	0.163	0.153	0.021	0.029	0.018
thalamus	0.022	0.022	0.012	0.009	0.006	0.005
midbrain	0.013	0.012	0.008	0.004	0.004	0.003
averaged brain w/o CSF	0.297	0.291	0.351	0.028	0.029	0.024
head	1.812	1.819	2.320	0.042	0.043	0.040

4   Conclusions

A detailed numerical evaluation for EEG electrodes artifacts is described in this paper. Versions of two basic exposure scenarios are evaluated: i) flat layered phantom with one electrode attached and ii) realistic head model with 32 electrodes attached. Results conclude in significant alternations in EM power absorption, E-field and SAR distributions, due to the co-polarization between the leads and the E-field. Concerning the realistic scenario, the use of 32 electrodes and their leads results in an 11% increase (12.3% with electric contact) of the head $psSAR_{10g}$, comparing to the reference simulation. Future work can be focused on altering the number of the electrodes and orientation of their leads on the realistic head. This study along with variation and uncertainty numerical evaluation completes the full numerical dosimetry assessment that should always precede a human provocation study.

References

1. Huber, R., Schuderer, J., Graf, T., Jütz, K., Borbély, A.A., Kuster, N., Acherman, P.: Radio frequency electromagnetic field exposure in humans: estimation of SAR distribution in the brain, effects on sleep and heart rate. Bioelectromagnetics 24, 262–276 (2003)
2. Angelone, L.M., Potthast, A., Segonne, F., Iwaki, S., Belliveau, J.W., Bonmassar, G.: Metallic electrodes and leads in simultaneous EEG-MRI: Specific Absorption Rate (SAR) simulation studies. Bioelectromagnetics 25, 285–295 (2004)

3. Hamblin, D.L., Anderson, V., McIntosh, R.L., McKenzie, R.J., Wood, A.W., Iskra, S., Croft, R.J.: EEG electrode caps can reduce SAR induced in the head by GSM 900 mobile phones. IEEE Trans. Biomed. Eng. 54, 914–920 (2007)
4. Murbach, M., Kuehn, S., Christopoulou, M., Christ, A., Achermann, P., Kuster, N.: Evaluation of Artifacts by EEG Electrodes during RF Exposures. In: BioElectromagnetics Annual Meeting (BioEM), Davos, Switzerland, June 14-19 (2009)
5. Schmid, G., Cecil, S., Goger, C., Trimmel, M., Kuster, N., Molla-Djafari, H.: New head exposure system for use in human provocation studies with EEG recording during GSM900 and UMTS-like exposure. Bioelectromagnetics 28, 636–647 (2007)
6. Kuster, N., Schuderer, J., Christ, A., Futter, P., Ebert, S.: Guidance for Exposure Design of Human Studies Addressing Health Risk Evaluations of Mobile Phones. Bioelectromagnetics 25, 524–529 (2004)
7. Murbach, M., Christopoulou, M., Crespo-Valero, P., Achermann, P., Kuster, N.: System to Study CNS Responses of ELF Modulation and Cortex versus Subcortical RF Exposures. Bioelectromagnetics (2012) (accepted for publication)
8. Taflove, A.: Computational Electromagnetics-The Finite Difference Time Domain Method. Artech House Publishers, Boston (1995)
9. Gabriel, S., Lau, R.W., Gabriel, C.: The dielectric properties of biological tissues: III. Parametric models for the dielectric spectrum of tissues. Phys. Med. Biol. 41, 2271–2293 (1996)
10. Farkas, L.G.: Anthropometry of the head and face, App A, 2nd edn., p. 244. Raven Press, New York (1994)
11. Christ, A., Kainz, W., Hahn, E.G., Honegger, K., Zefferer, M., Neufeld, E., Rascher, W., Janka, R., Bautz, W., Chen, J., Kiefer, B., Schmitt, P., Hollenbach, H.P., Shen, J., Oberle, M., Szczerba, D., Kam, A., Guag, J.W., Kuster, N.: The Virtual Family—development of surface-based anatomical models of two adults and two children for dosimetric simulation. Phys. Med. Biol. 55, N23–N38 (2010)
12. Rowan, A.J., Tolunsky, E.: Primer of EEG with a Mini-Atlas. Elsevier Science, United States of America (2003)
13. IEEE: Recommended Practice for Measurements & Computations of RF EM fields with Respect to Human Exposure to Such Fields. IEEE Standard C95.3-2002 (2002)

On Location-Based Services for Patient Empowerment, Guidance and Safety

Andreas K. Triantafyllidis[1,2], Vassilis G. Koutkias[1,2], Ioannis Moulos[1,2], and Nicos Maglaveras[1,2]

[1] Lab of Medical Informatics, Medical School, Aristotle University of Thessaloniki, P.O. Box 323, 54124, Thessaloniki, Greece
[2] Institute of Biomedical and Biomolecular Research, Center for Research and Technology, 6th Km. Charilaou-Thermi, P.O. Box 60361, 57001, Thessaloniki, Greece
{atriant,bikout,joemoul,nicmag}@med.auth.gr

Abstract. The importance of patients' active participation in healthcare delivery and disease management has been highlighted in various studies, especially for chronic patients. Overall, patient empowerment, guidance and safety constitute concrete goals in modern healthcare systems. To this end, mobile computing technologies can enable to realize such services in a pervasive manner, tailored to each patient's specific needs. The current work elaborates on the utilization of outdoor location information to deliver personalized services to patients, discriminating among reactive and proactive services according to their initiator and nature. In this regard, we present various use cases and relevant applications that have been developed by our group, exploiting this way the virtue and applicability introduced via the adoption of location-based services. The ultimate goal of this work is the development and establishment of an integrated framework for providing location-based healthcare information services targeting patient safety, empowerment and guidance.

Keywords: location-based services, mobile computing, patient safety, patient guidance, patient empowerment, personalization.

1 Introduction

Modern healthcare services delivery approaches emphasize on continuity and quality of care, personalization of services, and patients' active involvement in managing their health. Along these axes, patient empowerment, guidance and safety constitute concrete goals [1-2]. To this end, the utilization of mobile computing technologies, and the availability of location information in particular, lead to the development of healthcare services that are realized in a pervasive manner and meet the patients' specific needs and requirements in an "anytime-anywhere" fashion.

Location-based services in healthcare (both indoor and outdoor) have been recently explored in several works. For example, Kjeldskov et al. present a location-based service called 'GeoHealth' to support home healthcare workers, who attend patients at

K.S. Nikita et al. (Eds.): MobiHealth 2011, LNICST 83, pp. 416–422, 2012.
© Institute for Computer Sciences, Social Informatics and Telecommunications Engineering 2012

home within a large geographical area [3]. Boulos et al. elaborate on location-based services for patient independent living, that are provided via a light, wearable device [4], while Marco et al. target at the indoor localization of elder and disabled persons [5]. Location identification is crucial in health emergency scenarios where rapid response and action is needed. For example, in the work by Weixing et al. a mobile geospatial information system based on location-based services and wireless communications is presented aiming to enable rapid response to public health emergencies [6], while Vicente et al. investigate location-based access control policies in emergency situations [7]. Furthermore, location information is commonly employed and analyzed in conjunction with other attributes, e.g., various vital signs, such as the heart rate, and blood pressure, in order to monitor a subject's activity patterns [8]. Finally, mobile social networking systems have been presented [9], which employ location data for the delivery of various social networking services, mainly relying on the exchange of personal information.

In the current work, the primary focus lies on illustrating the use of outdoor location information as an important attribute in the design of pervasive health systems aiming to support the patients throughout their daily activities. Various use cases and applications that have been recently developed by our group are presented, aiming to illustrate the virtue and applicability of location-based healthcare services delivery.

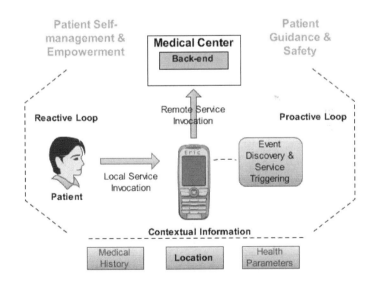

Fig. 1. The proposed conceptual framework for designing and providing location-based services for patient safety, guidance and empowerment

2   Methods

Figure 1 illustrates the overall conceptual framework of the current work. Using a mobile device, e.g., a smartphone or a Personal Digital Assistant (PDA), with outdoor location detection, communication and computational capabilities, the patient is enabled to "consume" personalized services, which are either installed locally at his/her mobile terminal or remote ones that are available at the back-end infrastructure of the medical center. Among such services are self-reporting of symptoms coupled with the location of interest, information delivery on demand concerning community-based activities in terms of proximity, generation and communication of alerts to the caregivers when the location of the patient may indicate harm, etc.

Such services are deployed using the patient's location as a primary attribute. However, location can be considered as only one dimension of the contextual information related to the patient, which can be used in conjunction with others that include his/her medical history and/or user profile, various health parameters that can be monitored via available sensing devices, such as the heart rate and the blood pressure, the time of the day, etc.

In the current work, location-based services are discriminated into two categories [10]: a) *Reactive services*, which are usually initiated by the users and require their constant interaction and attention being suitable for patient self-management and empowerment [11], and b) *Proactive services*, which are initiated either by the system or by the patient's family or caregivers and are typically executed in an event-driven manner, being particularly appropriate for patient guidance and safety. The services of the former type (employed in the *Reactive Loop* depicted in Fig. 1) are specifically targeted at chronic patients who are highly aware of their disease and wish to play a more active role in their disease management, gaining potential positive outcomes in their quality of life and well-being [12]. The services of the latter type (employed in the *Proactive Loop* depicted in Fig. 1) are adequate for emergency cases, where rapid response is required after the identification of a possibly hazardous situation within the *Event Discovery and Service Triggering* module.

In the following, we present various use cases and relevant applications that we recently developed corresponding to both reactive and proactive location-based services following the conceptual framework depicted in Fig. 1.

2.1   Reactive Location-Based Services

A characteristic reactive location-based service involves logging on a mobile phone various subjective elements concerning the patient's health status during his/her daily activities. This is particularly important for helping patients to understand their disease, providing them potentially with better insights into their self-management and treatment. Specifically, the patients are able to record various symptoms (e.g. dizziness, chest pain, shortness of breath, etc.), and associate their health status with their current location as illustrated in Fig. 2(a). Healthcare professionals may in turn receive this information via appropriate reports that are generated by the back-end framework, in order to identify and assess the potential health problems and hazards that patients may face and proceed with fine-tuning of their treatment and exercise plans.

Besides self-reporting, another reactive location-based service that we developed involves conveying information to the patients, which is generated by the members of relevant communities (e.g. young obese patients). Thus, the users can create events (e.g. walking, cycling, etc.) through their interaction with a map-based application, enabling them to share the relevant information with other members of the community. The aim of this service is to provide emotional support to the patients and/or increase their self-confidence towards achieving their personal health goals (e.g. losing weight). This information is provided by applying a filtering mechanism according to user-defined spatial criteria (e.g. physical distance from the events defined in the social networking platform).

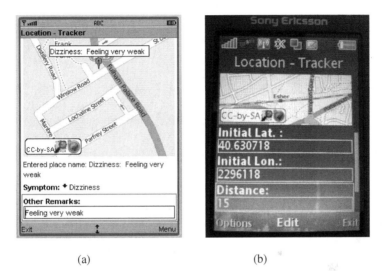

(a)   (b)

Fig. 2. (a) Sample reactive, location-based service: reporting a health symptom on a map, and (b) sample proactive location-based service: tracking the location of a person with Alzheimer's disease

2.2   Proactive Location-Based Services

As a typical proactive location-based service, we developed a tracking tool focusing on safety aspects (Fig. 2(b)), which relies on a GPS (Global Positioning System) enabled mobile phone. The tool is particularly useful for easily tracking patients suffering from diseases such as Alzheimer's disease. In a first step, a geographic area is specified (e.g. by the health professional or the patient's family), denoting the patient tracking boundaries. This area is determined by defining a point of interest (POI) with a certain longitude and latitude and specifying a radius, formulating this way a virtual circle (the Region of Interest - ROI). If the patient enters or leaves the specified ROI, he/she is provided with relevant recommendations or guidance information, while the caregivers are informed with appropriate alerts encapsulated in SMS (Short Message Service) messages, denoting the patient's current location.

In the case where the constant monitoring of the patient's health condition is required, various sensor-enhanced devices may be utilized in combination with location data, as additional information sources for defining the patient's context [13]. In the current work, we elaborated on continuous monitoring of the patients' health condition by employing a wearable multi-sensing device that incorporates sensors for monitoring the heart rate, the respiratory rate, the skin temperature and the activity. By analysing the acquired data, event-driven patterns in the form of personalized ECA (Event-Condition-Action) rules may be defined by the healthcare professional in order to identify an emergency situation (e.g. a high heart rate) (Fig. 3(a)). Thus, in the case of an emergency, a service can be triggered aimed at the detection of the user's location and the discovery of the nearest hospitals or pharmacies (Fig. 3(b)), notifying also the healthcare professionals for the occurred event.

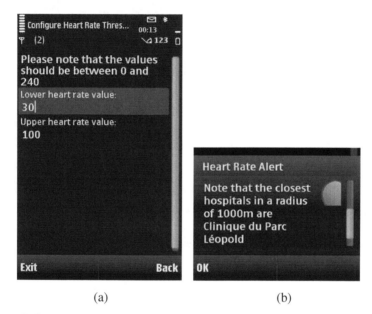

(a)   (b)

Fig. 3. (a) Definition of emergency situations related to heart rate, according to configured thresholds, and (b) the message obtained by triggering a proximity service for the discovery of the nearest hospitals to the patient's location

3 Implementation

Prototypes of the presented use cases and services have been implemented and tested in JavaME platform (Nokia N86 and Sony Ericsson C905 devices) as well as on Android-based smartphones (Motorola Milestone), in order to illustrate their technical feasibility and realization. JavaME provides Application Programming Interfaces (APIs) that are appropriate for dealing with the small memory footprint and the limited processing capabilities that are typically met in mobile phones. On the other hand, Android offers a Java-enabled platform that is particularly suitable for high-end smartphone devices.

As regards the location-based software modules developed, the Java Specification Request (JSR) 179 location API was used [14], in order to develop the required functionality, according to the user's geographic location. The Nutiteq's Mobile Maps API (http://www.nutiteq.com/mobile-map-api-sdk-guides/) together with the OpenStreetMap (http://www.openstreetmap.org/) were also employed for the provision of the necessary mapping capabilities, e.g. definition of POIs/ROIs and route handling. The SMS implementation relied on the Wireless Messaging API (WMA 2.0) JSR-205 [15].

Zephyr BioHarness (http://www.zephyr-technology.com/bioharness-bt) was employed as the multi-sensing device, while the communication between the mobile device and BioHarness was achieved by adopting the JSR 082 API for Bluetooth [16], which provides all the necessary methods for device discovery and the appropriate Bluetooth stream handling.

The communication between the back-end platform and the mobile terminal was realized by deploying a Service-Oriented Architecture (SOA) [17] based on Web services. For this reason, the JSR 172 API was employed [18], in order to provide the required Web service functionality based on the Simple Object Access Protocol (SOAP) and the Web Service Definition Language (WSDL). Moreover, aiming to flexibly implement and handle the required back-end operations, the Medical Center utilizes the Drupal platform (http://drupal.org/), which constitutes an open source Content Management System (CMS) based on the PHP server-side scripting language and offering various add-ons for robust service management.

4 Conclusion

Location-based services constitute a significant mean for healthcare service delivery particularly targeting patient empowerment, guidance and safety. Especially considering outdoor activities, modern mobile phone platforms may support efficiently the deployment of advanced location-based services through their (constantly advancing) sensing, communication and computational capabilities. This work discriminated location-based services into reactive and proactive ones, according to their initiator and nature. Reactive services are typically initiated by the users requiring their constant interaction and attention and being suitable for patient self-management and empowerment, while proactive services are initiated either by the system or by the patient's family or caregivers being particularly appropriate for patient guidance and safety scenarios. The presented use cases and applications for each service category highlighted their virtue and applicability in various scenarios of healthcare service delivery. The ultimate goal of this research is the development and establishment of an integrated framework for providing location-based services targeting patient safety, empowerment and guidance, through the incorporation of various personalized services such as the presented ones.

Acknowledgments. The research leading to these results has received funding from the Ambient Assisted Living (AAL) Joint Programme under Grant Agreement n° AAL-2008-1-147 – the REMOTE project (http://www.remote-project.eu/).

References

1. Lang, A., Edwards, N., Fleiszer, A.: Safety in home care: A broadened perspective of patient safety. Int. J. Quality Health Care 20(2), 130–135 (2008)
2. Koutkias, V.G., Chouvarda, I., Triantafyllidis, A., Malousi, A., Giaglis, G.D., Maglaveras, N.: A personalized framework for medication treatment management in chronic care. IEEE Transactions on Information Technology in Biomedicine 14(2), 464–472 (2010)
3. Kjeldskov, J., Christensen, C.M., Rasmussen, K.K.: GeoHealth: a location-based service for home healthcare workers. J. Locat. Based Serv. 4(1), 3–27 (2010)
4. Boulos, M.N., Rocha, A., Martins, A., Vicente, M.E., Bolz, A., Feld, R.: CAALYX: a new generation of location-based services in healthcare. Int. J. Health Geogr. 6(9) (2007)
5. Marco, A., Casas, R., Falco, J., Gracia, H., Artigas, J., Roy, A.: Location-based services for elderly and disabled people. Computer Communications 31, 1055–1066 (2008)
6. Weixing, W., Jianhua, G., Lihui, Z., Jinjin, Z., Fang, L., Cao, W.: Design and Implementation of Mobile GeoSpatial Information System for Public Health Emergency. In: 5th International Conference on Wireless Communications, Networking and Mobile Computing, September 24-26, pp. 1–4 (2009)
7. Vicente, C.R., Kirkpatrick, M., Ghinita, G., Bertino, E., Jensen, C.S.: Towards location-based access control in healthcare emergency response. In: 2nd International Workshop on Security and Privacy in GIS and LBS, pp. 22–26 (2009)
8. Duncan, J.S., Badland, H.M., Schofield, G.: Combining GPS with heart rate monitoring to measure physical activity in children: A feasibility study. Journal of Science and Medicine in Sport 12(5), 583–585 (2009)
9. Cheng, R., Zhuo, Y., Feng, X.: iZone: A Location-Based Mobile Social Networking System, Parallel Architectures. In: 3rd International Symposium on Parallel Architectures, Algorithms and Programming, pp. 33–38 (2010)
10. Bellavista, P., Kupper, A., Helal, S.: Location-Based Services: Back to the Future. IEEE Pervasive Computing 7, 85–89 (2008)
11. Warsi, A., Wang, P., LaValley, M., Avorn, J., Solomon, D.H.: Self-management education programs in chronic disease: a systematic review and methodological critique of the literature. Arch. Intern. Med. 164(15), 1641–1649 (2004)
12. Mosen, D.M., Schmittdiel, J., Hibbard, J., Sobel, D., Rem-mers, C., Bellows, J.: Is patient activation associated with out-comes of care for adults with chronic conditions? J. Ambul. Care Manage. 30(1), 21–29 (2007)
13. Konstantas, D.: An overview of wearable and implantable medical sensors. IMIA Yearbook 2(1), 66–69 (2007)
14. Java Community Process: JSR-000179 Location API for J2ME™, Final Release (2003)
15. Java Community Process: JSR-000205 Wireless Messaging API 2.0, Final Release (2004)
16. Java Community Process: JSR-000082 Java™ APIs for Bluetooth, Final Release (2002)
17. Singh, M.P., Huhns, M.N.: Service-oriented computing: Semantics, processes, agents. J. Wiley and Sons (2005)
18. Java Community Process: JSR-00172 J2ME™ Web Services Specification, Final Release (2004)

Author Index